Predicting Photosynthesis for Ecosystem Models

Volume II

Editors

J. D. Hesketh, Ph.D.
Crop Physiologist
Agricultural Research, Scientific and Education Administration
U.S. Department of Agriculture
University of Illinois
Urbana, Illinois

James W. Jones, Ph.D.
Associate Professor
Agricultural Engineering
University of Florida
Gainesville, Florida

CRC Press
Taylor & Francis Group
Boca Raton London New York

CRC Press is an imprint of the
Taylor & Francis Group, an informa business

First published 1980 by CRC Press
Taylor & Francis Group
6000 Broken Sound Parkway NW, Suite 300
Boca Raton, FL 33487-2742

Reissued 2018 by CRC Press

Library of Congress Cataloging in Publication Data
Main entry under title:

Predicting photosynthesis for ecosystem models.

Bibliography: p.
Includes indexes.
1. Photosynthesis—Simulation methods. I. Hesketh, John D. II. Jones, James Wigington, 1944-
QK882.P73 581.1'3342'0184 79-4047
ISBN 0-8493-5335-1 (v. 1)
ISBN 0-8493-5336-X (v. 2)

A Library of Congress record exists under LC control number: 79004047

Publisher's Note
The publisher has gone to great lengths to ensure the quality of this reprint but points out that some imperfections in the original copies may be apparent.

Disclaimer
The publisher has made every effort to trace copyright holders and welcomes correspondence from those they have been unable to contact.

ISBN 13: 978-1-315-89685-4 (hbk)
ISBN 13: 978-1-351-07595-4 (ebk)

Visit the Taylor & Francis Web site at http://www.taylorandfrancis.com and the
CRC Press Web site at http://www.crcpress.com

PREFACE

Research and development agencies in the United States have emphasized the need for methods of predicting ecosystem behavior, so as to minimize pollution (insecticide, herbicide, and fertilizer usage) and conserve energy (petroleum products) while maximizing productivity and profits. Some of this pressure may have evolved from the successful use of system analysis techniques by the National Aeronautics and Space Administration in their missions in the 1960s. The editors of these volumes collaborated in the late 1960s and early 1970s in a regional cotton project using such system analysis techniques while located at Mississippi and North Carolina State Universities with the U.S. Department of Agriculture. Dr. W. G. Duncan (Universities of Kentucky and Florida) and Dr. D. N. Baker (USDA, Mississippi State) played important roles in these collaborations. It was an exciting period involving several disciplines in an uniquely exploratory effort. For example, at one point we were interested in the role of the crop in regulating the density and physiological state of the boll weevil, an insect pest. Therefore, we studied the nitrogen and carbon compositions of various plant parts with respect to the nutritional requirements of the boll weevil larvae and adults. Diets high in protein induced reproduction while diets high in carbohydrates induced storage of fats and preparation for overwintering (diapause).

Photosynthate, as recently formed reductant in photosynthetic cells or as reduced nitrogen or carbon, plays a key role in supplying the energy needs for an ecosystem. In the context of predicting ecosystem behavior, photosynthesis must be predicted as it interacts with other plant physiological processes as well as with animal behavior, such as the effect of insects feeding on leaves. Likewise a study of other aspects of ecosystem behavior must take into account their interactions with photosynthesis activities. Therefore our coverage is broad, including descriptions of how one might integrate a photosynthetic model into an ecosystem predictor. The understanding and description of interactions among processes is vital to a successful ecosystem model.

We invited several full-time experimentalists unassociated with any prediction effort (one of us, JDH, is a full-time experimentalist) to review problems one frequently encounters when attempting to predict field photosynthesis. Some of these problems may only be resolved using a whole plant approach and studying photosynthesis in relation to other plant physiological processes. In analyzing the recent literature, it would seem that experimentalists in the various specialties of plant physiology have developed, independently, the same techniques as those using the multidisciplinary approach to study photosynthesis for ecosystem predictions. Photosynthesis is being studied in many laboratories as it relates to whole plant behavior. In our opinion, the experimentalists authoring chapters in these volumes had no problem identifying relevant subject matter for discussion. Indeed, a major limiting factor in ecosystem prediction efforts is the participation of experimentalists in reviewing the status of various biological problems involved. We recommend more of this in the near future. It is likely that participation will only come with the invitation to review a carefully defined problem.

We go into greater detail justifying this effort in the first chapter. There are now four methods for predicting photosynthesis, each with limitations and strengths. Their present usage depends upon the purpose of the prediction to be made, as well as improvements in associated techniques that seem to revolutionize the methodology. At this point in time, comparisons among methods are valuable, justifying a state-of-the-art review for all photosynthetic scientists.

We were faced with the need to develop general notation for all the chapters but the subject matter was so diverse, with so many traditions in notation, that we gave up. Perhaps our failure in this respect will focus on a solution to the problem. Rarely has so much material been brought together in one place.

We can't help but feel that progress in this general area of research will be rapid in the near future. We hope these volumes speed up the process.

J. D. Hesketh
J. W. Jones
July 1979

CONTRIBUTORS

Randall S. Alberte, Ph. D.
Assistant Professor
Department of Biology
University of Chicago
Chicago, Illinois

Larry G. Brown, Ph.D.
Associate Professor, Industrial Engineering
Mississippi State University
Mississippi State, Mississippi

J. Robert Cooke, Ph.D.
Director of Instruction
College of Agriculture and Life Sciences
Professor of Agricultural Engineering
Cornell University
Ithaca, New York

Gary L. Cunningham, Ph.D.
Professor of Biology
New Mexico State University
Las Cruces, New Mexico

C. D. Elmore, Ph.D.
Plant Physiologist
AR-SEA-USDA Southern Weed Science Laboratory
Stoneville, Mississippi

David M. Gates, Ph.D.
Professor of Botany
Director, Biological Station
University of Michigan
Ann Arbor, Michigan

Gene Guinn, Ph.D.
Plant Physiologist (Research Leader)
AR-SEA-USDA Western Cotton Research Laboratory
Phoenix, Arizona

Peter C. Harley, M.S.
Research Assistant
Biological Station
University of Michigan
Ann Arbor, Michigan

John D. Hesketh, Ph.D.
Crop Physiologist
AR-SEA-USDA
University of Illinois
Urbana, Illinois

James W. Jones, Ph.D.
Associate Professor
Department of Agricultural Engineering
University of Florida
Gainesville, Florida

Kenneth R. Knoerr, Ph.D.
Professor of Forest Meteorology
Department of Botany
Duke University
Durham, North Carolina

Harry R. Leffler, Ph.D.
Plant Physiologist
AR-SEA-USDA Cotton Physiology and Genetics Laboratory
Stoneville, Mississippi

Jackson R. Mauney, Ph.D.
Plant Physiologist
AR-SEA-USDA Western Cotton Research Laboratory
Phoenix, Arizona

Bobby L. McMichael, Ph.D.
Plant Physiologist
AR-SEA-USDA Plant and Soil Science Department
Texas Technical University
Lubbock, Texas

John M. Norman, Ph.D.
Professor of Agronomy
The University of Nebraska
Lincoln, Nebraska

David T. Patterson, Ph.D.
Plant Physiologist
AR-SEA-USDA Southern Weed
Science Laboratory
Stoneville, Mississippi

John W. Radin, Ph.D.
Plant Physiologist
AR-SEA-USDA Western Cotton
Research Laboratory
Phoenix, Arizona

Richard H. Rand, Sc.D.
Associate Professor
Department of Theoretical and Applied
Mechanics
Cornell University
Ithaca, New York

James F. Reynolds, Ph.D.
Assistant Professor
Department of Botany
North Carolina State University
Raleigh, North Carolina

Boyd R. Strain, Ph.D.
Professor of Botany
Duke University
Durham, North Carolina

John D. Tenhunen, Ph.D.
Research Associate
Biological Institute
University of Wurzburg
Wurzburg, West Germany

THE EDITORS

J. D. Hesketh was born in Maine and attended the University of Maine and Cornell University. He has worked in Connecticut, Arizona, Canberra (Australia), Mississippi, North Carolina, and Illinois with various governmental agencies. His early research was concerned with gas exchange aspects of the so-called C_3-C_4 phenomenon, as well as the effect of environment on crop photosynthesis. His interest in the latter has continued to date, as well as collaborations in various efforts to forecast crop yields.

J. W. Jones was born in Mississippi and received degrees from Texas Tech, Mississippi State, and North Carolina State Universities. He worked in Mississippi and in North Carolina with the U.S. Department of Agriculture and in Florida with the University of Florida. His early research focused on analysis of cotton production systems with emphasis on modeling cotton crop growth and yield as well as crop-pest interactions. Currently, his interests are in modeling crop growth processes as affected by water and other stresses. He has taught several courses related to systems analysis of agricultural and biological systems.

TABLE OF CONTENTS

VOLUME I

VOLUME II

Chapter 1

ANALYSIS OF CO_2 EXCHANGE ASSUMPTIONS: FEEDBACK CONTROL

G. Guinn and J. R. Mauney

TABLE OF CONTENTS

I. INTRODUCTION

Neales and Incoll[1] reviewed the evidence for end-product inhibition of photosynthesis in 1968. Despite a century of investigation by many workers, they concluded that the hypothesis had neither been proved nor disproved. At the time of their review, the evidence for feedback control was primarily circumstantial, based on observations that manipulations of sources and sinks and environmental conditions that caused (or were presumed to cause) accumulation of assimilates in leaves also decreased photosynthesis (as estimated by gas exchange or increases in dry weight). They concluded that a satisfactory proof of the hypothesis must depend upon the demonstration of (1) a negative correlation between photosynthetic rate and assimilate level in the leaf and (2) a mechanism.

At the time Neales and Incoll[1] wrote their review, the emphasis was on sugars; there was little evidence, or concern, that starch was involved in end-product inhibition of photosynthesis. Since that time, however, most of the evidence for feedback control has involved starch rather than sugars. Results with sugars have been mostly negative. Few authors have proposed mechanisms to explain end-product inhibition, and the bulk of the evidence is still circumstantial.

Few workers have investigated hormonal mechanisms for the feedback control of photosynthesis. The limited data available suggest that such mechanisms may operate in at least some plant species.

II. EXPERIMENTAL MANIPULATIONS

A. Source-Sink Relations

Although we recognize that some organs can change from sinks to sources (e.g., storage roots, stems), for the purpose of this review we shall consider photosynthetically active leaves as sources and other organs as sinks. The reader is referred to Warren Wilson[2] for a comprehensive discussion of concepts of sources and sinks.

Additional evidence has accumulated that sinks can affect the rate of apparent photosynthesis (AP) in source leaves, but not all workers agree that the effect is caused by accumulation of assimilates, or even that such an effect exists.

King et al.,[3] working with wheat (*Triticum aestivum* L.) plants pruned to a main shoot and two earless tillers, found that removal of the main shoot ear decreased AP as much as 50% within 3 to 15 hr. Darkening the lower leaves caused a partial recovery in AP of the flag leaf. Rate of AP was also decreased by subjecting intact plants to continuous, high-intensity light for "more than a week", but was restored by 24 hr of darkness or by spraying the ear with 3-(3,4-dichlorophenyl)-1,1-dimethyl urea (diuron) to inhibit photosynthesis in the ear. In contrast, Austin and Edrich[4] reported only small and inconsistent changes in AP after they removed the main shoot ear. Different light regimes may have caused the different results. King et al.[3] placed their plants under continuous illumination for 3 days before treatment, whereas Austin and Edrich[4] grew their plants in field plots with natural light. Continuous light probably caused large accumulations of carbohydrates; King et al.[3] reported that the extractable dry weight increased to 44.9% after 7 days of continuous light. Austin and Edrich,[4] however, did not find any increase in concentration of extractable carbohydrates under the conditions of their experiment.

Barlow and Boersma[5] chilled either roots or the apical meristem region of corn (*Zea mays* L.) for 7 hr to change sink demand. They measured AP of leaf five continuously during the 7-hr treatment and then measured soluble carbohydrate in leaves five, six, and seven. Lowering the root temperature from 27.5 to 15°C caused a steady decrease in AP until, at the end of the 7-hr treatment, it was only 53% of the initial rate. Soluble carbohydrate contents of leaves five, six, and seven were 27%, 47%, and 45% higher than that of corresponding control leaves. They estimated that only about a third of the decrease in AP was due to decreased leaf-water potential. Chilling the shoot meristem region from 27.5 to 6°C increased the soluble carbohydrate content of the source leaf 23% and decreased its AP 24%. In the latter case there should have been no change in water potential of the source leaf.

Chalmers et al.[6] measured AP at various levels in a peach (*Prunus persica* L. Batsch) tree during a growing season, and found that AP increased as the rate of fruit growth increased, and then decreased after the peaches were harvested. Changes in AP with sink (fruit) demand were much greater in the middle part of the canopy that contained most of the fruit than at the base or the top of the tree. They assumed that accumulation of assimilate in leaves before rapid fruit growth and after harvest limited AP. A relatively high specific leaf weight (SLW) of upper leaves, even during the period of rapid fruit growth, indicated that assimilate was not depleted in those leaves and may have remained at inhibitory levels. Tests with ^{14}C-labeled assimilate indicated that assimilate accumulated in leaves after the peaches were harvested; time required for half the label to disappear from the leaves doubled after harvest, and AP decreased.

The presence of fruit also affected the rate of AP in eggplant (*Solanum melongena* L.) leaves. Claussen[7] found that rate of AP started decreasing within 5 hr after the start of illumination. The decrease was greater for plants without than for plants with fruit, and the difference was greatest during the stage of rapid fruit growth. After 12 to 16 hr of illumination, plants with fruit showed 60% higher rates of AP than plants without fruits (defruited plants). Although he did not speculate why the presence of fruit prevented the large decline in AP during the day, the results suggest end-product inhibition in the absence of fruit.

Flinn[8] compared rates of AP in leaflets of pea (*Pisum sativum* L.) at fruited, defruited, and vegetative nodes. Changes in rates of AP in leaflets that subtended an intact developing fruit were correlated with changing growth rates, and thus changing sink demand, of the developing fruit. Rates of AP were lower and less variable in leaflets at nodes where the fruit was removed at anthesis and in leaflets at vegetative nodes.

Habeshaw[9] altered sink demand in sugar beet (*Beta vulgaris*) by cooling the roots from 17 to 9°C for 2 days. This treatment decreased the translocation rate out of all but the two youngest leaves and, likewise, decreased AP in all but the youngest leaf. The nearer the leaves were to the root sinks, the greater the effects on translocation and AP rates. In a separate experiment, Habeshaw[9] cooled the roots 10°C below air temperature and sampled leaves for carbohydrate analyses at the end of the light period (14 hr) of the second day and at the end of the dark period (10 hr) of the second day. Results provided little support for the hypothesis that sugars or starch inhibited AP when roots were cooled. Starch content of leaves was increased by root cooling, but was low in all cases (14 to 37 $\mu g/cm^2$). Root cooling apparently caused a slight decrease in sucrose content of leaves, but the results were not statistically significant. Perhaps root cooling decreased the conversion of reducing sugars to sucrose. Root cooling almost doubled the reducing sugar content of leaves at the end of the dark period, but had no effect at the end of the light period. An explanation of these variable results is not readily apparent. Habeshaw's statement that, "Cooling the root tends to reduce the sucrose levels in the leaf, presumably because of lower translocation rates," does not provide a satisfactory explanation.

Lenz[10] reviewed work on the effects of fruit on photosynthesis and respiration of *Citrus madurensis* Lour. and *Solanum melongena* L., and discussed possible mechanisms by which fruit might stimulate photosynthesis. He reported that the presence of fruit increased AP, and that heating the fruit to 32°C further increased AP. No differences in dark respiration rates were found, but rates of photorespiration were lower in fruiting plants than in nonfruiting plants. Leaves of nonfruiting plants accumulated more sugars and starch and had higher gaseous diffusive resistance than leaves of fruiting plants. He discussed some possible mechanisms by which the presence of fruit might stimulate photosynthesis, including end-product inhibition by carbohydrates and stimulation by hormones produced in fruits and translocated to leaves.

Thorne and Koller[11] used shade to alter the source-sink relationships of potted soybean (*Glycine max* L. Merr.) plants. The one unshaded leaf (the source leaf) on otherwise shaded plants lost starch, gained phosphate, and showed increased RuDP-carboxylase activity and increased AP, when compared to a leaf in the same position on unshaded control plants. Because sucrose content of the source leaf increased as AP increased, they concluded that sucrose was not a mass-action regulator of photosynthesis. Although starch content decreased ten-fold in the source leaf of shaded plants, they questioned its role in the observed increase in AP. The increase in RuDP-carboxylase activity caused them to speculate that shading caused a source-sink hormonal interaction which was responsible for the increased AP.

Removal of all axillary buds of cotton (*Gossypium hirsutum* L.) prevented the development of reproductive sinks, but did not affect AP of young, fully expanded leaves.[12] As the leaves aged, however, they showed a slightly more rapid decline in rate of AP on debudded than on control plants.

Geiger[13] removed all source leaves except one, or removed the growing point and axillary buds to alter the balance between sources and sinks of bean plants (*Phaseolus vulgaris* L.). He also cooled the petiole of the source leaf to 2°C to decrease translocation of assimilate out of that leaf. None of the treatments affected the rate of AP during the 2-day measurement periods. Geiger summarized most of the work to that time in tabular form and concluded that, in those cases where sink demand affected photosynthesis, the control mechanism was probably more complex than simple feedback control by assimilates. Likewise, Carmi and Koller[14] found no evidence for feedback inhibition of AP in bean. They altered demand by removing the top and all buds and altered translocation by girdling the petioles with hot water. Although starch granules accumulated in leaf mesophyll cells after petioles were girdled, the treatments did not affect the rate of AP.

Liu et al.[15] did find evidence for a correlation between AP and rate of translocation of assimilates out of bean leaves. The cultivar Michelite-62 had higher AP and translocation rates than Red Kidney. Only 38% of the label remained in the treated leaf of Michelite-62, compared to 60% in Red Kidney, 8 hr after a pulse of $^{14}CO_2$. The results supported their hypothesis that translocation is one of the important factors controlling varietal differences in photosynthetic efficiency in *Phaseolus vulgaris.*

B. Temperature and Photoperiod

In addition to cooling the sinks, some workers have cooled entire plants to cause assimilates to accumulate. Various photoperiods have also been used, either as a separate treatment or as an additional variable.

Cool nights inhibit growth and cause starch to accumulate in some plants, such as pangolagrass (*Digitaria decumbens* Stent). Chatterton et al.[16] reported that chloroplasts of pangolagrass accumulated starch during the day unless the plants were tillering. The starch disappeared during warm nights but not during cool nights. Nontillering plants subjected to cool nights accumulated starch and their rate of AP decreased, presumably because of the starch. Tillering plants accumulated little starch during the day and cool nights did not decrease their AP the following day. Thus, cool nights per se did not decrease AP in this experiment.

Correlations do not establish causal relationships. Wilson[17] subjected *Lolium perenne* L. to temperature regimes of 15°C day and 10°C night, or to 25°C day and 20°C night, and then determined rates of AP. He also examined leaf sections with an electron microscope. Cooler temperatures caused many large starch grains to accumulate in chloroplasts of all mesophyll cells examined. The 15°C day and 10°C night temperature regime also depressed subsequent AP. However, Wilson[17] also noted that the lower temperatures increased cell size, and he attributed the differences in rates of AP to differences in size of mesophyll cells, rather than to accumulation of starch in chloroplasts. He based this conclusion on results of earlier work, which indicated higher rates of AP of small cells,[18] and that AP was negatively correlated with cell size in leaves grown at a constant 15°C.[19]

A recent report by Carter and Garrard[20] also shows how results of tests in which starch content is manipulated by photoperiod and night temperature can be complicated by other effects. They subjected pangolagrass to either 24 hr or 28 hr of treatment in four combinations of light and temperature: (a) 12 hr at 30°C in light, followed by 12 hr at 30°C in dark; (b) 12 hr at 30°C in light, followed by 12 hr at 10°C in dark; (c) 4 hr at 30°C in light, followed by 12 hr at 30°C in dark (to deplete assimilate), and then 12 hr at 10°C in dark; and (d) 4 hr at 30°C in light, followed by 24 hr at 30°C in dark. Subsequent measurements of AP at the end of the dark period indicated that treatment (a) had the highest rate but, surprisingly, treatment (d) had the lowest. Rates of AP in mg $CO_2/dm^2/hr$ were 40.0, 25.8, 29.7, and 12.7 for treatments (a) through (d), respectively. They postulated that treatment (d), which caused extreme depletion of carbohydrate, depressed photosynthesis because precursors were insufficient for maximum rates of carbon fixation. In contrast, exposure of sugar beet to continuous darkness for 56 hr, and then to 1 hr in light (temperature not specified), resulted in a 3.3-fold higher AP rate than that of plants exposed to 56 hr of continuous light at 12 klux.[9] The temperature of sugar beet plants in the dark was probably below 17°C, because Habeshaw[9] indicated an average air temperature of 17°C. Therefore, depletion may not have been as severe in sugar beet[9] as it was in pangolagrass,[20] even though time in the dark was longer. Furthermore, the different species may have responded differently.

Austin[21] used continuous light, CO_2 enrichment, or low temperature to obtain very high sugar contents in discs or intact leaves of sugar beet. Rates of AP, estimated by increases in dry weight of the discs, were depressed when sugar content exceeded about 30% of the residual dry weight. Because sugars do not accumulate to such high levels under normal conditions, he concluded that photosynthesis of sugar beet leaves is not normally limited by accumulation of sugars, but may be regulated by other mechanisms. Milford and Pearman[22] also used long photoperiods to cause sugar beet leaves to accumulate carbohydrates. Leaves that were given a 48-hr dark pretreatment initially showed higher rates of AP than leaves that were given a 48-hr light pretreatment. The difference disappeared after 30 hr of continuous illumination, because the rate of AP in the dark-pretreated leaves dropped to the level of the light-pretreated leaves as they accumulated starch and sugars. Their data showed a closer relationship between starch and AP than between sugars and AP. Although their data indicated an inhibition of AP by starch, they concluded that carbohydrates were unlikely to accumulate to inhibitory levels under field conditions.

Different species apparently respond differently to light and temperature treatments. Rate of AP of pangolagrass began to decline after only 4 hr in the light, and continued to decline during the following 8 hr as starch and sugars accumulated.[23] Conversely, after a lag period of 4 hr, AP increased with time in the dark, as sugar and starch content of the leaves decreased. (These plants were illuminated only during photosynthesis measurements.) In a somewhat similar experiment with *Paspalum dilatatum* Poir., Forde et al.[24] found no evidence for feedback inhibition of photosynthesis. The maximum starch content they measured, however, was only 11.6% (dry weight) and may have been too low to cause inhibition. Changing the day/night temperature regime from 24/19°C to 17/10°C decreased photosynthetic efficiency of sugar cane (*Saccharum officinarum* L.).[25] Returning the plants to 24/19°C day/night temperatures caused rapid recovery of transpiration rate, but slower and only partial recovery of photosynthetic efficiency. When Waldron et al.[25] detached sugarcane leaves and illuminated them for 12 hr, they observed an increase in sucrose content and a decrease in AP. However, AP showed a substantial recovery after 12 hr of darkness, although there was no decline in sugars. They attributed the decline in AP of detached leaves to water deficit, and concluded that sugars did not repress AP in sugarcane leaves in their experiments.

Moldau and Karolin[26] used different photoperiods to regulate the size of the reserve pool of carbohydrates in dwarf bean (*Phaseolus vulgaris* L.). Increasing the photoperiod from 13 to 22 hr of light per day doubled the size of the reserve pool and suppressed photosynthesis. In contrast to other results[11,27,28] the suppression of photosynthesis was caused solely by stomatal closure.

C. Carbon Dioxide Enrichment

The low concentration of atmospheric CO_2 frequently limits rate of photosynthesis in C_3 plants.[29] Therefore, raising the CO_2 concentration increases the rate of net photosynthesis and, in many plants, causes starch to accumulate. The burning of massive amounts of fossil fuels is causing the CO_2 content of the atmosphere to increase and, according to some projections,[30] it could reach 600 vpm by the year 2035. If the accumulation of assimilate does inhibit photosynthesis, feedback inhibition could prevent plants from maintaining the high rates of photosynthesis that should be possible in a CO_2-enriched atmosphere. Thus, the question of feedback inhibition may assume increasing significance as the CO_2 content of the atmosphere increases.

Several workers have used CO_2 enrichment to cause accumulation of assimilates. Hofstra and Hesketh[27] exposed two soybean cultivars to CO_2 enrichment (800 to 1000 vpm) under four different temperature regimes. Rate of AP was negatively correlated with mesophyll resistance, starch content of leaves, and SLW. Sugar content of leaves, on the other hand, tended to be positively correlated with photosynthesis. They found no correlation between AP and stomatal resistance, indicating that the inhibition of AP at high starch levels was not a stomatal effect.

Mauney et al.[31] grew four species of plants in normal (330 vpm) and high (630 vpm) concentrations of CO_2. The CO_2 enrichment caused large increases in starch content of leaves of cotton and soybean, but caused only small increases in starch content of sunflower (*Helianthus annuus* L.) and sorghum (*Sorghum bicolor* L.). Rate of AP of sorghum, a C_4 plant, was relatively insensitive to CO_2 enrichment, and its leaves always contained low concentrations of starch. Starch content of cotton leaves, on the other hand, increased to levels as high as 50% of their dry weight after a few days in high CO_2. When cotton plants were transferred from normal to high CO_2, AP increased 45% in the first 2 hr after transfer, but then gradually declined as starch accumulated, until AP was only 15% greater for leaves in high CO_2 than for leaves in normal CO_2. When these plants, high in starch, were returned to normal air, AP dropped below that of plants that had remained in normal air. Significant (1% level) negative correlations were obtained between starch contents and rates of AP measured in normal air (Figure 1). Rates of AP returned to normal in about 3 days.

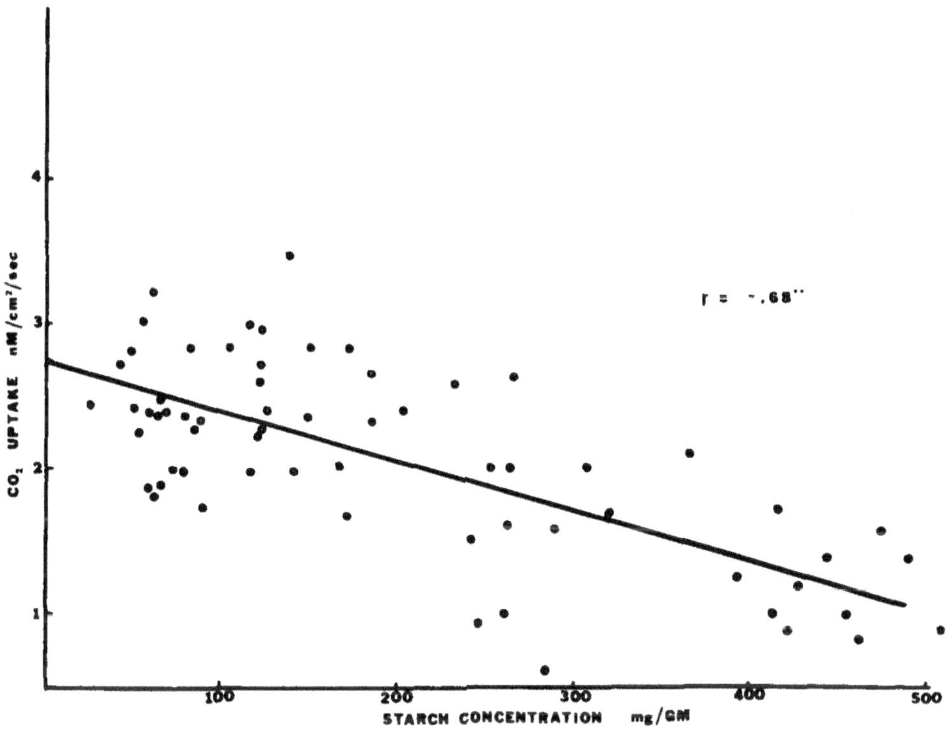

FIGURE 1. Correlation of leaf starch concentration and CO_2 uptake in leaves of greenhouse-grown cotton. Plants were grown and measured at 330 vpm CO_2 or grown in 630 vpm and measured at 330 vpm. (From Mauney, J. R., Guinn, G., Fry, K. E., and Hesketh, J. D., Correlation of photosynthetic carbon dioxide uptake and carbohydrate accumulation in four crop species, *Photosynthetica*, in press. With permission.)

Nafziger and Koller[28] subjected soybean plants to 50, 300, and 2000 vpm CO_2 for 12.5 hr, and then measured AP at 300 vpm CO_2. The CO_2 enrichment increased the starch, but not the sugar, content of leaves. Rate of AP decreased with increasing starch. Most of the difference in AP appeared to be due to increased mesophyll resistance which increased more than stomatal resistance as starch content of leaves increased. This effect was apparent only at starch levels above 1 mg/cm^2 (about 200 mg/g dry weight) which, as they pointed out, may explain why some workers failed to observe inhibition of AP by starch.

Ito[32] reported that starch accumulated to high levels in tomato (*Lycopersicon esculentum* L.) leaves (in the afternoon) in a CO_2-enriched atmosphere, and starch content showed a high negative correlation with rate of AP. Fruit removal decreased AP more in 1065 vpm CO_2 than it did in 300 vpm CO_2, possibly because of the additive effects of fruit removal and CO_2 enrichment on the accumulation of assimilates. Madsen[33-35] grew tomatoes in different concentrations of CO_2, from ambient to 3200 vpm. The CO_2 enrichment caused large increases in starch content of leaves and severe deformation of chloroplasts.[33] The rate of AP increased with increasing CO_2 only up to about 1000 vpm, and then decreased at higher concentrations of CO_2, possibly because of chloroplast deformation.[34] Likewise, yields increased with increasing CO_2 up to 1000 vpm, but then decreased at higher CO_2 concentrations; the yield at 3200 vpm was less than that in normal air.[35]

D. Application of Sugars

A few experiments have been conducted in which carbohydrate status of tissue was manipulated by direct addition of sugars. Moore et al.[36] placed petioles of *Sinapis* cotyledons in 60 m*M* sucrose solution or in water. Carbohydrate content increased 28-fold in the sucrose solution and 8-fold in water during 7 days in the light. Although carbohydrate was supplied as sucrose, most of the increase was in the form of starch and reducing sugars. Fixation of $^{14}CO_2$ remained constant in cotyledons cultured in water, but decreased by 50% during the first 2 days in cotyledons cultured in sucrose. Osmotic effects may have been a factor in this test. Chlorophyll content remained constant in controls, but fell rapidly after 4 days in sucrose, which probably decreased AP. Habeshaw[9] painted the lower surface of sugar beet leaves at twelve 1-hr intervals with either water or 2.5% sucrose solution in the dark. He then fed a pulse of $^{14}CO_2$ in the light, and subsequently determined fixation rates and distribution of radioactivity. The sucrose treatment depressed CO_2 fixation, and caused a higher fraction of the radioactive assimilate to remain in the chloroplasts.

Natr and Ludlow[37] and Natr et al.[38] floated barley (*Hordeum vulgare* L.) leaf segments on a 5% glucose solution in dim light, or on distilled water at low- and high-light intensities. A given increase in dry matter, due to glucose, caused a stronger inhibition of photosynthesis than did a comparable increase in dry matter that resulted from photosynthesis. Glucose also decreased transpiration and increased r_s and r_m. They concluded that the differential inhibition was not due to starch, because leaves that were floated on water in bright light contained more starch than leaves floated on glucose in dim light. Waldron et al.[25] immersed sugar cane leaf discs in water or in 0.4*M* solutions of sucrose, glucose, or mannitol. Rates of AP of the leaf discs declined about 24% after 22 hr in continuous light, but declined no more in the sugar solutions than in water.

When phosphorus-deficient leaves of subterranean clover (*Trifolium subterraneum* L.) were kept for 3 hr in 3 or 6 m*M* fructose-1,6-diphosphate or glucose-6-phosphate, rate of AP increased within 24 hr.[39] Fructose at 6 m*M* had no effect on AP. Thus, in these phosphorus-deficient leaves, application of phosphorylated sugars increased, rather than decreased, the rate of AP. (See Bowes and Ogren[40] and Buchanan and Schürmann[41] for an opposite effect with nondeficient systems.)

E. No Manipulations

Several workers have measured changes in AP with time, during a day or during a growing season, and have attempted to relate differences in AP to changes in SLW, carbohydrates, or sink demand. Chatterton[42] found a close negative correlation between diurnal variations in SLW and AP in alfalfa (*Medicago sativa* L.) leaves, which he interpreted as evidence for end-product inhibition of photosynthesis. Upmeyer and Koller[43] reported that AP declined late in the photoperiod, concomitant with a decline in rate of starch accumulation and a slight increase in sugar content of soybean leaves. They suggested that the decline in AP may have been caused by the increase in soluble carbohydrate, but their data are not convincing on this point because of the small change in soluble carbohydrate. In a later paper, Thorne and Koller[11] concluded that sucrose was not a mass-action regulator of photosynthesis in soybean leaves, because sucrose content of leaves increased, rather than decreased, with increasing photosynthesis.

Little and Loach[44] measured changes in concentrations of sugars and starch and AP during a growing season in fir (*Abies balsamea* (L.) Mill.) needles. On the basis of a positive, rather than a negative, correlation between carbohydrate (mostly starch) and AP, they concluded that photosynthate accumulation does not influence rate of photosynthesis in fir under normal growing conditions.

Fraser and Bidwell[45] measured gross or true photosynthesis, photorespiration, transpiration, r_s, and r_m in bean (*Phaseolus vulgaris* L.) leaves during ontogeny of the plant. Both true and net photosynthesis increased with the onset of flowering, except in the first trifoliate leaf, where an increase in photorespiration prevented an increase in net photosynthesis. There also appeared to be a transient increase in photosynthesis of leaves at the emergence of each subsequent leaf, which is a sink during early development and emergence. They suggested that flowering had a greater effect on photosynthesis than leaf emergence, because flowering is a stronger and longer-lasting sink. Although they did not determine the mechanism for the stimulation, they mentioned the work of Bidwell and others, which indicated that auxins had a stimulatory effect on AP. See Section III.B.1. Indoleacetic Acid (IAA). Fraser and Bidwell[45] found that r_m was more highly correlated with photosynthesis than was r_s. This correlation would, however, be compatible with more than one possible mechanism.

III. POSSIBLE MECHANISMS

Most of the papers published since the review by Neales and Incoll[1] have dealt with evidence, or lack of it, for feedback control, and few have dealt with possible mechanisms. Control of photosynthesis by sink demand could be achieved, not only by end-product inhibition (e.g., by sugars or starch), but also by hormonal effects. We shall briefly consider possible mechanisms for both types of control.

A. End-Product Inhibition

1. Sugars

Certain phosphorylated sugars inhibit RuDP-carboxylase activity by competing with ribulose-1,5-diphosphate for binding sites on the enzyme.[40,41] Fructose-1,6-diphosphate was the most effective inhibitor, but glucose-1,6-diphosphate, glucose-1-phosphate, glucose-6-phosphate, 3-phosphoglycerate, and 2-phosphoglycolate were also inhibitory. Sucrose, glucose, and fructose had no appreciable inhibitory effect.[40] Thus, if sugars accumulate in the above phosphorylated forms, they could inhibit AP by inhibiting RuDP-carboxylase activity.

Accumulation of nonphosphorylated sugars may affect photosynthesis by sequestering orthophosphate,[46,47] thereby inducing a deficiency of phosphate. Phosphate deficiency has been shown to depress AP.[48] Only glucose and mannose were reported to sequester phosphate, but other sugars, such as fructose and sucrose, may also have an effect, or may be converted to glucose. The sequestration of phosphate by sugars could explain the inhibition of AP by applied sugar solutions.[36,38]

Because sucrose, unlike the hexoses, is not synthesized inside chloroplasts,[49] it is more difficult to visualize mechanisms by which sucrose would inhibit photosynthesis. One possible mechanism is through inhibition of sucrose phosphatase activity,[50] which might cause accumulation of phosphorylated sugars in addition to sucrose phosphate and, thereby, inhibit RuDP-carboxylase activity.[40,41]

2. Starch

Relatively little attention was given to starch in possible end-product inhibition of photosynthesis before the review by Neales and Incoll.[1] Since then, however, the proportion of research concerned with starch has increased.

Starch is synthesized inside chloroplasts,[49] and most of the proposed mechanisms for feedback inhibition by starch are based on physical, rather than biochemical, effects. The proposed mechanisms include: binding of magnesium ions,[51] interference with light transmission,[32] increased distance for CO_2 diffusion,[28,32] and physical damage to chloroplasts.[33,52,53] These effects are not mutually exclusive.

Klemme and Jacobi[51] reported evidence that amylose, but not amylopectin, binds Mg^{++} ions. Because of the key role of magnesium as an activator or cofactor of many enzymes in photosynthesis, a magnesium deficiency should inhibit AP.

Although interference with light transmission has been proposed,[32] Nafziger and Koller[28] discounted this possibility because they determined AP at a three-fold higher irradiance than that at which the plants had been growing, and still found an inverse correlation between starch and AP in soybean.

Increased pathway for diffusion of CO_2 and O_2 around large starch grains provides an attractive explanation for feedback inhibition by starch. RuDP carboxylase also has oxygenase activity.[54,55] Therefore, any increase in the diffusion pathway for CO_2 toward, and O_2 away from, the reaction sites would decrease net CO_2 fixation by decreasing the ratio of CO_2 to O_2 at the reaction sites.[55] Inhibition by this mechanism should be less in an atmosphere that is high in CO_2 and/or low in O_2. Mauney et al.[31] obtained a higher negative correlation between starch content and AP in cotton leaves when they measured AP in 330 vpm CO_2 than when they measured AP in 630 vpm CO_2. Even though starch grains should interfere with CO_2 diffusion in both atmospheres, the concentration of CO_2 should be higher at the reactive sites in an atmosphere of 630 vpm CO_2 than it would in one of 330 vpm CO_2. Thus, CO_2 concentration should be less limiting at 630 vpm, and a lower correlation coefficient should be observed than at 330 vpm CO_2. Because the CO_2 compensation concentration was not affected by starch content in their experiment, Nafziger and Koller[28] suggested that starch increased r_m by increasing the diffusion-path length, or by decreasing cytoplasmic streaming, rather than by a direct biochemical effect.

Large starch grains may cause physical damage to chloroplasts. Noble and Craig[52] correlated rates of AP with ultrastructural changes in soybean chloroplasts. At high-light intensity they observed a progressive enlargement of chloroplasts due to increasing size and number of starch grains. By late afternoon, these chloroplasts showed a deterioration of thylakoid membranes and a virtual absence of intact microbodies. These changes correlated with a decline in AP. Smith and Struckmeyer[56] reported that chloroplasts could not readily be identified in alfalfa leaves in which starch had accumulated to 40.3% (dry weight) at low temperature. On the basis of a change in packing volume and leakage of chlorophyll, West[53] postulated that chloroplasts of pangolagrass and *Pennisetum typhoides* (Burm. f. Stapf and C. E. Hubb.) were damaged by starch grains that accumulated with cool nights.

B. Hormonal Control

Some evidence indicates that sinks, such as roots and fruits, may regulate AP in source leaves by hormonal signals.

1. Indoleacetic Acid

Bidwell and Turner,[57] Bidwell et al.,[58] and Tamas and coworkers[59,60] obtained evidence that IAA stimulated photophosphorylation and CO_2 fixation in isolated chloroplasts and intact leaves. The short time required for a measurable stimulation of photophosphorylation (5 min) indicated that the effect was direct, rather than through stimulation of enzyme synthesis. Also, the optimum concentration (10^{-7} *M*) was a physiological, rather than an abnormally high, concentration. Therefore, it seems possible that IAA produced in developing fruits could stimulate AP in source leaves if the IAA is transported from the fruits to the leaves. However, King et al.[3] were unable to find evidence of such stimulation in wheat. They applied tri-iodobenzoic acid (TIBA) as an injection or in a lanolin ring. This treatment blocked the movement of applied IAA, but did not affect AP of flag leaves.

2. *Cytokinins*

Wareing et al.[61] reported that partial defoliation 3 days before measurement stimulated AP, compared to partial defoliation immediately before measurement. The increased AP was evidently due to increased RuDP-carboxylase activity in bean and to increased phosphoenolpyruvate-carboxylase activity in maize. This increased enzyme activity was accompanied by a general increase in protein content of remaining leaves after partial defoliation. Root pruning prevented the increase in AP by partial defoliation, but application of kinetin partially compensated for root pruning. Spraying the leaves with kinetin stimulated AP, both in root-pruned and in control plants. A considerable body of evidence indicates that cytokinins are produced in roots.[62] Wareing et al.[61] concluded that, "partial defoliation not only increases the relative demand for photosynthates on the remaining leaves, but it also increases the photosynthetic efficiency of the latter by reducing the competition between leaves for mineral nutrients and possibly also for specific hormonal factors, such as cytokinins, supplied by the roots."

3. *Gibberellins*

Spraying with GA_3 stimulated RuDP-carboxylase activity and AP in peas, and the effect lasted at least 4 days.[63] It also increased RuDP-carboxylase activity and slightly increased AP in a nonflowering mutant of red clover (*Trifolium pratense* L.).[64] In both cases, the authors concluded that GA caused a general stimulation of protein synthesis, rather than a specific increase in RuDP carboxylase.[63,64] Unlike IAA, rather high concentrations of GA were used (10 and 100 ppm with red clover and $7.5 \times 10^{-4} M$ with peas). The site of GA synthesis has not been established with certainty, but roots appear to be an important source.[62] Therefore, cytokinins and gibberellins both appear to stimulate AP via a general stimulation of enzyme synthesis, and both are supplied by roots. Therefore, root sinks could regulate photosynthesis in source leaves through cytokinins, gibberellins, or both.

4. *Abscisic and Phaseic Acids*

Fruiting apple trees had higher rates of AP than trees without fruits.[65] A subsequent test[66] indicated that one reason for the higher AP was greater stomatal aperture in leaves on fruiting trees. Hansen[66] suggested that kinetin and other hormones may have affected stomatal aperture, but he did not measure hormonal concentrations nor suggest why they should be different. Defruiting may have affected the abscisic acid (ABA) and phaseic acid (PA) contents of his apple leaves. Loveys and Kriedemann[67] and Kriedemann et al.[68] reported that fruit removal and stem girdling increased the concentrations of ABA and, especially, PA in grape (*Vitis vinifera* L.) leaves. Stomatal resistance increased from 1.41 to 7.14 s/cm within 7 days after fruit removal and stem girdling. In addition to stomatal effects, Kriedemann et al.[68,69] found that PA had a strong nonstomatal inhibitory effect on photosynthesis. Thus, in contrast to IAA, cytokinins, and GA, all of which stimulated photosynthesis, ABA and PA inhibited it. Why fruit removal and stem girdling should cause increases in ABA and PA content of leaves is an intriguing question. If accumulation of sugars were shown to stimulate synthesis of these inhibitory hormones, another possible mechanism for end product inhibition of photosynthesis would be provided.

IV. DISCUSSION

Demonstration of end-product inhibition of photosynthesis is difficult. Demonstration of a negative correlation between photosynthesis and concentration of assimilates is complicated by the fact that assimilates are products of photosynthesis. Therefore, higher rates of photosynthesis should cause the production of more assimilates and result in a positive, rather than a negative, correlation. Thus, failure to demonstrate a negative correlation does not prove the absence of end-product inhibition when accumulation of assimilates and measurement of AP are not separated in time.

The use of low temperatures to alter sink demand or assimilate concentration can cause changes in leaf water potential, particularly if the roots are cooled. Root temperature affects permeability and rate of water uptake.[70] When roots are chilled, they may not be able to absorb water fast enough to maintain leaf turgor. Therefore, stomatal aperture and photosynthesis may be decreased. The ABA content increases rapidly in stressed leaves,[71] and its presence may delay stomatal opening after roots are returned to normal temperature. Because PA is a metabolite of ABA, PA may also increase during a period of water deficit in leaves. The stomatal effects of ABA, and the nonstomatal inhibition of photosynthesis by PA,[67-69] could cause inhibition of AP for some time after temperatures are returned to normal. Crookston et al.[72] found that exposure of bean (*Phaseolus vulgaris* L.) plants to one cold (5°C) night increased starch content and depressed AP, but they concluded that photosynthesis was depressed by temporary water stress rather than by starch. Drake and Salisbury[73] reported that low temperature pretreatment influenced stomatal aperture in *Xanthium strumarium* L. leaves. The depression of AP reported by Carter and Garrard,[20] after subjecting pangolagrass to a 12-hr night at 10°C, may have been at least partly due to the effects of water deficit on stomatal aperture. A 10°C night caused a 26% decrease in AP when assimilate was depleted before the 10°C treatment, and a 36% decrease in AP when it was not depleted. Therefore, most of the depression in AP was due to temperature effects not related to accumulation of assimilate. Cooling the sinks can also affect their rate of production of hormones (cytokinins and gibberellins in roots, IAA in fruits). Thus, a decrease in AP after sinks are cooled does not prove end-product inhibition of photosynthesis, even when assimilate content of source leaves increases. Likewise, the removal of sinks or source leaves can alter the hormonal balance of remaining source leaves as well as their assimilate content.

The use of varying concentrations of CO_2 to alter starch content may affect photosynthesis via stomatal effects, since it is known that CO_2 enrichment causes partial stomatal closure. Nafziger and Koller[28] and Mauney et al.[31] attempted to circumvent this effect by allowing 30 min and 2 hr, respectively, for stomatal adjustment, after transferring plants to different concentrations of CO_2 before they determined AP.

Different species of plants respond differently to treatments designed to demonstrate end-product inhibition of photosynthesis. Some species accumulate starch readily, whereas others do not. For example, sugar beet, sugar cane, and sorghum do not accumulate much starch, and show no evidence of end-product inhibition by starch. Other species, such as alfalfa, cotton, pangolagrass, soybean, and tomato, do accumulate starch and do show evidence of end-product inhibition. Alfalfa and pangolagrass can apparently accumulate inhibitory amounts under natural conditions. Cotton, soybean, and tomato, on the other hand, require CO_2 enrichment to show statistically significant evidence of end-product inhibition. Limited evidence[28,31] indicates that a threshold value of about 200 mg of starch per gram dry weight of leaves is required before AP shows any evidence of end-product inhibition in soybean and cotton. Our results[31,74] indicate that prime cotton leaves in full sunlight and normal air increased from a low of about 50 to a high of about 150 mg of starch per gram dry weight of leaves during each day. The CO_2 enrichment increased the rate of starch synthesis but not breakdown; so starch accumulated. Starch content of older, shaded leaves on field-grown plants was much lower.[75] Thus, only young leaves in full sunlight are likely to accumulate enough starch to inhibit AP, and for only a portion of each day. If predicted increases in CO_2 content of the atmosphere occur,[30] starch may accumulate to superoptimal levels in several species. However, estimating the percentage of leaf area affected and the number of hours each day that starch is at an inhibitory level for each leaf will not be easy.

The following are suggestions for overcoming some of the problems of interpretation of experimental results on end-product inhibition:

1. Experiments should be conducted in such a way as to separate, in time, the accumulation of assimilates and the measurement of AP.
2. Experiments should be designed to minimize changes in water relations and hormonal status. To the extent that such confounding effects occur, they should be recognized and measured, if possible.
3. Results should not be extrapolated from one species to another.
4. Results should not be extrapolated from young leaves in full sunlight near the end of a long photoperiod to all leaves during the entire day.

V. CONCLUSIONS

Results of some of the experiments reviewed leave little doubt that rates of photosynthesis can be controlled by the balance between supply of, and demand for, photosynthates in some species. Some of the results indicate control by end-product inhibition (principally by starch), whereas others indicate hormonal control. Stimulation of photosynthesis by developing fruits could result from either type of control. It is conceivable that both types of regulation might occur in the same plant, either independently or as primary and secondary events. The questions that remain to be answered are (1) Is photosynthesis in all crop plants regulated by the balance between supply and demand? and (2) What mechanisms serve to effect such regulation? Much more work is needed on hormonal control, both as a primary control mechanism and as a possible complicating factor in experiments designed to test for end-product inhibition. Considering the possible effects and complex interactions of IAA, cytokinins, and GA (stimulatory); ABA and PA (inhibitory); and assimilates such as sugars and starch (inhibitory); it is not surprising that results appear contradictory and confusing.

Regardless of the kinds of controls, it is important for the modeler to have an estimate of the variation in rates of photosynthesis that can occur in the species with which he works, and the conditions that can cause variations. It is obvious that much more information is needed before numerical values can be assigned to all important crop species and to all conditions.

VI. SYMBOLS

ABA — abscisic acid

AP — apparent photosynthesis measured by CO_2 uptake, O_2 release, $^{14}CO_2$ incorporation, or increases in dry weight

Assimilate (n) — products of photosynthetic carbon fixation

Assimilate (v) — the process of photosynthetic carbon fixation

IAA — indoleacetic acid

PA — phaseic acid

r_m — mesophyll resistance

r_s — stomatal resistance

RuDP — ribulose-1,5-diphosphate

SLW — specific leaf weight, e.g., g/dm^2
Sink — a plant part that is a net importer of photosynthate
Source — a plant part that is a net exporter of photosynthate
vpm — volumes per million, e.g., ul/l

REFERENCES

1. **Neales, T. F. and Incoll, L. D.,** The control of leaf photosynthesis rate by the level of assimilate concentration in the leaf: a review of the hypothesis, *Bot. Rev.*, 34, 107, 1968.
2. **Warren Wilson, J.,** Control of crop processes, in *Crop Processes in Controlled Environments,* Rees, A. R., Cockshull, K. E., Hand, D. W., and Hurd, R. G., Eds., Academic Press, London, 1972, 7.
3. **King, R. W., Wardlaw, I. F., and Evans, L. T.,** Effect of assimilate utilization on photosynthetic rate in wheat, *Planta*, 77, 261, 1967.
4. **Austin, R. B. and Edrich, J.,** Effects of ear removal on photosynthesis, carbohydrate accumulation and on the distribution of assimilated ^{14}C in wheat, *Ann. Bot. (London)*, 39, 141, 1975.
5. **Barlow, E. W. R. and Boersma, L.,** Interaction between leaf elongation, photosynthesis, and carbohydrate levels of water-stressed corn seedlings, *Agron. J.*, 68, 923, 1976.
6. **Chalmers, D. J., Canterfore, R. L., Jerie, P. H., Jones, T. R., and Ugalde, T. D.,** Photosynthesis in relation to growth and distribution of fruit in peach trees, *Aust. J. Plant Physiol.*, 2, 635, 1975.
7. **Claussen, W.,** Einfluss der Frucht auf Netto-Assimilationsleistung und Netto-Photosyntheseraten der Aubergine (*Solanum melongena* L.), *Gartenbauwissenschaft*, 42, 61, 1977.
8. **Flinn, A. M.,** Regulation of leaflet photosynthesis by developing fruit in the pea, *Physiol. Plant.*, 31, 275, 1974.
9. **Habeshaw, D.,** Translocation and the control of photosynthesis in sugar beet, *Planta*, 110, 213, 1973.
10. **Lenz, V. F.,** Einfluß der Frucht auf Photosynthese und Atmung, *Z. Pflanzenernachr. Bodenkd.*, 140, 51, 1977.
11. **Thorne, J. H. and Koller, H. R.,** Influence of assimilate demand on photosynthesis, diffusive resistances, translocation, and carbohydrate levels of soybean leaves, *Plant Physiol.*, 54, 201, 1974.
12. **Nagarajah, S.,** Effect of debudding on photosynthesis in leaves of cotton, *Physiol. Plant.*, 33, 28, 1975.
13. **Geiger, D. R.,** Effects of translocation and assimilate demand on photosynthesis, *Can. J. Bot.*, 54, 2337, 1976.
14. **Carmi, A. and Koller, D.,** Endogenous regulation of photosynthetic rate in primary leaves of bean (*Phaseolus vulgaris* L.), *Ann. Bot.*, 41, 59, 1977.
15. **Liu, P., Wallace, D. H., and Ozbun, J. L.,** Influence of translocation on photosynthetic efficiency of *Phaseolus vulgaris* L., *Plant Physiol.*, 52, 412, 1973.
16. **Chatterton, N. J., Carlson, G. E., Hungerford, W. E., and Lee, D. R.,** Effect of tillering and cool nights on photosynthesis and chloroplast starch in Pangola, *Crop Sci.*, 12, 206, 1972.
17. **Wilson, D.,** Starch and apparent rate of photosynthesis of leaves of *Lolium perenne* grown at different temperatures, *Planta*, 91, 274, 1970.
18. **Wilson, D. and Cooper, J. P.,** Effect of light intensity and CO_2 on apparent photosynthesis and its relationship with leaf anatomy in genotypes of *Lolium perenne* L., *New Phytol.*, 68, 627, 1969.
19. **Wilson, D. and Cooper, J. P.,** Apparent photosynthesis and leaf characters in relation to leaf position and age, among contrasting *Lolium* genotypes, *New Phytol.*, 68, 645, 1969.
20. **Carter, J. L. and Garrard, L. A.,** Effects of 10 C-night temperature and leaf assimilate levels on photosynthesis of *Digitaria decumbens* Stent., *Soil Crop Sci. Soc. Fla. Proc.*, 35, 3, 1976.
21. **Austin, R. B.,** The relationship between dry matter increment and sugar concentrations in beetroot leaves, *Photosynthetica*, 6, 123, 1972.
22. **Milford, G. F. J. and Pearman, I.,** The relationship between photosynthesis and the concentrations of carbohydrates in the leaves of sugar beet, *Photosynthetica*, 9, 78, 1975.
23. **Garrard, L. A. and Carter, J. L.,** Leaf assimilate level and photosynthesis of *Digitaria decumbens* Stent., *Soil Crop Sci. Soc. Fla. Proc.*, 35, 6, 1976.
24. **Forde, B. J., Whitehead, H. C. M., and Rowley, J. A.,** Effect of light intensity and temperature on photosynthetic rate, leaf starch content and ultrastructure of *Paspalum dilatatum*, *Aust. J. Plant Physiol.*, 2, 185, 1975.

25. **Waldron, J. C., Glasziou, K. T., and Bull, T. A.**, The physiology of sugar-cane. IX. Factors affecting photosynthesis and sugar storage, *Aust. J. Biol. Sci.*, 20, 1043, 1967.
26. **Moldau, H. and Karolin, A.**, Effect of the reserve pool on the relationship between respiration and photosynthesis, *Photosynthetica*, 11, 38, 1977.
27. **Hofstra, G. and Hesketh, J. D.**, The effects of temperature and CO_2 enrichment on photosynthesis in soybean, in *Environment and Biological Control of Photosynthesis*, W. Junk, the Hague, 1975, 71.
28. **Nafziger, E. D. and Koller, H. R.**, Influence of leaf starch concentration on CO_2 assimilation in soybean, *Plant Physiol.*, 57, 560, 1976.
29. **Akita, S. and Tanaka, I.**, Studies on the mechanism of differences in photosynthesis among species IV. The differential response in dry matter production between C_3 and C_4 species to atmospheric carbon dioxide enrichment, *Proc. Crop Sci. Soc. Japan*, 42, 288, 1973.
30. **Baes, C. F., Jr., Goeller, H. E., Olson, J. S., and Rotty, R. M.**, Carbon dioxide and climate: the uncontrolled experiment, *Am. Scientist*, 65, 310, 1977.
31. **Mauney, J. R., Guinn, G., Fry, K. E., and Hesketh, J. D.**, Correlation of photosynthetic carbon dioxide uptake and carbohydrate accumulation in four crop species, *Photosynthetica*, in press.
32. **Ito, T.**, Plant growth and physiology of vegetable plants as influenced by carbon dioxide environment, Transactions of the Faculty of Horticulture No. 7, Chiba University, No. 7, Japan, 1973.
33. **Madens, E.**, Cytological changes due to the effect of carbon dioxide concentration on the accumulation of starch in chloroplasts of tomato leaves, Royal Veterinary and Agricultural University Yearbook, Copenhagen, 1971, 191.
34. **Madsen, E.**, The effect of carbon dioxide concentration on the photosynthetic rate in tomato leaves, Royal Veterinary and Agricultural University Yearbook, Copenhagen, 1971, 195.
35. **Madsen, E.**, Effect of CO_2 concentration on growth and fruit production of tomato plants, *Acta Agric. Scand.*, 24, 242, 1974.
36. **Moore, K. G., Illsley, A., and Lovell, P. H.**, Effects of sucrose on petiolar carbohydrate accumulation and photosynthesis in excised *Sinapis* cotyledons, *J. Exp. Bot.*, 25, 887, 1974.
37. **Natr, L. and Ludlow, M. M.**, Influence of glucose absorption and photosynthate accumulation on gas exchange of barley leaf segments, *Photosynthetica*, 4, 288, 1970.
38. **Natr, L., Watson, B. T., and Weatherley, P. E.**, Glucose absorption, carbohydrate accumulation, presence of starch, and rate of photosynthesis in barley leaf segments, *Ann. Bot.*, 38, 589, 1974.
39. **Bouma, D.**, Effects of some metabolic phosphorus compounds on rates of photosynthesis of detached phosphorus-deficient subterranean clover leaves, *J. Exp. Bot.*, 26, 52, 1975.
40. **Bowes, G. and Ogren, W. L.**, Oxygen inhibition and other properties of soybean ribulose-1,5,-diphosphate carboxylase, *J. Biol. Chem.*, 247, 2171, 1972.
41. **Buchanan, B. B., and Schürmann, P.**, Regulation of ribulose-1,5-diphosphate carboxylase in the photosynthetic assimilation of carbon dioxide, *J. Biol. Chem.*, 248, 4956, 1973.
42. **Chatterton, N. J.**, Product inhibition of photosynthesis in alfalfa leaves as related to specific leaf weight, *Crop Sci.*, 13, 284, 1973.
43. **Upmeyer, D. J. and Koller, H. R.**, Diurnal trends in net photosynthetic rate and carbohydrate levels of soybean leaves, *Plant Physiol.*, 51, 871, 1973.
44. **Little, C. H. A. and Loach, K.**, Effect of changes in carbohydrate concentration on the rate of net photosynthesis in mature leaves of *Abies balsamea*, *Can. J. Bot.*, 51, 751, 1973.
45. **Fraser, D. E. and Bidwell, R. G. S.**, Photosynthesis and photorespiration during the ontogeny of the bean plant, *Can. J. Bot.*, 52, 2561, 1974.
46. **Herold, A., Lewis, D. H., and Walker, D. A.**, Sequestration of cytoplasmic orthophosphate by mannose and its differential effect on photosynthetic starch synthesis in C_3 and C_4 species, *New Phytol.*, 76, 397, 1976.
47. **Sheu-hwa, C., Lewis, D. H., and Walker, D. A.**, Stimulation of photosynthetic starch formation by sequestration of cytoplasmic orthophosphate, *New Phytol.*, 74, 383, 1975.
48. **Terry, N. and Ulrich, A.**, Effects of phosphorus deficiency on the photosynthesis and respiration of leaves of sugar beet, *Plant Physiol.*, 51, 43, 1973.
49. **Kelly, G. T., Latzko, E., and Gibbs, M.**, Regulatory aspects of photosynthetic carbon metabolism, *Annu. Rev. Plant Physiol.*, 27, 181, 1976.
50. **Hawker, J. S.**, Inhibition of sucrose phosphatase by sucrose, *Biochem. J.*, 102, 401, 1967.
51. **Klemme, B. and Jacobi, G.**, Der Einfluβ von Stärke auf di Aktivitat der Pyrophosphatase aus isolierten Chloroplasten, *Planta*, 120, 155, 1974.
52. **Noble, R. D. and Crang, R. E.**, Photoduration effects on photosynthesis and photosynthetic apparatus in soybean, *Plant Physiol.*, 51(S), 28(155), 1973.
53. **West, S. H.**, Carbohydrate metabolism and photosynthesis of tropical grasses subjected to low temperatures, in *Plant Response to Climatic Factors*, Proc. Uppsala Symp., 1970, 165.

54. **Jensen, R. G. and Bahr, J. T.**, Ribulose 1,5-bisphosphate carboxylase-oxygenase, *Annu. Rev. Plant Physiol.*, 28, 379, 1977.
55. **Raven, J. A.**, Ribulose bisphosphate carboxylase activity in terrestrial plants: significance of O_2 and CO_2 diffusion, *Current Adv. Plant Sci.*, 9, 579, 1977.
56. **Smith, D. and Struckmeyer, B. E.**, Gross morphology and starch accumulation in leaves of alfalfa plants grown at high and low temperatures, *Crop Sci.*, 14, 433, 1974.
57. **Bidwell, R. G. S. and Turner, W. B.**, Effect of growth regulators on CO_2 assimilation in leaves, and its correlation with the bud break response in photosynthesis, *Plant Physiol.*, 41, 267, 1966.
58. **Bidwell, R. G. S., (Turner) Levin, W. B., and Tamas, I. A.**, The effects of auxin on photosynthesis and respiration, in *Biochemistry and Physiology of Plant Growth Substances*, Wightman, F. and Setterfield, G., Eds., Runge Press, Ottawa, 1968, 361.
59. **Tamas, I. A., Atkins, B. D., Ware, S. M., and Bidwell, R. G. S.**, Indoleacetic acid stimulation of phosphorylation and bicarbonate fixation by chloroplast preparations in light, *Can. J. Bot.*, 50, 1523, 1972.
60. **Tamas, I. A., Schwartz, J. W., Hagin, J. M., and Simmonds, R.**, Hormonal control of photosynthesis in isolated chloroplasts, in *Mechanisms of Regulation of Plant Growth*, Bieleski, R. L., Ferguson, A. R., and Cresswell, M. M., Eds., Bulletin 12, The Royal Society of New Zealand, Wellington, 1974, 261.
61. **Wareing, P. F., Khalifa, M. M., and Treharne, K. T.**, Rate-limiting processes in photosynthesis at saturating light intensities, *Nature*, 220, 453, 1968.
62. **Torrey, J. G.**, Root hormones and plant growth, *Annu. Rev. Plant Physiol.*, 27, 435, 1976.
63. **Broughton, W. J., Hellmuth, E. O., and Yeung, D.**, Role of glucose in development of the gibberellin response in peas, *Biochim. Biophys. Acta*, 222, 491, 1970.
64. **Treharne, K. J. and Stoddart, J. L.**, Effects of gibberellin on photosynthesis in red clover (*Trifolium pratense* L.), *Nature*, 220, 457, 1968.
65. **Hansen, P.**, ^{14}C-Studies on apple trees. VI. The influence of the fruit on photosynthesis of the leaves, and the relative photosynthetic yields of fruits and leaves, *Physiol. Plant.*, 23, 805, 1970.
66. **Hansen, P.**, The effect of fruiting upon transpiration rate and stomatal opening in apple leaves, *Physiol. Plant.*, 25, 181, 1971.
67. **Loveys, B. R. and Kriedemann, P. E.**, Internal control of stomatal physiology and photosynthesis. I. Stomatal regulation and associated changes in endogenous levels of abscisic and phaseic acids, *Aust. J. Plant Physiol.*, 1, 407, 1974.
68. **Kriedemann, P. E., Loveys, B. R., Possingham, J. V., and Satoh, M.**, Sink effects on stomatal phyisology and photosynthesis, in *Transport and Transfer Processes in Plants*, Wardlaw, I. F. and Passioura, J. B., Eds., Academic Press, New York, 1976, 401.
69. **Kriedemann, P. E., Loveys, B. R., and Downton, J. S.**, Internal control of stomatal physiology and photosynthesis, II. Photosynthetic responses to phaseic acid, *Aust. J. Plant Physiol.*, 2, 553, 1975.
70. **Kramer, P. J.**, *Plant and Soil Water Relationships*, McGraw-Hill, New York, 1949, Chap. 9.
71. **Hiron, R. W. P. and Wright, S. T. C.**, The role of endogenous abscisic acid in the response of plants to stress, *J. Exp. Bot.*, 24, 769, 1973.
72. **Crookston, R. K., O'Toole, J., Lee, R., Ozbun, J. L., and Wallace, D. H.**, Photosynthetic depression in beans after exposure to cold for one night, *Crop Sci.*, 14, 457, 1974.
73. **Drake, B. G. and Salisbury, F. B.**, Aftereffects of low and high temperature pretreatment on leaf resistance, transpiration, and leaf temperature in *Xanthium*, *Plant Physiol.*, 50, 572, 1972.
74. Unpublished results.
75. **Guinn, G.**, Unpublished.

Chapter 2

MODELING C_3 LEAF RESPIRATION IN THE LIGHT

J. D. Tenhunen, J. D. Hesketh, and P. C. Harley

TABLE OF CONTENTS

I. CHARACTERISTICS OF PHOTORESPIRATION AS DETECTED IN THE NET PHOTOSYNTHESIS RESPONSE

The "discovery" of photorespiration or CO_2 release in the light from a source other than mitochondrial respiration is relatively recent, and details of the process have been elucidated only in the past decade. It is now apparent that photorespiration played a significant role in the Warburg-Emerson controversy over quantum yields, and, in Rabinowitch's words, "the possibility of such an effect is a nightmare oppressing all who are concerned with the exact measurement of photosynthesis."[1] Early evidence of CO_2 release in the light, differing from mitochondrial or dark respiration, was obtained by Decker,[2] whose CO_2-response curves for *Mimulus*, at a range of temperatures, are shown in Figure 1. According to his data, respiration in the light in CO_2-free air (which is actually a net photosynthesis rate) was greater than dark respiration in either ambient or CO_2-free air. Decker[3] had earlier observed the inexplicable phenomenon of a burst of CO_2 release immediately after a period in the light, a phenomenon now readily explained as a result of photorespiration. Before Decker's work, which helped to draw attention to the photorespiratory phenomenon, there already existed a large but unrelated data base, much of which is discussed in Rabinowitch[1] under the key phrases of photorespiration, light respiration, C_4 metabolism, and O_2 effects.

In the early 1960s, it was discovered that high rates of CO_2 evolution into CO_2-free air were characteristic only of certain species.[4,5] Thus, two distinct groups of plants were described, and designated C_3 or C_4, according to the number of C atoms in the initial product of carboxylation.[6,7] These two groups were found to differ in several respects. The C_4 plants, which failed to release CO_2 into CO_2-free air or exhibit a burst of CO_2 after a light period, had much lower light and CO_2-compensation points than C_3 plants and much higher photosynthesis rates at ambient CO_2 and O_2 concentrations (e.g., maize in Figure 2). All these characteristics suggested that C_4 plants apparently lacked light-dependent respiration or photorespiration, particularly at high temperatures. Differences in leaf anatomy were found to correlate with these differences in gas-exchange characteristics.[8]

Net photosynthesis of C_3 plants, when measured at low O_2 concentrations (approximately 2%), was found to behave like that of C_4 plants, i.e., photorespiration was inhibited.[9-15] As is shown in Figure 3, CO_2 evolution into CO_2-free air is stimulated by O_2 in C_3 plants. But in addition, when net photosynthesis in C_3 plants was measured as a function of increasing O_2 concentration, the initial slopes of the CO_2-response curves were found to decrease, contributing to the inhibition of net photosynthesis under ambient conditions. Analysis of CO_2-response curves based on CO_2 concentrations outside (C_{CO_2}) and inside the leaf adjacent to the cell walls (C_{cw})[9-20] indicated that the O_2 effect on the initial slope was responsible for between 60 and 80% of the total inhibition of net photosynthesis in ambient air, as compared with net photosynthesis in O_2-depleted air.

Thus, it was these comparisons and the C_3-C_4 concept that served to bring together the large body of unrelated data mentioned by Rabinowitch,[1] and synthesize a concept of the behavior of photorespiration. Furthermore, the discovery of C_4 photosynthesis, and the realization that photorespiration might be a serious limiting factor to growth, stimulated increasing study of the biochemical nature of the process and its subsequent linking to the metabolism of glycolate.[21] In addition, hypotheses were developed concerning the relative advantages of C_3 and C_4 metabolism, particularly with respect to water-use efficiency.[22] Although C_3 and C_4 species of the same genus can be hybridized, yielding progeny intermediate in photosynthetic behavior,[23-27] and although some

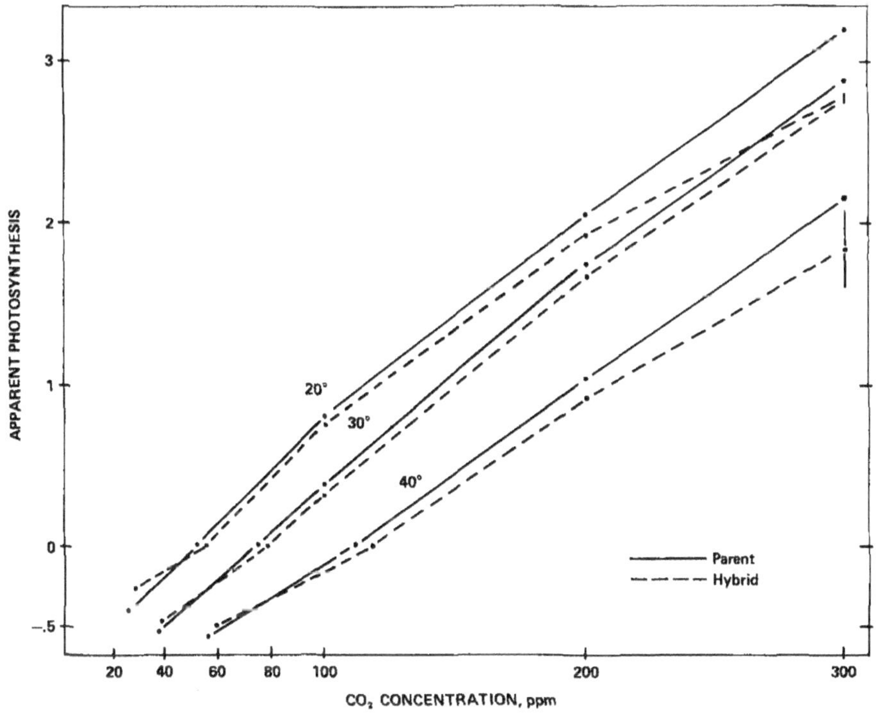

FIGURE 1. Effects of air temperature and of CO_2 concentration on apparent photosynthesis of intact leaves of *Mimulus*. Illumination was 2000 fc. Ordinal unit: $\mu g\ CO_2/dm^2/sec$. (Elevated light respiration is indicated by extrapolating the responses to zero carbon dioxide and comparing this value to dark respiration. Dark respiration was measured in $\mu g\ CO_2/dm^2/sec$ to be 0.15 to 0.22 at 20°C, 0.29 to 0.39 at 30°C, and 0.46 to 0.63 at 40°C. (From Decker, J., *Plant Physiol.*, 34, 103, 1959. With permission.)

tropical grasses are apparently intermediate in behavior in nature,[28, 29] it has not yet proven possible to genetically manipulate to advantage the characteristics of photorespiration in crop plants or to breed photorespiration out of C_3 plants.

Early attempts at modeling photorespiration focused on the change in the initial slope of the CO_2-response curves at different O_2 concentrations, alluded to above. Forrester et al.[9] and Tregunna et al.[10] denoted the initial slope as the carboxylation efficiency (CE), and reported its logarithmic relation to O_2 concentration. Since, at low CO_2 concentrations the photosynthesis response is a linear function of CO_2, the following equation was applied:[9]

$$J_{CO_2} = CE_o\, e^{-k[O]}\, (C_{CO_2} - u[O]) \qquad (1)$$

where k was the slope of the line relating ln CE to O_2 concentration, u was the slope of the CO_2 compensation concentration vs. O_2 concentration, and CE_o was the carboxylation efficiency at low O_2 concentration. While Equation 1 adequately describes net photosynthesis at low CO_2 concentrations, it is desirable to utilize parameters which are more meaningful physiologically. Only then can one assess the meaning of a change in parameter values. Attempts have been made to relate the parameters of Equation 1 to biochemical kinetic parameters,[14] but this was made difficult because the J_{CO_2} vs.

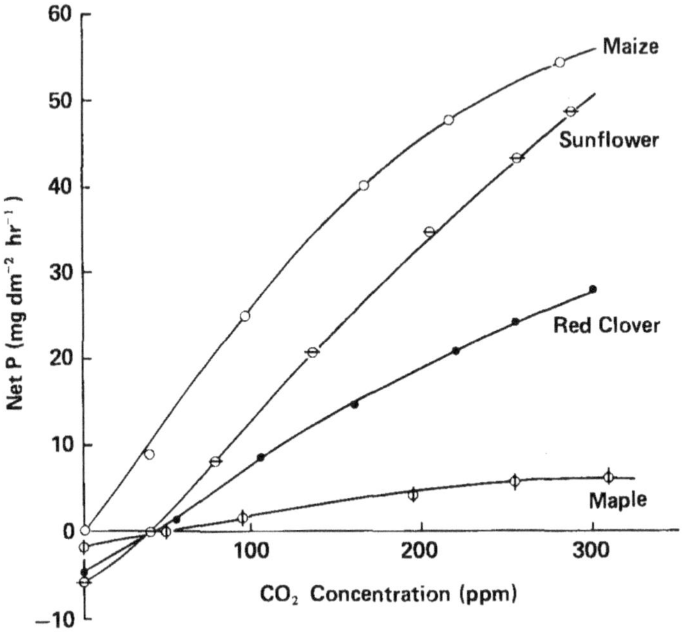

FIGURE 2. Relation to CO_2 concentration of the net phtosynthesis rate (net P) of 4 species at 2.0 ly/min, illustrating the greater photosynthesis rates and lower CO_2 compensation point in C_4 plants (maize) versus C_3 plants (sunflower, red clover, and maple). (From Hesketh, J., *Crop Sci.*, 3, 493, 1963. With permission.)

C_{cw} data obtained initially were distributed over a much too narrow range of CO_2 concentrations to reveal the effects of kinetic parameters. Ku and Edwards[19,20] continued the effort and obtained new data in the nonlinear portion of the CO_2-response curves. They attempted to apply a theoretical analysis based on biochemical kinetic theory. The proper analysis to be used, however, is still being debated, as will be discussed below.

The development of the Bowes-Ogren hypothesis in the early 1970s has had a dramatic impact on photosynthesis modeling. According to this hypothesis, which forms the basis of the modeling effort described below, RuDP carboxylase-oxygenase is capable of catalyzing both the carboxylation and the oxygenation of RuDP, yielding, in the first case, two molecules of phosphoglycerate and, in the second, one molecule of phosphoglycerate and one 2-C molecule of glycolate. The glycolate is subsequently metabolized via the glycolate pathway, an intermediate in that pathway being decarboxylated. Among those who support the idea that this is the major source of glycolate in the cell, the oxygenation of RuDP, as well as the decarboxylation, may be implied by the somewhat imprecise term of photorespiration. Based on this hypothesis, the overall effect of photorespiration on net photosynthesis may be attributed to: (1) a decrease in the concentration of acceptor-enzyme complex (EA in Figure 13 of Chapter 6, Volume I) available to CO_2, (2) an increase in the amount of reductant required for recycling phosphoglyceraldehyde and some of the glycolate, and for refixation of photorespired CO_2, and (3) diffusion of photorespired CO_2 out of the leaf. Earlier models of net photosynthesis that ignored the role of these factors are now obsolete. However, mathematical expression of the photorespiration function of a leaf is neither simple nor straightforward. The exact interpretation and underlying assumptions in the Bowes-Ogren hypothesis must still be resolved. It must still be determined whether the hypothesis validly describes photorespiration in the leaf.

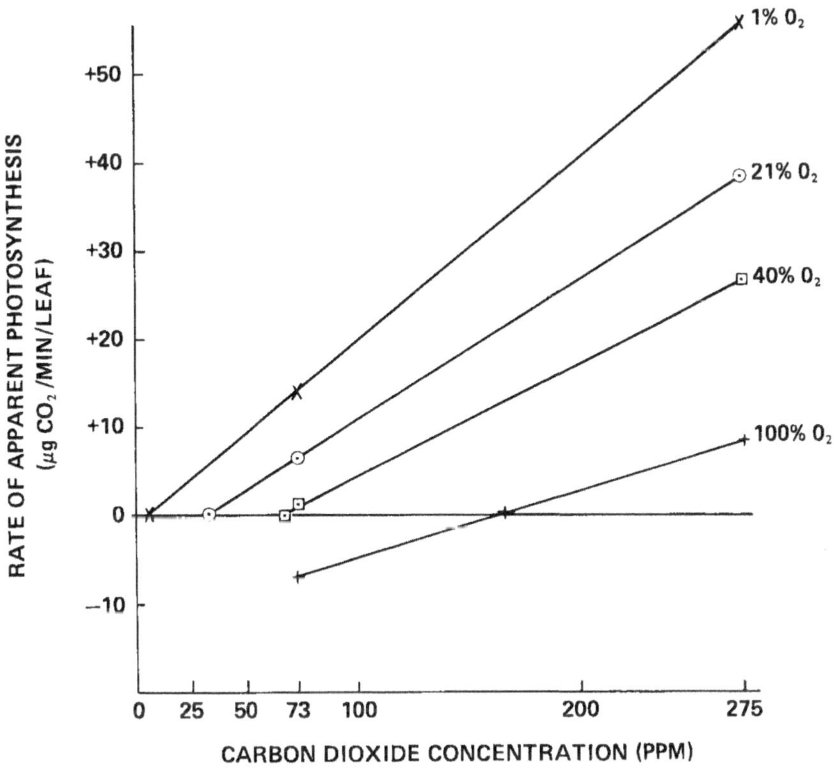

FIGURE 3. Effect of CO_2 on the rate of apparent photosynthesis at various O_2 concentrations in detached soybean (C_3) leaves. (From Forrester, M., Krotkov, G., and Nelson, C., *Plant Physiol.*, 41, 422, 1966. With permission.)

The Bowes-Ogren hypothesis has important implications for those seeking to regulate plant productivity by reducing the effects of photorespiration, either by altering the pathways of glycolate metabolism or by some means breeding out photorespiration. Those who support the Bowes-Ogren hypothesis have linked the formation of glycolate to the oxygenation reaction of RuDP carboxylase-oxygenase.[30-35] Others[36, 37] maintain that multiple pathways of glycolate formation exist, and that there are control points other than that at RuDP carboxylase-oxygenase. The supporters of this view feel that genetic modification of these controls may result in plants more efficient in net photosynthesis.[38] Those who support the idea of competitive inhibition at RuDP carboxylase-oxygenase emphasize the intimate coupling of photosynthesis and photorespiration,[39, 40] and suggest that modifications of photorespiration, apart from photosynthesis, may not be possible. Decarboxylation of glycolate seems to be responsible for the CO_2 evolved in photorespiration, but the reactions involved in this decarboxylation and the location of these reactions are not yet resolved.[39]

Many of the early researchers of photorespiration and the C_3-C_4 phenomenon had difficulty in publishing their findings, and, once published, their work was slow to gain acceptance. Contributing experimentalists include J. Decker,[2] M. Forrester, B. Tregunna, G. Krotkov, and C. Nelson,[9, 10] H. Kortshak, C. Hartt and G. Burr,[6] I. Zelitch,[21] M. Hatch and R. Slack,[7] and Y. Murata and J. Iyama.[41] Direct comparisons of gas exchange and carbon fixation of C_3 and C_4 plants were necessary before any synthesis of ideas took place. Some reviewers seemed to imply that scientists have been studying the C_3-C_4 phenomenon for years, when, in fact, the first papers on the subject appeared in the early 1960s. The C_3-C_4 phenomenon was certainly one of the most interesting photosynthetic discoveries of the past century, and the dedication of the pioneers in the face of ridicule was remarkable.

II. MEASUREMENT OF PHOTORESPIRATION

More complete understanding of the photorespiratory process is made exceedingly difficult by the inability to measure the process directly, despite the utilization of a number of imaginative techniques.[42, 43] Measurements of gas exchange outside the leaf do not take into account photorespired CO_2, which is recycled internally. Thus, for example, photorespiration may exist in the bundle sheath cells of C_4 plants, but be hidden by an efficient recycling process.[44] Some techniques, utilizing radioactively labeled CO_2, attempt to circumvent this problem, but it remains impossible to measure the CO_2 concentration at the site of fixation (C_i).[45] Hence, it remains impossible to distinguish the direct effects of the competing reactions at RuDP carboxylase-oxygenase from those of diffusion ($R^{mes}_{CO_2}$, $R^{chl}_{CO_2}$, and R'_{CO_2}, see Chapter 6, Volume I). In analyzing the various techniques for measuring photorespiration, or in studying data obtained by such techniques, therefore, we should keep the following points in mind:

1. Labeled CO_2 or O_2 diffuses in and out of the leaf independently of the concentration of its unlabeled counterpart. Thus, the initial fixation rate of $^{14}CO_2$ in air of 21% O_2 at steady state, with respect to the total CO_2 concentration external to the leaf, approaches that of nonlabeled CO_2 in air of 1% O_2. The former rate, however, immediately begins a gradual decline to the normal fixation rate of CO_2 in air of 21% O_2, indicating that labeled $^{14}CO_2$ is rapidly relased in the decarboxylation step of the glycolate pathway, thereby slowing the flux of $^{14}CO_2$ into the leaf.[46] The initial fixation rate of $^{14}CO_2$ has been assumed to reveal the true photosynthesis rate (P_g). When this estimate of P_g is compared with the rate of net photosynthesis measured simultaneously, one obtains a concurrent rate of CO_2 evolution attributed to photorespiration.
2. According to the Bowes-Ogren hypothesis, the photorespired CO_2 is derived from RuDP, i.e., recent assimilate contributes the substrate of photorespiration. This has been the subject of several studies.[47-50] Estimates of photorespiratory activity, based on $^{14}CO_2$ uptake, must be obtained very rapidly if they are to be valid. No quantitative study has established just how rapidly evolved $^{14}CO_2$ will interfere with uptake, or determined the effect of stomatal and mesophyll resistances and filling of the air space and substrate pools on such determinations.
3. Large quantities of CO_2 diffuse out of the leaf into CO_2-free air, suggesting to some that the diffusion resistance between the site of photorespiratory CO_2 release and the intercellular air spaces is less than the diffusion resistance between the photorespiration site and the site of fixation. As discussed in Section III, however, other views are also compatible with this observation. To resolve the question, the accuracy of the $^{14}CO_2$ technique, discussed in 1. and 2., for estimating P_g and Q_p, must be established, and the magnitude of recycling of CO_2 must be assessed in some manner. Relying solely on data obtained outside the leaf will provide no new insight.

Thus, gas-exchange methods for estimating photorespiratory fluxes include measuring: (1) the rate of CO_2 evolution into CO_2-free air or the rate of $^{14}CO_2$ evolution into $^{14}CO_2$-free air of ambient unlabeled CO_2 concentration, (2) the burst of CO_2 or $^{14}CO_2$ evolution immediately after the termination of a light period, (3) the enhancement of photosynthesis in O_2-depleted air, (4) the uptake of $^{14}CO_2$ vs. CO_2 after first applying labeled C, and (5) the CO_2 compensation concentration in the light. All of these measurements are indicators of the photorespiratory phenomenon, but none measures the absolute rate of the process itself. All of these techniques have been discussed by Ludlow and Jarvis.[51] Rate of glycolate synthesis and use in photorespiration may give a better indication of this rate.[42, 43] As demonstrated in the following sections, it is very important to the understanding of photorespiration and to the development of a model of photorespiration that a second means, in addition to leaf CO_2 exchange, be established to observe the behavior of simultaneous photosynthesis and photorespiration.

The theory and predicted rates presented in the following sections should be considered initial steps in attempting to model photorespiration, based on competition at RuDP carboxylase-oxygenase and diffusion theory. As such, they may be valuable in establishing limits, and in characterizing the behavior of a system described by a particular model. From predicted rates, the importance of establishing the reliability of the $^{14}CO_2$ technique discussed above is apparent. Any model will only approximate reality when all assumptions, explicit and implicit in the model, are correct. At present, however, we believe the approach which follows will, at the very least, serve to clarify our thinking about complex interactions between processes occurring in the leaf.

III. A BASIC MODEL OF PHOTORESPIRATION

A. Characteristics of the Model

In Chapter 6, Volume I, Section II.C., a model is presented that is considered a basic model of simultaneous photosynthesis and photorespiration. Similar models have been derived by Charles-Edwards and Ludwig[52] and Peisker.[53] We recommend that the material in Sections II.C, D, E, and G of Chapter 6, Volume I be studied thoroughly before proceeding with the following sections of this chapter, much of which concerns problems confronted in the basic model and variations in the basic model worth considering further. Symbolism used in this chapter is identical to that used in Chapter 6, Volume I, and questions related to many concepts described here may be clarified by reference to that chapter.

At any constant light intensity and leaf temperature, the rate of net photosynthesis is a function of CO_2 and O_2 concentrations at the fixation site. The CO_2 concentration at the fixation site (C_i) depends on the CO_2 concentration in the intercellular air space, the mesophyll diffusion resistance between the air space and the fixation site, the flux of photosynthesis, and the flux of photorespiratory CO_2 to the fixation site. For the moment, we will consider the mitochondrial respiration flux in the light (Q_m) equal to zero. Several important dependencies are combined, such that three equations are obtained that must be in balance when the rate of net photosynthesis is at steady state. At this point, we should consider exactly how this balance is achieved, and how the predicted rates of net photosynthesis are obtained in the iteration procedure described by Tenhunen et al.[54]

At a set O_2 concentration ([O]), the following iterative estimation of CO_2 concentration at the chloroplast balances these equations. The parameter values of P^{max}_g, Q^{max}_p, K', K'_O, and $R^{mes}_{CO_2}$ are known from previous analysis, e.g., Table 1. An initial value of C_i is chosen, and used in Equation 2 to calculate the rate of photorespiration (Q_p); see Equation 45 in Chapter 6, Volume I for definitions.

$$Q_p = \frac{Q_p^{max}}{1 + \frac{K'_O}{[O]}\,(1 + C_i / K')} \tag{2}$$

TABLE 1

	Sunflower		Wheat		Units of Expression
Case number	1	2	3	4	
Leaf temperature	25	25	30	35	°C
P_g^{max}	2.80	3.46	3.92	4.95	n mol CO_2 cm^{-2} sec^{-1}
Q_p^{max}	0.92	3.46	3.92	4.95	n mol CO_2 cm^{-2} sec^{-1}
K′	0.0183	0.493	0.201	1.61	n mol CO_2 cm^{-3}
K'_O	0.001	0.0823	0.0294	0.273	(atmosphere, V/V)
$R^{mes}_{CO_2}$	1.84	2.27	1.94	1.28	sec cm^{-1}
Residual Respiration Correction	0.12	0.0	0.05	0.32	n moles CO_2 cm^2 sec

Note: Parameter values determined according to the kinetic analysis of Tenhunen et al.[54] discussed in Chapter 6, Volume I. Case 1 is for data from sunflower observed by Ludwig[12] at 25°C and light intensity of 116 Watts m^{-2}; 400 to 700 nm. Cases 2 through 4 for data from wheat observed by Ku and Edwards[19,20] at three temperatures and light intensity of 0.15 μEinsteins cm^{-2} sec^{-1} PhAR. Data analyzed are average rates versus estimated internal air space CO_2 concentrations. Ku and Edwards data analyzed under the restriction $P_g^{max} = Q_p^{max}$. Predicted responses for case 1 shown in Figure 11, for cases 2 through 4 shown in Figures 3 and 4 of Chapter 6, Volume I.

Then, Q_p is used to calculate the photosynthesis rate with Equation 3; see Equation 50 in Chapter 6, Volume I for definitions.

$$J^*_{CO_2} = \frac{a - (a^2 - b)^{1/2}}{2\,R^{mes}_{CO_2}} \qquad (3)$$

where

$$a = \left(C_{cw} + K'\beta + R^{mes}_{CO_2}\,(P_g^{max} - Q_p) - Q_p\,\frac{(R^{mes}_{CO_2})}{1+M}\right)$$

$$b = 4R^{mes}_{CO_2}\left[\left(C_{cw} - Q_p\,\frac{R^{mes}_{CO_2}}{1+M}\right)(P_g^{max} - Q_p) - Q_p K'\beta\right]$$

$$\beta = (1 + [O]\,/\,K'_O)$$

and where M = 100, C_{cw} is calculated based on water exchange, and [O] is measured outside the leaf; see discussion of Equations 45 and 50 of Chapter 6, Volume I. The values of Q_p and $J^*_{CO_2}$ are then used in the Fick's Law equation to calculate a new C_i compatible with these fluxes.

$$C_i = C_{cw} - J^*_{CO_2}\,R^{mes}_{CO_2} - Q_p\,\frac{(R^{mes}_{CO_2})}{1+M} \qquad (4)$$

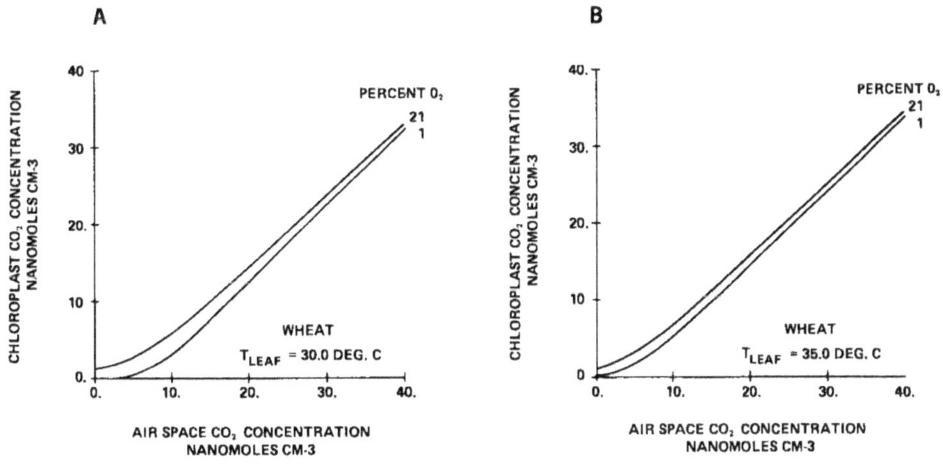

FIGURE 4. Predicted relationship of the iteratively determined chloroplast CO_2 concentration (C_i) to air space CO_2 concentration (C_{cw}) at 1 and 21% oxygen. Case A corresponds to case 3 of Table 1 (30°C responses of Figure 3 in Chapter 6, Volume I) case B correponds to case 4 of Table 1 (35°C responses of Figure 3 in Chapter 6, Volume I. Photorespiratory site assumed adjacent to the photosynthesis site.

The second estimate of C_i is compared to the first, and used again in Equation 2, until there is a change in C_i between iteration steps that is acceptably small. Calculations of this sort are possible by hand, but are extremely tedious.

In this fashion, the predicted responses of net photosynthesis vs. C_{cw} at different O_2 concentrations shown in Figures 3 and 4 of Chapter 6, Volume I are obtained. The parameter values used are those determined (Table 1) by the kinetic analysis of Tenhunen et al.,[54] which has been discussed in Section II.C. of Chapter 6, Volume I. For further referral, case numbers are assigned to the various problem solutions in Table 1. Iteratively determined C_i values were previously reported[54] for case 1 of Table 1. The C_i values for cases 3 and 4 of Table 1, derived from Ku and Edwards' data, at 30 and 35°C, are presented in Figure 4 and correspond to the predicted photosynthesis responses in Figure 3, Chapter 6, Volume I.

The same iterative procedure can be used to calculate net photosynthesis based on external CO_2 concentration (C_{CO_2}), substituting C_{CO_2} for C_{cw} and $R^{total}_{CO_2}$ for $R^{mes}_{CO_2}$ in Equations 2 through 4, where $R^{total}_{CO_2} \simeq R^{mes}_{CO_2} + R^{st}_{CO_2}$. The stomatal resistance, $R^{st}_{CO_2}$, must be known as a function of external CO_2 concentration and O_2 concentration. Assuming that stomatal resistance of sunflower is relatively independent of these concentrations in certain concentration ranges, as suggested by Ludwig and Canvin,[11] net photosynthesis, as a function of C_{CO_2} for four different O_2 concentrations, was calculated, as shown in Figure 5(A), and compared to observations made by Ludwig.[12] The parameter values are presented in the figure caption to avoid confusing these estimates (based on external CO_2 concentration) with those recorded in Table 1, which are based on estimated internal air-space CO_2 concentrations.

Several criteria were presented in Section II.C of Chapter 6, Volume I that must be satisfied by a successful description of photosynthesis and photorespiration. From observations of net CO_2 exchange summarized, in part, in the literature review above, and also in the previous chapter, an adequate model must certainly account for saturation of net photosynthesis and suppression of O_2 inhibition at high CO_2 concentration. From Figure 5(A), one observes that O_2 inhibition is suppressed only at very high CO_2 concentrations, which are predicted by the model to be above 3000 parts per million (ppm) CO_2 external to the leaf. This agrees well with unpublished data obtained by Canvin[75] from sunflower (shown in Figure 5(B)).

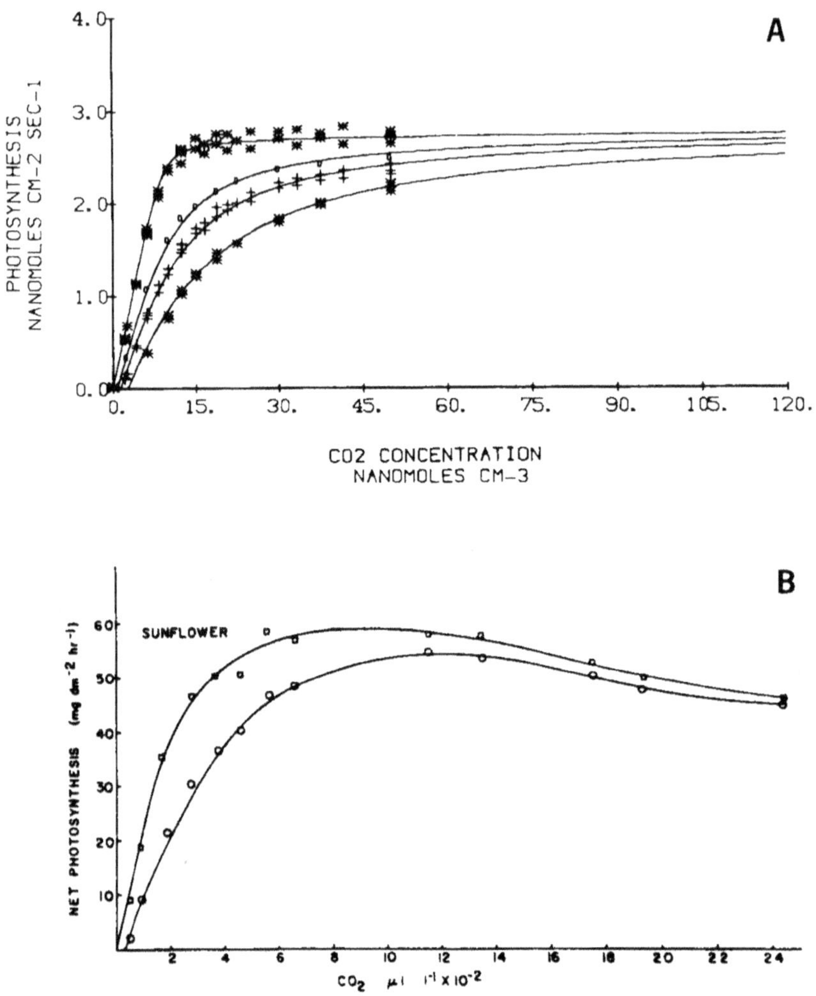

FIGURE 5. Net photosynthesis responses to air space CO_2 concentration external to the leaf for sunflower. Data in Figure 5(A) obtained by Ludwig[12] at 1% (upper * symbols), 12% (o symbols), 21% (+ symbols), and 40% (lower * symbols) oxygen and corrected for residual respiration of 0.03 nmol/cm²/sec, according to assumption 4, discussed in Section III.B. Solid lines in Figure 5(A) are predicted by the iterative method outlined in the text; with P_g^{max} = 2.72 nmol/cm²/sec, Q_p^{max} = 0.967 nmol/cm²/sec, K' = 0.163 nmol/cm³, K'_o = 0.00087 atmosphere O_2 (decimal fraction partial pressure), and $R^{total}_{CO_2}$ = 3.36 sec/cm The sum of mesophyll plus stomatal and boundary layer resistances is $R^{total}_{CO_2}$. Data in 5(B) obtained by Canvin[76] at 2% (□ symbols) and 21% oxygen (o symbols) from zero to 2400 ppm CO_2. Leaf temperatures 25°C; light intensity in 5(A): 116 W m² (400 to 700 nm), and in 5(B): 0.08 μEinsteins/cm²/sec. The analysis of data in Figure 5(A) assumes no effect of CO_2 and O_2 on stomatal resistance which, judging from Figure 5(B), may introduce some error. Therefore, the analysis of these data based on estimated air space CO_2 concentration as reported in Table 1 is preferable. 300 ppm = 12 nmol/cm³; 120 nmol/cm³ = 3000 ppm = 30 μl/l × 10⁻².

From Figure 5, and Figures 3 and 4 of Chapter 6, Volume I, it is seen that the model predicts a decrease in net photosynthesis as O_2 concentration is increased above 1%, and accurately predicts the response of net photosynthesis to CO_2 and O_2 concentrations. Estimates of CO_2 evolution into CO_2-free air (zero C_{cw}) are also compatible with observation. The predicted decrease in the initial slope of the CO_2 response curve with

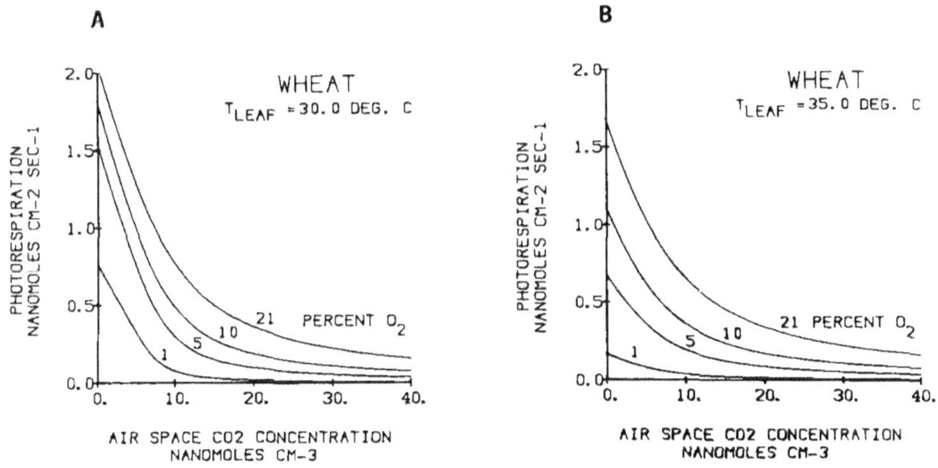

FIGURE 6. Predicted relationship of photorespiration (Q_p) to air space CO_2 concentration (C_{cw}) at 1, 5, 10, and 21% oxygen. Case A corresponds to case 3 of Table 1, case A of Figure 4, and 30°C case of Figure 3 in Chapter 6. Case B corresponds to case 4 of Table 1, case B of Figure 4, and 35°C case of Figure 3 in Chapter 6. Photorespiratory site assumed adjacent to photosynthesis site.

increasing O_2 concentration is similar to that described by Forrester et al.[7] (Equation 1), and was discussed in detail by Tenhunen et al.[54] In Sections II.E and II.G of Chapter 6, Volume I, it was demonstrated that this model will always result in a linear increase of the CO_2 compensation point with increasing O_2 concentration, and that the slope of this dependency is determined by the ratio $Q_p^{max}K'/P_g^{max}K'_O$. Compensation points are discussed further in Section V of this chapter.

To this point, we have been concerned primarily with photorespiration as it affects net photosynthesis. It is also important to examine the characteristics of the photorespiratory process itself, as described by the model, even though no direct comparison can yet be made between predicted and observed photorespiration rates. Photorespiration rates are obtained in the iterative solutions of Equations 2 through 4, described above. These rates depend on CO_2 and O_2 concentration at the fixation site. The dependencies of Q_p on air-space CO_2 concentration at four O_2 concentrations are shown in Figure 6, as predicted with the parameter values in Table 1, determined in analysis of the data of Ku and Edwards at 30 and 35°C (cases 3 and 4). These two cases are chosen for presentation here because they have not been presented previously, and can be compared with the responses based on analysis of Ludwig's sunflower data.[54] They are further interesting because the affinity constants for O_2 (K'_O) are about 3% and 27%, respectively, whereas K'_O from Ludwig's data is much lower, about equal to 0.1% O_2. From these three cases, we find that certain characteristics of photorespiration, as envisioned in this model, are independent of the value of K'_O for values of K'_O near zero to well above ambient O_2 concentration. Thus, while more work is needed to establish values of K'_O (Section IV), qualitative characteristics of these metabolic processes, regardless of the value of K'_O, can be discussed.

As shown in Figure 6, even with a low affinity for O_2 (K'_O = 0.27 atm at 35°C), predicted rates of photorespiration are substantial. The rate of photorespiration is most strongly dependent on C_{cw} at low concentrations. If a model is to be based on the Bowes-Ogren hypothesis, which implies that control of the rate of photorespiration involves oxygenation of RuDP at RuDP carboxylase-oxygenase, then the competitive effect of CO_2 in Equation 2 on the photorespiration rate must be included (see also Section II.C of Chapter 6, Volume I). As discussed later in Section III.B and Section

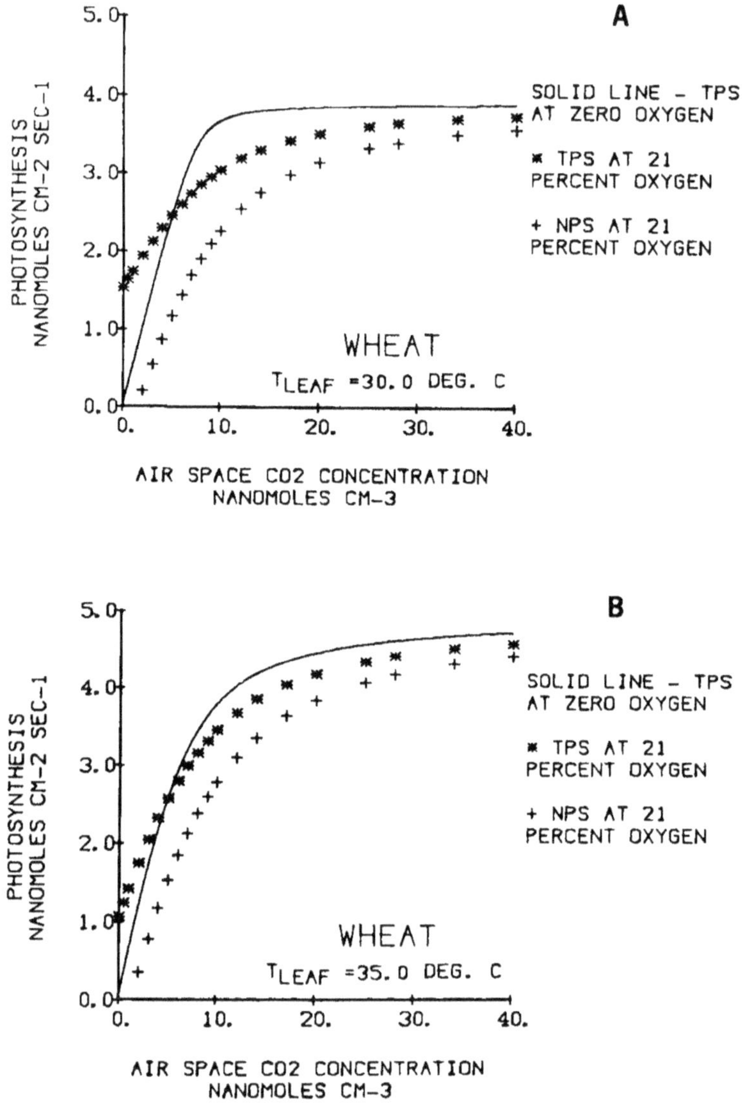

FIGURE 7. Simulated contributions to the net photosynthesis response of wheat at 21% oxygen according to Equations 2 through 4. Case A corresponds to case A in Figures 4 and 6 and case 3 in Table 1. Case B corresponds to case B in Figures 4 and 6 and case 4 in Table 1. Solid line is the true or total photosynthesis rate at zero oxygen or with no photorespiration occurring. Asterisks (*) indicate the true photosynthesis rate predicted at 21% oxygen (net photosynthesis plus photorespiration) and demonstrate the stimulation at low CO_2 concentration effected by photorespiration. Plus symbols (+) indicate the predicted net photosynthesis rate at 21% oxygen which differs from true photosynthesis due to diffusion of CO_2 from the photorespiratory flux out of the leaf. Photorespiratory site is assumed adjacent to photosynthesis site.

V, the rates shown in Figure 6 are probably overestimates. A crucial assumption discussed below is that the photorespiratory site is functionally closer to the fixation site than to the intercellular air space. A result of this assumption is that photorespiration will increase CO_2 concentration at the fixation site, and stimulate refixation of photorespired CO_2. This is shown in Figures 7 and 8. This is not the case when the site of

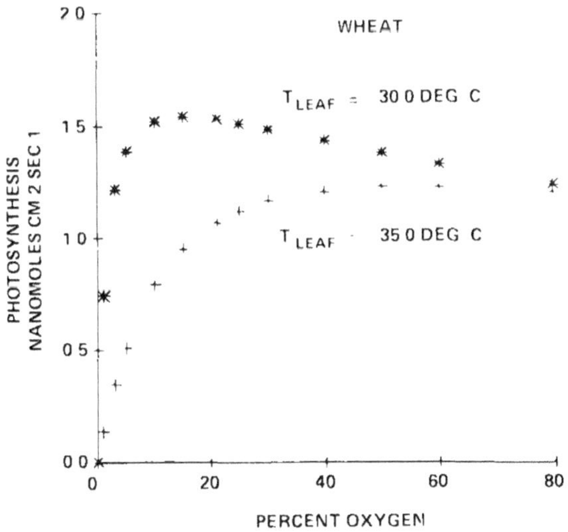

FIGURE 8. Predicted rate of true photosynthesis (net photosynthesis plus photorespiration) at zero CO_2 concentration in the intercellular air space as a function of oxygen concentration in the air. Asterisks (*) correspond to case A in Figures 4, 6, and 7 and case 3 in Table 1. Plus symbols (+) correspond to case B in Figures 4, 6, and 7 and case 4 in Table 1. Photorespiratory site assumed adjacent to photosynthesis site.

photorespiration is located further from the fixation site, and predicted photorespiration rates decrease. Recycling of CO_2 evolved in mitochondrial respiration or decarboxylation of glycolate produced via pathways other than the oxygenation of RuDP can also shift the O_2 and CO_2 dependencies of photorespiration considerably. Unless the photorespiratory function in Equation 2 is altered, there will always be a decrease in photorespiration as air-space CO_2 concentration increases. With the photorespiratory site close to the fixation site, net photosynthesis rate at zero C_{cw} and 1% O_2 is essentially zero, rather than negative, as suggested by the predicted photorespiration rates, because of recycling of CO_2. The chloroplast CO_2 concentration (C_i) remains relatively high, due to photorespiration and diffusion resistance to the escape of CO_2. The rate at which materials are cycled through the Calvin cycle (Q_p + $J^*_{CO_2}$) is stimulated, rather than inhibited, by photorespiration at low C_{cw}. The amount of stimulation can be determined by comparing the response expected for true photosynthesis when photorespiration is eliminated to the response expected when photorespiration occurs. The results of such a comparison for the Ku and Edwards wheat data at 30 and 35°C are illustrated in Figure 7. In Figure 7, the solid lines are the true photosynthesis C_{cw} responses obtained when no photorespiratory CO_2 is available for recycling, and all CO_2 fixed must diffuse in from the air space. This response is closely approximated by data obtained at 1% O_2, because, as indicated in Figure 6, the rate of photorespiration is low under these conditions. As O_2 concentration increases, the true photosynthesis response is increasingly modified. At high C_{cw}, the true photosynthesis rate is inhibited by O_2, because of the increase in the effective K′ for CO_2, due to competition at the carboxylase and due to diffusion of photorespiratory CO_2 out of the leaf. The effect of this diffusion out is to decrease net photosynthesis ($J^*_{CO_2}$). At low C_{cw}, on the other hand, photorespiratory CO_2 is available for recycling, and results in a stimulation of true photosynthesis.

Figure 7 demonstrates the C_{cw} dependent effect of photorespiration on true photosynthesis at 21% O_2. Where the asterisks are above the solid line in Figure 7, true photosynthesis is stimulated, and where below, it is inhibited. The plus symbols in Figure 7 indicate the net photosynthesis rate. The difference between the plus symbols and the asterisks is indirectly a measure of how much CO_2 is being contributed to true photosynthesis by recycling of photorespiratory CO_2. When net photosynthesis is zero, the entire rate is supported by photorespiration. When net photosynthesis is negative, photorespiration provides CO_2 for diffusion out of the leaf, as well as maintenance of the true photosynthesis rate.

Comparison of the results shown at 30 and 35°C demonstrates that the photorespiratory effect on true photosynthesis is greater if K'_O is low and photorespiration rates are relatively high. The effect is nevertheless significant, even when K'_O is above ambient O_2 concentration. Particularly interesting is the fact that even with a high K'_O, the rate of true photosynthesis at 21% oxygen can never decrease below 20 or 30% of maximum. Thus, a leaf structured according to the model uses photoproducts more effectively at low C_{cw} than a leaf with no photorespiration. It would be tempting to conclude that photorespiration protects the leaf reaction centers from possibly damaging effects of excess photoproducts during periods of water stress and high irradiance when stomata are closed, as suggested by Osmond and Björkman.[55] Clearly, it is important to verify independently the assumptions of this basic model.

The rate of true photosynthesis (rate of Calvin cycle cycling), at zero CO_2 concentration in the intercellular air space, is one measure of this possible protective capacity of photorespiration. From Figure 6, apparently the rate of photorespiration and, thus, the stimulation of true photosynthesis at zero C_{cw} is strongly dependent on O_2 concentration. The O_2 dependency for the predicted rate of true photosynthesis at zero C_{cw} is shown in Figure 8 for wheat, according to the analysis of Ku and Edwards' data at 30 and 35°C. Whether K'_O is relatively high or low, stimulation of true photosynthesis is near maximal at the ambient concentration of 21%. The magnitude of K'_O influences how rapidly the increase to maximal stimulation takes place. Even with extremely low K'_O values, as determined for the data of Ludwig (case 1, Table 1), a similar behavior is observed, but the plateau is reached more rapidly. It is conceivable that C_4 plants have evolved means (trapping mechanisms) to reclaim the net loss of C from the photorespiratory flux and diffusion out of the leaf.

A much different picture is obtained if photorespiratory CO_2 is evolved more distantly from the fixation site. As the site of CO_2 evolution is moved further and further away (M decreases), photorespired CO_2 contributes less to the internal CO_2 concentration and only to the net flux. At low CO_2 concentration, essentially all CO_2 is lost to the air space. In this case, photosynthesis is predicted to have a much lower affinity for CO_2 ($K_C \simeq 0.25$ to 0.5 nmol/cm^3) and photorespiration to be a more wasteful process. Even so, CO_2 at the compensation concentration is recycled when $P_g = Q_p$. No stimulation is seen at zero C_{cw}, but the leaf will only experience this condition when artificially imposed during experimentation. The rate of recycling is the rate of photorespiration at the compensation point, which was shown in Chapter 6, Volume I, Equation 62 to be a Michaelis-Menton function, with respect to O_2 concentration. This function resembles the function shown in Figure 8, except that the maximum rate is lower. Recycling at the compensation concentration, especially at high temperatures, is measured by the $^{14}CO_2$ technique to be at least 1 nmol/cm^2/sec, which is considerable.

A model of the photorespiratory process is inescapably complex. It has been apparent for some time that behavior of CO_2 fixation, according to the Bowes-Ogren hypothesis within a system where diffusion resistances are significant, is not readily explainable. Quantification may thus reveal rather surprising interrelationships between the CO_2 fluxes within the leaf. While the magnitude of photorespiratory fluxes and effects may change as more is learned, we feel that many of the characteristics of the processes discussd will remain part of the basic description. We can expect further surprises, because we are figuratively moving about in a complex multidimensional space as we proceed in our comparison of predictions of the the theoretical model to observation.

Equations 2 through 4 above provide a model of CO_2 and O_2 dependencies of simultaneous photosynthesis and photorespiration in a C_3 leaf. As demonstrated in Chapter 6, Volume I, the complexity becomes even greater when the effects of light and temperature are included in the model. Following sections consider possible improvements of the basic model, and problems that have been encountered in its development.

B. Review of the Model Assumptions

It is appropriate here to examine more closely the assumptions upon which the photorespiratory model is based. From such an examination, possible new directions for model development, as well as potential problems and limitations, are identified. It may be helpful to the reader to review Section II.C of Chapter 6, Volume I to better appreciate the role of these assumptions in the theoretical development of the model.

1. It was assumed that $R^{mes}{}_{CO_2}$ is a constant diffusion resistance unaffected by CO_2 or O_2 concentrations. The basis for this assumption is that it explains the linear increase in net photosynthesis rate as a function of air-space CO_2 concentration at 1% O_2, or where $O_p \simeq 0$.[56, 57] The mesophyll resistance has been discussed as a separate problem previously,[58-60] and there are many reasons why this assumption may be unjustifiable. Until further research improves our understanding of the details of CO_2 transfer to the chloroplast, however, the assumption is appropriate and necessary to maintain the delicate balance between extreme complexity and oversimplification in a model of whole-leaf photosynthesis.
2. More importantly, the numerical value obtained for $R^{mes}{}_{CO_2}$ results from the assumption that the resistance, R_4, in Figure 3(A) of Chapter 6, Volume I is much smaller than R_2 and R_3. Setting M equal to 100 in Equation 3 yields this same result (see discussion of Figure 3(A) in Chapter 6, Volume I). In real terms, this means that the site of photorespiration is closely associated with the site of fixation, much more closely than with the intercellular air space. This is supported by the close physical proximity of the chloroplasts, peroxysomes, and mitochondria, which appear to be involved in photorespiratory CO_2 evolution.[61] Evidence now suggests that the functioning of these organelles may be interlocked.[62, 63] Others feel that the magnitude of CO_2 evolution at low C_{cw} suggests that the resistance between the air space and the photorespiratory site is not much greater than the resistance between this site and the fixation site. Either assumption allows description of the net photosynthesis responses observed. The magnitude of CO_2 evolution, in one case, is the result of very high rates of photorespiration with recycling of CO_2, even at zero C_{cw}, and, in the other case, the result of less photorespiration and less recycling. The true behavior is unclear, but may be between these extremes. Based on the $^{14}CO_2$ technique, Canvin[76] suggests refixation of no more than 20 to 30% of the photorespiratory flux below the compensation point. Modifications of the model indicate that R_4 need not be greater than R_2 to obtain such behavior (M = 1). If more information could be obtained on the magnitude of recycling, it is possible that the relative magnitudes of R_2, R_3, and R_4 could be defined.

3. Basic to the model is the assumption that at steady state, overall kinetics of fixation at a set temperature and light intensity are determined by a competitive interaction of CO_2 and O_2 at RuDP carboxylase-oxygenase. This has been supported by comparison of the properties of RuDP-1,5-carboxylase-oxygenase in vitro to properties of whole-leaf net photosynthesis.[14,30] It is supported by incorporation of O^{18} into the glycolate pool.[32] Finally, a model based on this assumption will explain many phenomena of whole-leaf photosynthesis that have otherwise been unexplained. As yet unpublished work on a model of the type described by Tenhunen et al.,[54, 64, 65] coupled with temperature dependencies for K′ and K'_O, similar to those described by Laing et al.,[14] for the affinity constants of RuDP carboxylase-oxygenase in vitro, indicates that such a model successfully reproduces the following: (1) a linear dependency of the CO_2-compensation point on O_2 concentration, (2) a shift in this O_2 dependency with temperature, similar to that described in wheat by Jolliffe and Tregunna,[66] (3) a reduction of the initial slope of light responses of net photosynthesis with increase in O_2 concentration, as described by Ehleringer and Björkman[67] (see also Section II.D of Chapter 6, Volume I), (4) a reduction in the initial slope of the light response curve of net photosynthesis with an increase in temperature, as described by Ehleringer and Björkman for an ambient atmosphere external to the leaf, (5) a shift in the temperature optimum of net photosynthesis at ambient CO_2 and O_2 to higher temperatures as light intensity is increased, and (6) a CO_2-compensation point at a single temperature unaffected by light intensity (see also Section II.G of Chapter 6, Volume I). The model also provides a basis for shifts in the temperature optimum observed in plants grown under different temperature regimes (see also Section III of Chapter 6, Volume I). Discussion in Section V. indicates that many effects, attributable to competitive effects of CO_2 on photorespiration, are retained, when modification is made to include a constant flux of CO_2 from the mitochondria or alternate pathways of glycolate formation leading to subsequent decarboxylation.

In discussion of Figure 6, we stated that the photorespiratory function in Equation 2 will always result in photorespiration decreasing with an increase in CO_2 concentration. Results with the $^{14}CO_2$ technique indicated that photorespiration is relatively unaffected by CO_2 concentration. At most, there may be a 20 to 25% decrease between 0 and 400 ppm CO_2 at higher temperatures. Modifications of the basic model, in which the photorespiratory site is placed further from the fixation site ($M = 1$), and in which a constant flux of CO_2 is contributed by other reactions (see Section V.), flatten the CO_2 response of photorespiration, but still result in a decrease of photorespiration 30 to 40% over the same concentration range. The response can be flattened even more by increasing the magnitude contributed in the constant term. On the other hand, if extreme modification of Equation 2 is required before theory is compatible with observation, it indicates that the competitive relationship is less important in regulating the rate of photorespiration than postulated in this model. Control may be located at some step insensitive to CO_2, like glycolate oxidase, or may indicate that contributions to the glycolate flux are homeostatically regulated. An alternative relationship may then be appropriate in Equation 2. Thus, it is extremely important to reevaluate the $^{14}CO_2$ technique to determine the reliability of photorespiration rates obtained by this method. It is imperative to determine whether any error may be inherent in such measurements that varies with CO_2 concentration.

4. It has been assumed that mitochondrial respiration in the light is unaffected by light, CO_2, and O_2 concentration. Woo and Osmund[62] have hypothesized that

the peroxysomes and mitochondria may be functionally interlocked in the light, allowing the possibility that environmental variables that affect photorespiration may indirectly affect the mitochondrial respiratory process. It has also been assumed that the rate of mitochondrial respiration in the light can be estimated by extrapolation of the 1% O_2-CO_2-response curve to zero CO_2 concentration. This estimate of residual respiration has been used to correct the observed net photosynthesis data.[54] The corrected data set is then treated as though $Q_m = 0$. Implied is that one observes only the effect of photorespiration on net photosynthesis after this correction. Such a treatment of mitochondrial respiration in the light is clearly inadequate, because the effect of mitochondrial respiration on the recycling process and chloroplast CO_2 concentration is ignored. As discussed further in Section V., all respiratory fluxes should be included in Equation 2, and be considered in the iterative solution discussed above. We will demonstrate that failure to do so will very significantly shift the value of K′, and change the magnitude of predicted photorespiration.

5. The value of Q_p^{max} has, in some analyses, been assumed equal to P_g^{max}. Based on the kinetic equations of the Peisker model (Section II.G of Chapter 6, Volume I), it is not unreasonable to assume that Q_p^{max} is proportional to P_g^{max}. We could make Q_p^{max} a different fixed proportion of P_g^{max} or independent of P_g^{max}. The predicted magnitude of photorespiration, however, does not change, because in fitting the model to the data, compensatory change is made in the parameter K'_O. The restriction is useful because a parameter is eliminated. Numerical values of K'_O are of interest to the whole-plant modeler only as they vary in nature and indicate changes in photorespiratory properties. In this application, the simplification presents no problem. The biochemist, on the other hand, would like to relate the whole leaf affinity for O_2 to biochemical processes, like oxygenation of RuDP by RuDP carboxylase-oxygenase. There is no reason that the K′ or K'_O value determined for intact leaves must agree with affinity constants determined for isolated carboxylase. The biochemist would, nevertheless, prefer an unbiased estimate of K'_O to make such comparisons. In the context of the extreme complexity encountered in analyzing the leaf system, an unbiased estimate of K'_O and K′ may be unattainable. For now, the assumption is considered a useful working hypothesis.

Parameter estimation in this model is dependent on a family of CO_2-response curves at different O_2 concentrations. On this basis, there is no unique solution. Assumptions must be made to establish the relative magnitudes of R_2, R_3, and R_4, and to define the behavior of photorespiration as a function of CO_2 and O_2 concentrations. Independent tests of the model assumptions and predictions are needed. A model allows attention to be focused on assumptions, and may, as a result, help resolve difficulties. Familiarity with a particular model, coupled with knowledge derived from further research at the cellular and subcellular level, may lead to improvements in the model that will allow description of the actual behavior of photorespiration.

IV. LEAF-AFFINITY CONSTANTS FOR CO_2 and O_2

Results thus far obtained in analyses with the basic model (summarized in Table 1) demonstrate considerable variation in the values of K′ and K'_O. These analyses suggest affinity constants for CO_2 for the whole leaf (K′) of 0.1 to 1.0 μM dissolved CO_2, based on the data of Ku and Edwards,[20] or even lower, if based on the data of Ludwig.[12] Values obtained for the affinity constant for O_2 (K'_O) are 3 to 27% O_2, from the data of Ku and Edwards, and lower for the data of Ludwig. These are also extremely low, compared to values commonly referred to in the plant physiology literature. Numbers of these magnitudes assigned to the affinity constants of the leaf can be expected to spark considerable controversy.

The lowest value of the affinity constant for CO_2 of RuDP carboxylase-oxygenase that has been reported is about 5 to 10 μM dissolved CO_2. Bowes et al.[68] report a Km(CO_2) for the carboxylase of 7 μM. Bahr and Jensen[69] report a similar value from spinach and tobacco. These authors stated that their Km(CO_2) values are compatible with whole leaf photosynthesis. What has not been clearly emphasized is that the value for the affinity constant for CO_2 in the whole leaf is a function of the model used in obtaining that value. Thus, Servaites and Ogren[70] recently reported a value of Km(CO_2) for soybean mesophyll cells of 20 μM dissolved CO_2 at 25°C, pH 7.8, and 2% O_2. This value of 20 μM is obtained by fitting a transformation of the Michaelis-Menton equation to the observed data. The photosynthesis model in this case is, therefore, the Michaelis-Menton equation.

If a model for photosynthesis is used that includes a diffusion resistance for CO_2, like that discussed for Equations 41 and 42 of Chapter 6, Volume I, an estimate of about 8 μM dissolved CO_2 is obtained for K′ from the same data. It is unclear that the term Km(CO_2) should be applied to this cellular system. Based on the second model, Jones and Slatyer[57] reported values for the affinity constant for whole leaves of cotton as low as 0.23 μM. Using the Jones and Slatyer[57] analysis, the authors obtained similar values from response curves of red kidney-bean leaves (*P. vulgaris*) and from the 1% O_2- CO_2- response data of Ku and Edwards, and Ludwig.

Still a third estimate may be obtained if the model imposed is based on competitive interaction of CO_2 and O_2 at RuDP carboxylase-oxygenase at the end of a diffusion path, as in the basic model considered here. In our experience, estimates for K′ are then somewhat lower whenever the resistance R_4, between the photorespiratory and fixation sites, is negligibly small. The value of K′ is the same order of magnitude as determined with the Jones and Slatyer analysis, when R_4 is larger and the fixation site is less influenced by photorespiration. Considerations discussed in Section V demonstrate that modification of the respiration function in the theoretical analysis also influences estimates of the leaf affinity constant for CO_2. It is easy to understand these changes in K′, because each modification of the model alters the relationship of net photosynthesis to CO_2 concentration at the fixation site; either the site is moved or additional CO_2 fluxes are included. These same problems are confronted in estimating K'_O, not because the concentration of O_2 at the reaction site varies, but because a particular rate of respiration, dependent in a competitive manner on CO_2, is needed to account for observed net photosynthesis rates. The values of the affinity constants depend on the theoretical underpinnings of the analysis. Until some degree of unanimity is reached as to the proper means of analysis, great care must be taken when comparing K values. In particular, the enzymologist must be careful to recognize the nature of whole-leaf affinity constants, when comparing them to Km values obtained by traditional enzymological techniques.

One must also investigate why the estimates of K′ and K'_O (reported in Table 1), obtained with the basic model, vary. Is there real significance to this variation? Lommen et al.[56] demonstrated, that with photorespiration equal to zero, the influence of K′ on the net photosynthesis vs. C_{cw}-response curve is most striking in the curvature region. Similarly, in this model with photorespiration, K′ and K'_O determine the rate of curvature of the CO_2-response curves, and also the proximity of response curves obtained at two O_2 concentrations.

The 1 and 21% oxygen C_{cw}-response curves of net photosynthesis were used as ref-

TABLE 2

Variations in K′ and K′$_O$ Studied by Simulation to Determine the Effect of These Parameters on Positioning and Shape of the 1% and 21% Oxygen C_{ew} Response Curves of Photosynthesis

Changing K Mangitude→

Changing Index Value ↓		
	case A, IV = 3 K′ = 0.493 K′$_O$ = 0.1646	case B, IV = 3 K′ = 0.0493 K′$_O$ = 0.01646
case C, IV = 6 K′ = 0.986 K′$_O$ = 0.1646	case D, IV = 6 K′ = 0.493 K′$_O$ = 0.0823	case E, IV = 6 K′ = 0.0493 K′$_O$ = 0.00823
	case F, IV = 30 K′ = 0.493 K′$_O$ = 0.01646	

→ shown in Figure 9(A)

↓ shown in Figure 9(B)

Note: In Equations 2 through 4, $R_{CO_2}^{mes}$ was constant at 2.27 sec cm^{-1} and P_g^{max} = Q_p^{max} was constant at 3.325 n mol CO_2 cm^{-2} sec^{-1}. K′ in n mol CO_2 cm^{-3}, K′$_O$ in atmospheres or partial pressure (V/V). Results are shown in Figure 9. IV is an index value for the ratio K′/K′$_O$ ignoring dimensions. Q_m = 0.

erences to examine graphically the influence of changes in K′ and K′$_O$. The parameters K′ and K′$_O$ may vary in magnitude and relative to each other. Simulated C_{ew}-response curves at 1 and 21% O_2 were obtained for the variations in K′ and K′$_O$ shown in Table 2. Parameter values were otherwise based on values obtained in the analysis of Ku and Edwards' data at 25°C (case 2, Table 1): R^{mes} CO_2 was constant at 2.27 sec/cm, P_g^{max} = Q_p^{max} was constant at 3.325 n*M* cm^2 sec, and Q_m = 0. Moving across Table 2 from left to right, there is no change in the ratio K′/K′$_O$. This ratio (ignoring dimensions) is indicated as an index value IV, following the case letter in Table 2. Moving across the table from left to right, values of both parameters decrease. Moving from top to bottom in the table, the value of K′ is held constant, and the ratio K′/K′$_O$ increases because K′$_O$ decreases.

The alterations in response determined from these simulations are summarized in Figure 9(A) and (B). In Figure 9(A), one observes that for a constant IV, the position of the 21%-response curve is constant. As the magnitude of K′ and K′$_O$ change, however, the curvature of the 1%-response curve changes markedly. The change in curvature observed as K′ decreases is the same as that described by Lommen et al.[56] for the CO_2 response in the model in which photorespiration is eliminated. As a result of the change in the curvature, the proximity of the 1% response to the 21% response in the curvature region also changes.

In Figure 9(B), one observes that for a constant K′, the shape of the 1% response is defined. As the IV increases, however (K′$_O$ decreases), the 21%-response curve is rapidly positioned further and further away from the 1% response. The 1% response is shifted, as well, because more photorespiration is predicted at 1% O_2 as K′$_O$ decreases.

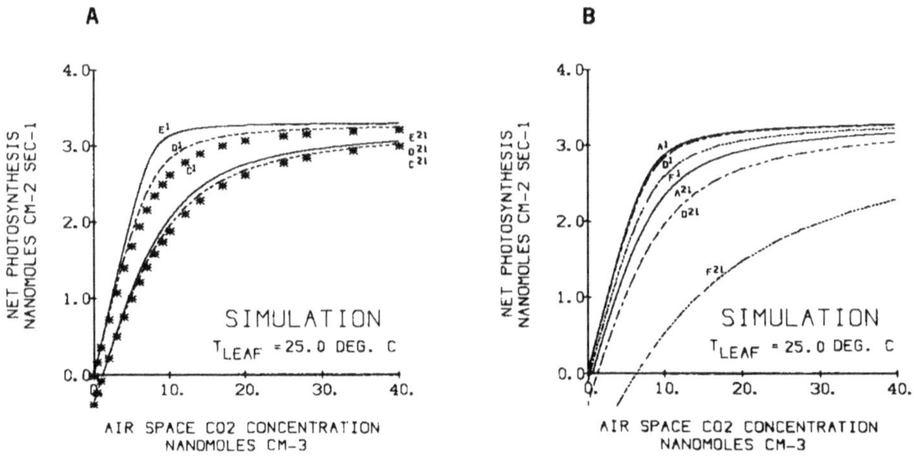

FIGURE 9. Simulated C_{cw}-response curves of net photosynthesis at 1 and 21% oxygen. K′ and K'_o set at the values given in Table 2 for the case letters corresponding to labels on the curves. Superscripts indicate the % oxygen that applies for a particular curve. In Figure 9(A), the ratio K'/K'_o is constant as K′ and K'_o vary. In figure 9(B), K′ is constant as K'_o varies.

Analysis of Ludwig's data resulted in an IV of 18 and K′ of 0.018 nmol cm^{-3}, while analysis of the 25°C data of Ku and Edwards (represented best by case D in Table 2) resulted in a IV of 6 and a much larger K′. For an IV of 18, we would expect the 1 and 21% O_2 responses to be more widely spaced than for an IV of 6. The low value of K′ obtained in the Ludwig analysis will offset any lowering of the 1% response curve that would be caused by a low K'_o and photorespiration occurring at 1% O_2. A direct comparison of the observed data is possible, because the mesophyll resistance (RCO_2^{mes}) for the two data sets is the same, and because standardizing the data (dividing observed rates by P_g^{max}) does not change the parameter values obtained in analysis. The comparison, shown in Figure 10, supports both of the above conclusions. The differences in distribution of the data in the curvature region can be deduced from the differences in values predicted for K′ and K'_o. Sunflower seems to be less affected by oxygen in the curvature region than does wheat. There is little difference in the initial slopes of the response curves where RCO_2^{mes} dominates the response.

The meaning of differences in K′ and K'_o in real terms, and the amount of this type of variation to be found in nature, is unknown. Again we must reemphasize that basic data of this type are lacking. Further data to high C_{cw} concentrations will allow us to separate problems introduced by shortcomings of the model from problems introduced by experimental measurement and those resulting from natural variation in plant material. In the meantime, the data that we do have can be used to study modifications of the basic mdel.

V. LEAF MITOCHONDRIAL RESPIRATION IN THE LIGHT

A. A Modified Photorespiratory Function

The estimates obtained for the affinity constants of the leaf for carbon dioxide and oxygen are a function of the model used in obtaining those estimates. The following modifications of the basic model draw further attention to this important point. Aspects of the model that require further study are those that most affect the concentration of CO_2 at the fixation site. We have previously assumed that CO_2 responses corrected for residual respiration (Q_m) provided data sets that were free of the influence

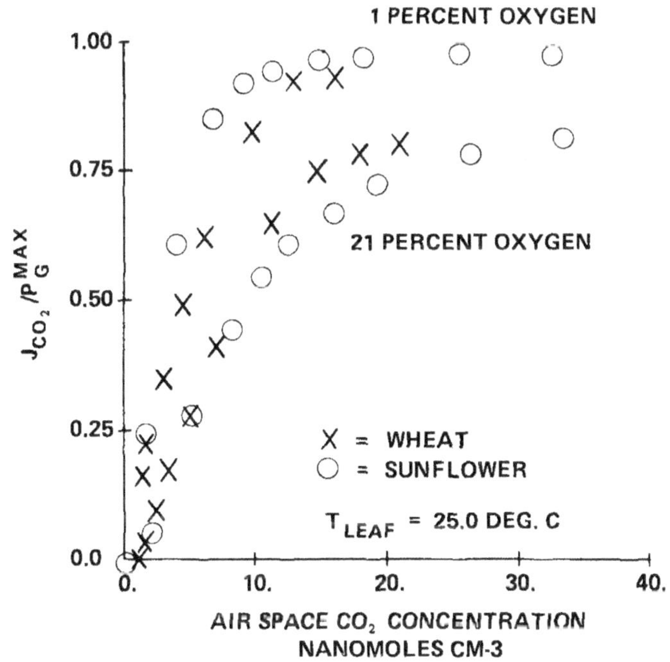

FIGURE 10. Observed net photosynthesis rate vs. air space CO_2 concentration at 1 and 21% oxygen from wheat [20] and sunflower.[12] Data are transformed by dividing by the apparent P_g^{max}. The results demonstrate the real difference in the curvature region responsible for differing estimates of K′ and K'_o from these data.

of leaf mitochondrial respiration in the light. The following suggests that this is not true and that parameter values must ultimately be derived from uncorrected (unbiased) data.

Several studies[71, 72] have suggested that leaf mitochondrial respiration continues in the light at rates comparable to dark respiration rates. These respiration rates are, in general, greater than those found by extrapolation of the 1% C_{cw}-response curve to zero CO_2 concentration. The treatment in the model of Q_m was modified to consider a constant contribution to the photorespiratory flux Q_p. Analysis was undertaken of the uncorrected net photosynthesis vs. C_{cw} response data of Ludwig shown in Figure 5 (A). The constant flux Q^*_m could also include decarboxylation of the glycolate formed via reactions other than oxygenation of RuDP.[36, 37] Such a flux could not be separated from the mitochondrial flux based on leaf-gas exchange. The iterative procedure for obtaining C_i is identical to that described in Section III, except that the following equations replace Equation 2, and JCO_2 (the net photosynthesis rate) replaces J^*CO_2 in Equation 3.

$$Q_p^* = \frac{Q_p^{max}}{1 + \frac{K_O'}{[O]}(1 + C_i / K')} + Q_m^* \qquad (5)$$

$$Q_p^* = Q_p + Q_m^* \qquad (6)$$

TABLE 3

Parameter Values Determined According to the Kinetic Analysis of Tenhunen et al.[54] Discussed in Chapter 6, Volume I.

	Q_m = 0.12 extrapolated and added to observed data	Q^*_m = 0.27 added as constant contribution to Q_p, see Equations 5 and 6.	n mol $CO_2/cm^2/sec$
Case number	1	2	
P_g^{max}	2.80	3.02	n mol CO_2 cm^{-2} sec^{-1}
Q_p^{max}	0.92	0.677	n mol CO_2 cm^{-2} sec^{-1}
K′	0.0183	0.177	n mol CO_2 cm^{-3}
K'_O	0.001	0.0058	(atmosphere, V/V)
$R_{CO_2}^{mes}$	1.84	2.16	sec cm^{-1}

Note: Case 1 is for the data from sunflower (case 1 of Table 1 based on air space CO_2 concentration) observed by Ludwig[12] at 25°C, corrected for residual respiration obtained by extrapolation of the 1% oxygen C_{cw}-response curve to zero C_{cw}. Case 2 considers the same data uncorrected and includes a constant contribution (Q^*_m) to the photorespiratory flux of 0.27 n mol CO_2 cm^{-2} sec^{-1}. The modified treatment is discussed in Section V. Predicted responses and observed data presented in Figure 11.

where Q^*_m is the mitochondrial respiration rate in the light plus any decarboxylation of glycolate contributed by alternative pathways ($nmol/cm^{-2}/sec^{-1}$) and Q^*_p is the total respiration in the light ($nmol/cm^{-2}/sec^{-1}$). Parameter values determined with this model are presented in Table 3. Case 1 of Table 3 is the result obtained previously,[54] where a residual respiration correction of Q_m = 0.12 $nmol/cm^{-2}/sec^{-1}$ was added to all observations before analysis. Case 2 is the result obtained when Q^*_m = 0.27 nmol/cm^{-2}/sec^{-1} in Equation 5.

Predicted values of net photosynthesis vs. C_{cw} at 1 and 21% oxygen, based on the two sets of parameter values, are compared to the observed net photosynthesis data in Figure 11. From Figure 11, it is apparent that either model successfully predicts net photosynthesis rates. The modified model is perhaps more satisfying, because it is not necessary to resort to artificial manipulations of the observed data, and because the rates predicted are clearly net photosynthesis rates rather than corrected rates of obscure meaning; Q^*_m was estimated by iteration. This estimate of 0.27 $nmol/cm^2/sec$ is higher than reported for dark respiration of these leaves by Ludwig.

In Table 3, several changes in parameter values are detected; P_g^{max} increases. This is expected, because at CO_2 saturation, Q^*_m must be added to the net photosynthesis flux to obtain P_g^{max}; Q^*_m results in higher C_i values at low C_{cw}, and alters the predicted dependency of P_g on C_i. The affinity constant for CO_2 is thus raised an order of magnitude. Less photorespiration is required to account for O_2 effects observed in the net photosynthesis data, leading to an increase in K'_O and decrease in Q_p^{max}. As discussed previously, K′ and K'_O change, because the modification made in the model imposes new restrictions and new possibilities.

B. Respiration Components and Behavior of the Modified Model

In earlier sections and previous work, predictions of the rate of photorespiration as a function of CO_2, O_2, and light intensity were made with the basic model. These predictions were made to ascertain characteristic behavior of a model in which the steady-state rates of photosynthesis and photorespiration are controlled according to the kinetics of RuDP carboxylase-oxygenase implied in the Bowes-Ogren hypothesis. The predictions were useful, because they helped examine the compatability of photo-

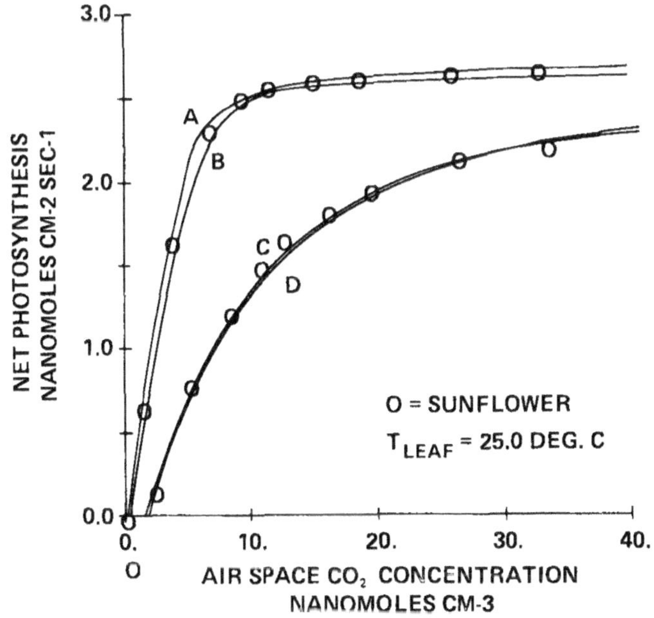

FIGURE 11. Comparison of the predicted responses to the observed data of Ludwig[12] of net photosynthesis rate vs. air space CO_2 concentration at 1 and 21% oxygen for the two models described in the text. Parameter values are given in Table 3. Case 1 of Table 3 which involves the basic model described in Section III and a residual respiration correction predicts curves A and C. Case 2 of Table 3, based on the modified model described in Section V, predicts curves B and D.

respiration rates, predicted by the model, with those measured by the $^{14}CO_2$ technique,[45-49] discussed in Section III.B. Here we will examine the components of respiration in the light predicted by the modified model described above. This comparison demonstrates further that certain characteristic behavior of the basic model is unaltered in the modified model, even though the magnitude of fluxes changes.

Predicted respiration in the light for both models (cases 1 and 2 of Table 3) is separated into components in Figures 12(A) and (B). The results obtained at 1% O_2 are presented in Figure 12(A), and the results at 21% O_2 are presented in Figure 12B. The plus symbols are rates of photorespiration (Q_p) predicted for the basic model (case 1). The asterisks are rates of photorespiration (Q_p) predicted for the modified model (case 2). Including the flux Q^*_m results in much lower predicted rates of Q_p at low C_{cw}, Q_p in both models decreases to similar values at high C_{cw} concentration.

The total respiration (Q^*_p), according to the modified model, is presented as the solid line in Figure 12 which is located 0.27 nM/cm^3/sec above the asterisks. The total respiration at $C_{cw} = 0$ is the same for both models. Modifying the model as described in Equation 5 not only dramatically increases respiration at high CO_2 concentrations, but also counteracts the necessity for respiration from the competitive term in Equation 5. Thus, if Q^*_m becomes large, the respiration attributed to oxygenation of RuDP becomes small. As discussed in Section III.B., it is important to determine whether the flat CO_2 dependency for photorespiration, suggested by Ludwig and Canvin,[11] is correct. If it is, then photosynthesis may be regulated by the kinetics at RuDP carboxylase-oxygenase, while photorespiration is not.

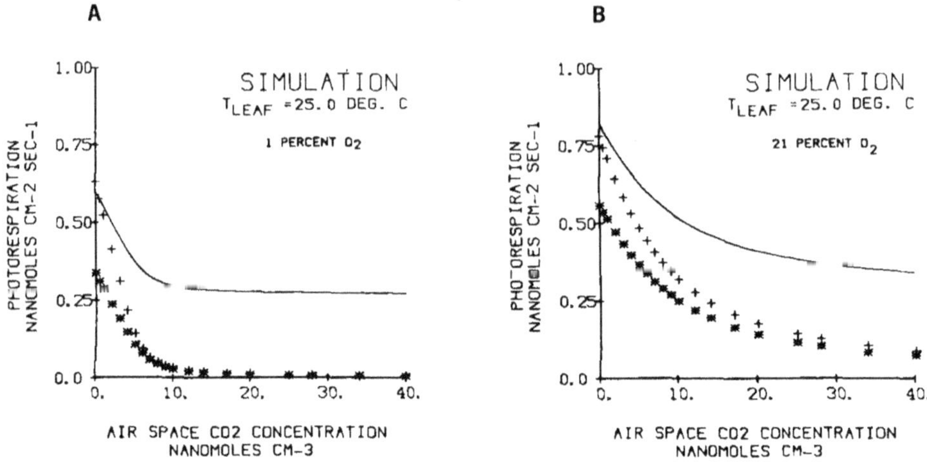

FIGURE 12. Comparison of predicted photorespiration rates at 1 and 21% oxygen according to the two models described in the text. Net photosynthesis responses are shown for the two models in Figure 11. Parameter values are given in Table 3. The plus (+) symbols are rates of photorespiration predicted for the basic model, where the entire photorespiratory flux results from oxygenation of RuDP. The asterisk (*) symbols are rates of photorespiration resulting from oxygenation of RuDP in the modified model, and the solid lines are rates of total respiration in the light in the modified model. The solid lines are obtained by adding to the asterisk (*) symbols; a constant respiratory flux of 0.27 nmol/cm²/sec attributed to mitochondrial respiration or decarboxylation of glycolate produced via alternative pathways.

It was determined in Equation 116 of Chapter 6, Volume I that the compensation point with $Q_m = 0$ is determined by the following relationship:

$$\Gamma = \frac{Q_p^{max} K'}{P_g^{max} K_O'}[O] \tag{7}$$

This equation is obtained by setting the equation describing total photosynthesis (Equation 8) equal to the equation describing photorespiration and solving for C_i, which is the compensation concentration Γ. A similar procedure using the modified Equation 5 in place of Equation 2 results in an alternate expression describing the oxygen dependence of the compensation point. For total photosynthesis,

$$P_g = \frac{P_g^{max}}{1 + \frac{K'(1 + [O] / K_O')}{C_i}} = \frac{P_g^{max} C_i}{C_i + K' + \frac{K'}{K_O'}[O]} \tag{8}$$

At the compensation point:

$$P_g = Q_p^* = Q_p + Q_m^* \tag{9}$$

Substituting from Equations 8 and 5:

$$\frac{P_g^{max}\, C_i}{C_i + K' + \frac{K'}{K'_O}[O]} = \frac{Q_p^{max}\, \frac{K'}{K'_O}[O]}{C_i + K' + \frac{K'}{K'_O}[O]} + Q_m^* \qquad (10)$$

Multiplying by the expression in the denominator:

$$P_g^{max}\, C_i = Q_p^{max}\, \frac{K'}{K'_O}[O] + Q_m^* C_i + Q_m^* K' + \frac{K'}{K'_O} Q_m^* [O] \qquad (11)$$

Regrouping terms and dividing by $(P_g^{max} - Q^*_m)$ to solve for C_i:

$$C_i = \frac{(Q_p^{max}\frac{K'}{K'_O} + Q_m^*\frac{K'}{K'_O})}{(P_g^{max} - Q_m^*)}[O] + \frac{K'\, Q_m^*}{(P_g^{max} - Q_m^*)} \qquad (12)$$

Thus, the compensation concentration in the modified model (Equation 12) is still a linear function of O_2 concentration. The value of P_g^{max}, as discussed with respect to the results presented in Table 3, is equal to the observed maximum rate plus "mitochondrial" respiration in the light (Q^*_m). If Q^*_m were about the same as the respiration rate in the dark, then K' could be calculated from the C_i intercept of the relationship $C_i = f([O])$ in Equation 12. If Q^*_m includes a constant rate of glycolate decarboxylation, this is not possible.

One further interesting point should be mentioned. In these simulations in which a mitochondrial respiration flux is included, it is possible to calculate, from the simulations, the residual respiration[64, 65] that would be estimated by extrapolation of the predicted 1% O_2-C_{cw}-response curve of net photosynthesis to zero C_{cw}. One finds that this residual is generally 0.50 to 0.25 of Q^*_m. If light response curves are simulated at saturating CO_2, however, as in Section II.D of Chapter 6, Volume I, the residual estimated at 1% O_2 is identical to the flux included as Q^*_m in the model. One must keep in mind that the results obtained from simulated light responses include the assumption that Q^*_m is unaffected by light intensity, which may not be the case due to indirect regulatory effects.[62]

VI. SUMMARY AND CONCLUSIONS

Considerable progress has been made in describing the net photosynthesis response of leaves as a function of CO_2 and O_2 concentrations, light intensity, and leaf temperature. The continuing effort must be to base this description, and that for respiratory fluxes, on parameters that are physiologically meaningful. The present effort depends heavily on the assumption that competition between CO_2 and O_2 at the active site of RuDP-1,5-carboxylase-oxygenase plays a major role in controlling the steady-state rates of photosynthesis and photorespiration. Such a model will explain many phenomena of whole-leaf net photosynthesis. Rates predicted for photorespiration by the present model are understandably controversial. Nevertheless, it is important to study the predicted rates and the characteristics of the models from which they are obtained. In this manner, clues may be found that will help explain what has remained an enigma. Variations in assumptions of the model may lead progressively to a description of the respiratory processes of the illuminated leaf.

The material presented in this chapter demonstrates how tightly the system is interlocked. Parameter values are tied to the specific way in which biochemical studies are interpreted and translated into the mathematics of the model. Still, the efforts presented here suggest that further basic data must be obtained, and that these data will be of great value, even if the models are altered. Studies of leaf respiration in the light at the cell or organelle level of organization that would help clarify problems identified in this chapter are needed. As discussed in Chapter 6, Volume I and in Section IV of this chapter, data must be obtained which will reveal the amount of natural variation that occurs in present model parameters. Studies are required that reveal the effects of leaf development and environmental effects, due to position in canopies on model parameters, so that canopy respiration fluxes can eventually be assessed. The work of Ludwig, Saeki, and Evans[73] demonstrated how dramatically respiratory fluxes vary at different levels in the canopy. Angus and Wilson[74] have studied the variation in the maximum rate of photosynthesis in canopies of barley and wheat. They found this maximum to be extremely plastic. If the plasticity in all parameters could be assessed and the environmental control of these parameters determined, it is possible that the information could be used in breeding programs to develop canopies more efficient in photosynthesis and possibly production.

Aspects of assessing canopy photosynthesis and respiration that must be considered, in addition to a properly constructed leaf photosynthesis-photorespiration model, are the functional dependencies on depth in the canopy of atmospheric CO_2 concentration, incident light intensity, leaf temperature, stomatal and mesophyll resistances, and mitochondrial respiration and photorespiration. Effects of age and water status on these functional relationships must be determined. Studies must not direct attention to only one of these relationships, but should be designed to elaborate interrelationships among several functional dependencies. Simpler studies to provide basic data should, at the same time, be related to several other of these dependencies, measured as references.

VII. SYMBOLS

a	Expression defined in Equation 3.
b	Expression defined in Equation 3.
C_2	Phosphoglycolate, a product of the oxygenation reaction of RuDP carboxylase-oxygenase.
CE	The initial slope of the C_x response curve for photosynthetic CO_2 gas exchange under saturating light, often restricted to refer to the initial slope of the J_{CO_2}- vs. C_{cw}-response curve (units of conductance, distance time).
CE_o	The initial slope of the J_{CO_2}- vs. C_{cw}-response curve with no photorespiration occurring, or at 1% oxygen (units of conductance, distance/time).
C_x	CO_2 concentration of air next to the boundary layer of the leaf (x = CO_2), air next to cell walls inside the leaf (x = cw), and in liquid at the chloroplast site (x = i) (weight/volume or moles/volume).
IV	An index value for the ratio of K'/K'_O. See discussion for Table 2.
J_{CO_2}	The net photosynthetic CO_2-exchange rate between the leaf and the atmosphere (weight or moles CO_2 per unit area of one side of the leaf/time).

$J^*_{CO_2}$	The net photosynthetic CO_2-exchange rate between the leaf and the atmosphere corrected for Q_m (residual mitochondrial respiration), estimated by extrapolation to zero CO_2 concentration; same units as J_{CO_2}.
k	The slope of ln(CE) vs. [O].
K′	The chloroplast carbon dioxide concentration (photorespiration ≃ 0) at which $P_g = P_g^{max}/2$; the leaf affinity constant for CO_2.
K'_O	The oxygen concentration with $C_i \simeq 0$ at which $Q_p = Q_p^{max}/2$ (oxygen concentration units); the leaf affinity constant for O_2.
Km(CO_2)	The affinity constant of RuDP carboxylase-oxygenase for CO_2 determined in vitro.
Km(O_2)	The affinity constant of RuDP carboxylase-oxygenase for O_2 determined in vitro.
M	A constant which measures the ratio of the diffusion resistance between C_{cw} and C_i to the diffusion resistance between the source of photorespiratory CO_2 and C_i; as M → ∞, the photorespiratory site is moved to, and becomes identical with, the fixation site (dimensionless).
[O]	Oxygen concentration (% or decimal fraction partial pressure — volume/volume).
P_g^{max}	The maximum photosynthetic rate at saturating CO_2 at the chloroplast reaction site; same units as J_{CO_2}; dependent on incident light intensity and leaf temperature.
ppm	Parts per million.
Q_m	The mitochondrial respiration rate in the light (residual respiration); same units as J_{CO_2}. This is estimated by extrapolation of light or CO_2-response curves of photosynthesis to zero light intensity or CO_2 concentration.
Q^*_m	The mitochondrial respiration rate in the light plus any glycolate decarboxylation resulting from reactions other than oxygenation of RuDP; same units as J_{CO_2}; assumed to be unaffected by light intensity and CO_2 and O_2 concentrations.
Q_p	The photorespiratory rate of CO_2 evolution at the photorespiratory site; same units as J_{CO_2}; dependent on [O], C_i, light intensity, and leaf temperature.
Q^*_p	The total rate of respiration in the light, including photorespiration due to oxygenation of RuDP (Q_p) and mitochondrial and glycolate decarboxylation from other pathways (Q^*_m).
Q_p^{max}	The maximum photorespiratory rate at saturating oxygen at the photorespiratory site; same units as J_{CO_2}; dependent on light intensity and leaf temperature.
R_1, R_2, R_3, R_4	Resistances to gas diffusion in the leaf defined in Figure 3(A) of Chapter 6, Volume I.
$R_{CO_2}^{chl}$	CO_2 flux resistance from the outside surface of the chloroplast membrane to the reaction site (time/distance).
$R_{CO_2}^{mes}$	CO_2 flux resistance associated with liquid phase from wet cell walls through membranes and cytoplasm to the chloroplast membranes (time/distance).
$R_{CO_2}^{st}$	The stomatal resistance to CO_2 flux (time/distance^{-1}).
$R_{CO_2}^{total}$	The total resistance for CO_2 flux to the site of fixation Components shown in Figure 1 of Chapter 6, Volume I (time/distance).
RuDP	CO_2 acceptor ribulose-1,5-diphosphate for CO_2 fixation dark reactions (moles/area).

u The slope of the linear dependency of compensation point Γ on oxygen concentration of the air (concentration/concentration).

β A modifier of the affinity constant for CO_2 = $(1 + [O]/K'_O)$.

Γ CO_2 compensation concentration at which $J_{CO_2} = 0$ (weight/volume or moles/volume).

REFERENCES

1. **Rabinowitch, E.**, *Photosynthesis and Related Processes*, Vol. 1, Interscience, New York, 1945.

1.a. **Rabinowitch, E.**, *Photosynthesis and Related Processes*, Vol. 2, Part 1, Interscience, New York, 1951.

1.b. **Rabinowitch, E.**, *Photosynthesis and Related Processes*, Vol. 2, Part 2, Interscience, New York, 1956.

2. **Decker, J.**, Some effects of temperature and carbon dioxide concentration on photosynthesis of *Mimulus*, *Plant Physiol.*, 34, 103, 1959.
3. **Decker, J.**, A rapid, postillumination deceleration of respiration in green leaves, *Plant Physiol.*, 30, 82, 1955.
4. **Hesketh, J.**, Limitations to photosynthesis responsible for differences among species, *Crop Sci.*, 3, 493, 1963.
5. **El-Sharkawy, M. and Hesketh, J.**, Photosynthesis among species in relation to characteristics of leaf anatomy and CO_2 diffusion resistances, *Crop Sci.*, 5, 517, 1965.
6. **Kortschak, H., Hartt, C., and Burr, G.**, Carbon dioxide fixation in sugar-cane leaves, *Plant Physiol.*, 40, 209, 1965.
7. **Hatch, M. and Slack, C.**, Photosynthesis by sugar cane leaves: A new carboxylation reaction and the pathway of sugar formation, *Biochem. J.*, 101, 103, 1966.
8. **Hesketh, J. and Baker, D.**, The relationship between leaf anatomy and photosynthetic CO_2 assimilation among and within species, in *Prediction and Measurement of Photosynthetic Productivity*, Setlik, I., Ed., Pudoc, Wagengingen, 1970, 317.
9. **Forrester, M., Krotkov, G., and Nelson, C.**, Effect of oxygen on photosynthesis, photorespiration and respiration in detached leaves. I. Soybean, *Plant Physiol.*, 41, 422, 1966.

9.a. **Forrester, M., Krotkov, G., and Nelson, C.**, Effect of oxygen on photosynthesis, photorespiration and respiration in detached leaves. II. Corn and other monocotyledons, *Plant Physiol.*, 41, 428, 1966.

10. **Tregunna, E., Krotkov, G., and Nelson, C.**, Effect of oxygen on the rate of photorespiration in detached tobacco leaves, *Physiol. Plant.*, 19, 723, 1966.
11. **Ludwig, L. and Canvin, D.**, The rate of photorespiration during photosynthesis and the relationship of the substrate of light respiration to the products of photosynthesis in sunflower leaves, *Plant Physiol.*, 48, 712, 1971.
12. **Ludwig, L.**, The relationship between photosynthesis and respiration, in *Crop Processes in Controlled Environments*, Rees, A., Cockshell, K., Hand, D., and Hard, K., Eds., Academic Press, London, 1972, 305.
13. **Kriedemann, P. and Canberford, R.**, The photosynthetic activity of pear leaves (*Pyrus communis* L.), *Physiol. Plant.*, 24, 197, 1971.
14. **Laing, W., Ogren, W., and Hageman, R.**, Regulation of soybean net photosynthetic CO_2 fixation by the interaction of CO_2, O_2 and ribulose-1,5-diphosphate carboxylase, *Plant Physiol.*, 54, 678, 1974.
15. **Peisker, M. and Apel, P.**, Influence of oxygen on photosynthesis and photorespiration in leaves of *Triticum aestivum* L., I. Relationship between oxygen concentration, CO_2 compensation point, and intracellular resistance to CO_2 uptake, *Photosynthetica*, 9, 16, 1975.
16. **Troughton, J. and Slatyer, R.**, Plant water status, leaf temperature, and the calculated mesophyll resistance to carbon dioxide of cotton leaves, *Aust. J. Biol. Sci.*, 22, 815, 1969.
17. **Lake, J. and Slatyer, R.**, Respiration of leaves during photosynthesis. III. Respiration rate and mesophyll resistance in turgid cotton leaves, with stomatal control eliminated, *Aust. J. Biol. Sci.*, 23, 529, 1970.
18. **Collatz, G.**, Influence of certain environmental factors on photosynthesis and photorespiration in *Simmondsia chinensis*, *Planta*, 134, 127, 1977.
19. **Ku, S. and Edwards, G. E.**, Oxygen inhibition of photosynthesis, I. Temperature dependence and relation to O_2/CO_2 solubility ratio, *Plant Physiol.*, 59, 986, 1977.

20. **Ku, S. and Edwards, G.**, Oxygen inhibition of photosynthesis, II. Kinetic characteristics as affected by temperature, *Plant Physiol.*, 59, 991, 1977.
21. **Zelitch, I.**, Organic acids and respiration in photosynthetic tissues, *Annu. Rev. Plant Physiol.*, 15, 121, 1964.
22. **Shantz, H. and Piemeisel, L.**, The water requirements of plants at Akron, Colorado, *J. Agric. Res.*, 34, 1093, 1972.
23. **Nobs, M., Björkman, O., and Pearcy, R.**, Hybrids between *Atriplex* species with and without β-carboxylation photosynthesis. Cytogenetic and morphological characteristics, *Carnegie Inst. Washington, Yearb.*, 69, 625, 1971.
24. **Boynton, J., Nobs, M., Bjorkman, O., and Pearcy, R.**, Hybrids between *Atriplex* species with and without β-carboxylation photosynthesis. Leaf anatomy and ultrastructure, *Carnegie Inst. Washington, Yearb.*, 69, 621, 1971.
25. **Pearcy, R. and Björkman, O.**, Hybrids between *Atriplex* species with and without β-carboxylation photosynthesis. Biochemical characteristics, *Carnegie Inst. Washington Yearb.*, 69, 632, 1971.
26. **Björkman, O., Pearcy, R., and Nobs, M.**, Hybrids between *Atriplex* species with and without β-carboxylation photosynthesis. Photosynthetic characteristics, *Carnegie Inst. Washington, Yearb.*, 69, 640, 1971.
27. **Hatch, M., Osmond, C., Troughton, J., and Björkman, O.**, Physiological and biochemical characteristics of C_3 and C_4 *Atriplex* species and hybrids in relation to the evolution of the C_4 pathway, *Carnegie Inst. Washington, Yearb.*, 71, 135, 1972.
28. **Brown, R. and Brown, W.**, Photosynthetic characteristics of *Panicum milioides*, a species with reduced photorespiration, *Crop Sci.*, 15, 681, 1975.
29. **Brown, R.**, Characteristics related to photosynthesis and photorespiration of *Panicum milioides*, in *CO_2 Metabolism and Plant Productivity*, Burris R. and Black, C., Eds., University Park Press, Baltimore, 1976, 311.
30. **Ogren, W., and Bowes, G.**, Ribulose diphosphate carboxylase regulates soybean photorespiration, *Nature New Biol.*, 230, 159, 1971.
31. **Bowes, G. and Ogren, W.**, Oxygen inhibition and other properties of soybean ribulose 1,5-diphosphate carboxylase, *J. Biol. Chem.*, 247, 2171, 1972.
32. **Lorimer, G., Krause, G., and Berry, J.**, The incorporation of (^{18}O) oxygen into glycolate by intact isolated chloroplasts, *FEBS Lett.*, 78, 199, 1977.
33. **Andrews, T., Lorimer, G., and Tolbert, N.**, Ribulose diphosphate oxygenase. I. Synthesis of phosphoglycolate by fraction-1 protein of leaves, *Biochemistry*, 12, 11, 1973.
34. **Lorimer, G., Andrews, T., and Tolbert, N.**, Ribulose diphosphate oxygenase. II. Further proof of reaction products and mechanisms of action, *Biochemistry*, 12, 18, 173.
35. **Tolbert, N. and Ryan, F.**, Glycolate biosynthesis and metabolism during photorespiration, in *CO_2 Metabolism and Plant Productivity*, Burris, R. and Black, C., Eds., University Park Press, Baltimore, 1976, 141.
36. **Zelitch, I.**, Pathways of Carbon fixation in green plants, *Annu. Rev. Bio. Chem.*, 44, 123, 1975.
37. **Kelly, G., Latzko, E., and Gibbs, M.**, Regulatory aspects of photosynthetic carbon metabolism, *Annu. Rev. Plant Physiol.*, 27, 181, 1976.
38. **Zelitch, I., Oliver, D., and Berlyn, M.**, Increasing photosynthetic carbon dioxide fixation by the biochemical and genetic regulation of photorespiration, in *Biological Solar Energy Conversion*, Mitsui, A., Miyachi, S., Pietro, A., and Tamura, S., Eds., Academic Press, New York, 1977, 231.
39. **Tolbert, N.**, Regulation of products of photosynthesis by photorespiration and reduction of carbon, in *Biological Solar Energy Conversion*, Mitsui, A., Miyachi, S., Pietro, A., and Tamura, S., Eds., Academic Press, New York, 1977, 243.
40. **Servaites, J. and Ogren, W.**, Chemical inhibition of the glycolate pathway in soybean leaf cells, *Plant Physiol.*, 60, 461, 1977.
41. **Murata, Y. and Iyama, J.**, Studies on the photosynthesis of forage crops. II. Influence of air temperature upon the photosynthesis of some forage and grain crops, *Proc. Crop Sci. Soc. Japan*, 31, 315, 1963.
42. **Raven, J.**, Endogenous inorganic carbon sources in plant photosynthesis. I. Occurrence of the dark respiratory pathways in illuminated green cells, *New Phytol.*, 71, 227, 1972.
43. **Raven, J.**, Endogenous inorganic carbon sources in plant photosynthesis. II. Comparison of total CO_2 production in the light with measured CO_2 evolution in the light, *New Phytol.*, 71, 995, 1972.
44. **Tolbert, N., Oeser, A., Yamazaki, R., Hageman, R. and Kisaki, T.**, A survey of plants for leaf peroxisomes, *Plant Physiol.*, 44, 135, 1969.
45. **Ludwig, L. and Canvin, D.**, An open gas-exchange system for the simultaneous measurement of the CO_2 and $^{14}CO_2$ fluxes from leaves, *Can. J. Bot.*, 49, 1299, 1971.
46. **Hew, C., Krotkov, G., and Canvin, D.**, Determination of the rate of CO_2 evolution by green leaves in light, *Plant Physiol.*, 44, 662, 1969.

47. **Ludwig, L.**, The Relationship between Photosynthesis and Respiration in Sunflower Leaves, Ph.D. thesis, Queen's University, Kingston, Ontario, Canada, 1968.
48. **Ludwig, L. and Krotkov, G.**, The kinetics of labeling of the substrates for CO_2 evolution by sunflower leaves in the light, *Plant Physiol.*, 42, 547, 1967.
49. **Ludwig, L., Krotkov, G., and Canvin, D.**, The relationship of the products of photosynthesis to the substrates for CO_2 evolution in light and darkness, *Prog. Photosynthesis Res.*, 1, 494, 1969.
50. **Kisaki, T. and Tolbert, N.**, Glycine as a substrate of photorespiration, *Plant Cell Physiol.*, 11, 247, 1970.
51. **Ludlow, M. and Jarvis, P.**, Methods for measuring photorespiration in leaves, in *Plant Photosynthetic Production Manual of Methods*, Sestak, Z., Catsky, J., and Jarvis, P., Eds., W. Junk, The Hague, 1971, 294.
52. **Charles-Edwards, D. and Ludwig, L.**, A model for leaf photosynthesis by C_3 plant species, *Ann. Bot. (London)*, 38, 921, 1974.
53. **Peisker, M.**, Ein Modell der Sauerstoffabhängigkeit des photosynthetischen CO_2-Gaswechsels von C_3-Pflanzen, *Kulturpflanze*, 24, 221, 1976.
54. **Tenhunen, J., Weber, J., Filipek, L., and Gates, D.**, Development of a photosynthesis model with an emphasis on ecological applications. III. Carbon dioxide and oxygen dependencies, *Oecologia*, 30, 189, 1977.
55. **Osmond, C. and Björkman, O.**, Simultaneous measurements of oxygen effects on net photosynthesis and glycolate metabolism in C_3 and C_4 species, *Carnegie Inst. Washington, Yearb.*, 71, 141, 1971.
56. **Lommen, P., Schwintzer, C., Yocum, C., and Gates, D.**, A model describing photosynthesis in terms of gas diffusion and enzyme kinetics, *Planta*, 98, 195, 1971.
57. **Jones, H. and Slatyer, R.**, Estimation of the transport and carboxylation components of the intracellular limitation to leaf photosynthesis, *Plant Physiol.*, 50, 283, 1972.
58. **Yocum, C. and Lommen, P.**, Mesophyll resistances, in *Perspectives of Biophysical Ecology*, Gates, D. and Schmerl, R., Eds., Springer-Verlag, Berlin, 1975, 45.
59. **Prioul, J., Reyss, A., and Chartier, P.**, Relationships between carbon dioxide transfer resistances and some physiological and anatomical features, in *Environmental and Biological Control of Photosynthesis*, Marcelle, R., Ed., W. Junk, The Hague, 1975, 17.
60. **Jarvis, P.**, The estimation of resistances to carbon dioxide transfer, in *Plant Photosynthetic Production Manual of Methods*, Sestak, Z., Catsky, J., and Jarvis, P., Eds., W. Junk, The Hague, 1971, 566.
61. **Newcomb, E. and Frederick, S.**, Distribution and structure of plant microbodies (peroxisomes), in *Photosynthesis and Photorespiration*, Hatch, M., Osmund, C., and Slatyer, R., Eds., John Wiley & Sons, New York, 1971, 442.
62. **Woo, K. and Osmond, C.**, Participation of leaf mitochondria in the photorespiratory carbon oxidation cycle: glycine decarboxylation activity in leaf mitochondria from different species and its intramitochondrial location, Photosynthetic Organelles, *Plant Cell Physiol.*, Special Issue, 315, 1977.
63. **Woo, K. and Osmond, C.**, Glycine decarboxylation in mitochondria isolated from spinach leaves, *Aust. J. Plant Physiol.*, 3, 771, 1976.
64. **Tenhunen, J., Yocum, C., and Gates, D.**, Development of a photosynthesis model with an emphasis on ecological applications. I. Theory, *Oecologia*, 26, 89, 1976.
65. **Tenhunen, J., Weber, J., Yocum, C., and Gates, D.**, Development of a photosynthesis model with an emphasis on ecological applications. II. Analysis of a data set describing the P_M surface, *Oecologia*, 26, 101, 1976.
66. **Jolliffe, P. and Tregunna, E.**, Environmental regulation of the oxygen effect on apparent photosynthesis in wheat, *Can. J. Bot.*, 51, 841, 1973.
67. **Ehleringer, J. and Björkman, O.**, Quantum yields for CO_2 uptake in C_3 and C_4 plants, *Plant Physiol.*, 59, 86, 1977.
68. **Bowes, G., Ogren, W., and Hageman, R.**, pH Dependence of the Km(CO_2) of ribulose 1,5-diphosphate carboxylase, *Plant Physiol.*, 56, 630, 1975.
69. **Bahr, J. and Jensen, R.**, Ribulose diphosphate carboxylase from freshly ruptured spinach chloroplasts having an in vivo Km(CO_2), *Plant Physiol.*, 53, 39, 1974.
70. **Servaites, J. and Ogren, W.**, Rapid isolation of mesophyll cells from leaves of soybean for photosynthetic studies, *Plant Physiol.*, 59, 587, 1977.
71. **Marsh, H., Galmiche, J., and Gibbs, M.**, Effect of light on the tricarboxylic acid cycle in *Scendesmus*, *Plant Physiol.*, 40, 1013, 1965.
72. **Raven, J.**, Endogenous inorganic carbon sources in plant photosynthesis. I. Occurrence of the dark respiratory pathways in illuminated green cells, *New Phytol.*, 71, 227, 1972.
73. **Ludwig, L., Saeki, T., and Evans, L.**, Photosynthesis in artificial communities of cotton plants in relation to leaf area. I. Experiments with progressive defoliation of mature plants, *Aust. J. Biol. Sci.*, 18, 1103, 1965.

74. **Angus, J. and Wilson, J.**, Photosynthesis of barley and wheat leaves in relation to canopy models, *Photosynthetica*, 10, 367, 1976.
75. **Canvin, D.**, personal communication.
76. **Canvin, D.**, personal communication.

Chapter 3

INTERFACING LEAF AND CANOPY LIGHT INTERCEPTION MODELS

J. M. Norman

TABLE OF CONTENTS

I. INTRODUCTION

Light is essential to plant growth because it is the source of energy for the photosynthetic machinery that converts carbon dioxide and water to life-giving photosynthate; no one questions this fact. However, the precise nature of the influence of light on plant growth, beginning with the arrival of photons at the surface of the earth and ending with the production of some form of dry matter, is the subject of much research and uncertainty. Even though light may be essential, if it is present in sufficient abundance so that other factors are more important in determining the final outcome of the growth of a season, then from a prediction point of view, light may not be a particularly useful variable. Clearly, water, nutrients, carbon dioxide, and reasonable temperatures also are essential to plant survival. The point being made here is that light is only one of many important variables, and it must be considered as interacting with the other variables and not in isolation. In any system as complex as a growing plant, many interacting variables are involved, and the ultimate importance of any one of them is very difficult to ascertain without understanding all of them.

From a modeling perspective, the quantification of radiation effects on vegetation must be at a level of sophistication comparable to considerations given other important variables. If only light is considered, as in the case of predicting yield from intercepted radiation, then very simple empiricisms may be adequate. However, if physiological characteristics of individual leaves are to be incorporated, then radiation incident on these individual leaves should be considered with the resultant increase in complexity. Since our present-day level of understanding is sufficiently refined to deal with individual leaves in a canopy, this chapter will emphasize the leaf level of organization. This is not to imply that more simple empiricisms may not be useful for some purposes, but they already exist in abundance in the literature[1] and represent something less than our present level of understanding.

This chapter is formulated with the notion that light (or radiation in general, including near-IR and thermal wavelengths) is only one of many relevant variables; further, the importance of radiation can best be understood by revealing interrelationships with other variables. No attempt is made to review all the relevant literature on radiative transfer in vegetation because several reviews are available,[2,3] and numerous books contain good discussions,[4,5] particularly Ross[6] for an overview and Monteith[7] for detailed discussions specific to numerous plant communities. Nor is the aim to produce a comprehensive set of equations that describe the detailed interactions between radiation and vegetation; their variety is almost without limit, and many are contained in the reviews cited above or in other publications.[6,8-11] The objective is to discuss why radiation models take the form that they do and how they are interfaced with models describing other aspects of plant growth; in particular, photosynthesis.

II. GENERAL CHARACTERISTICS OF RADIATION EXTINCTION IN PLANT CANOPIES

The radiation incident on a canopy from the sun, sky, and soil is attenuated or extinguished as it passes through the vegetation because foliage elements (leaves, stems, flowers, etc.) absorb, reflect, and transmit radiation that they intercept. Clearly, the foliage is exposed to a range of irradiances, varying from near zero to full sunlight depending on the characteristics of the incident radiation and the size, position, orientation, and spectral characteristics of the foliage. Thus, many factors are involved in determining the penetration of radiation into vegetation; a brief list would include:

Radiation above the canopy
- Magnitude
- Angular distribution
- Wavelength distribution
- Sun angle

Spectral properties of foliage and ground surface

Canopy structure
- Foliage area index (leaves, stems, etc.)
- Foliage angular distribution
- Foliage position distribution
- Foliage size and shape

Foliage movement
- Wind, wilting, phototropism, etc.

In fact, there are so many variables, all time dependent, that the traditional approach is to construct a model based on the principles of physics and statistics and then test it under a reasonable range of conditions; even this is difficult because of the wide range of combinations of variables that are possible.

The characteristics of the incident radiation above the canopy are very important in determining the radiant energy available for interception by individual leaves. Considerable information is available on this topic in the book by Kondratyev,[12] the reviews by Davies and Idso[13] and Norman,[3] and the discussions by Monteith[5] and Ross.[6] Approximately one half of the solar radiation is in the visible and thus contributes to photosynthesis; most of the remaining radiation is in the near-IR, with only a few percent in the ultraviolet. Considering the relatively unpredictable effects that clouds, haze, water vapor, pollution, etc. have on visible and solar radiation, it is desirable to measure beam and diffuse radiation in the visible and solar wavelengths separately throughout the day. It is particularly important to measure visible radiation separately from solar radiation because many plant responses are specific to visible wavelengths. For careful radiation work, either the thermal sky radiation or net thermal radiation should be measured; however, it is more easily predicted than solar radiation.[13] For use in radiation models where beam and diffuse radiation are attenuated differently, the sun angle is calculated from astronomical equations,[14] and the diffuse angular distribution is assumed to be isotropic.

The spectral properties of the vegetation determine the contribution of scattered radiation that is intercepted by other foliage and incident upon a given leaf. On the average, leaves absorb about 85% of the incident visible radiation (10% reflected and 5% transmitted) and only 15% of incident near-IR radiation (50% reflected and 35% transmitted).[6] Because of this great difference in leaf spectral properties and because visible radiation is most useful for photosynthesis, it is desirable to model visible and near-IR radiation separately rather than the entire solar spectrum as one waveband. At thermal wavelengths, most foliage has absorbtivities (and thus emissivities) of 95%, with only 5% reflectance. Usually leaves are assumed to behave as black bodies.

The extinction of radiation as it penetrates into a canopy of vegetation is highly variable. To illustrate this variability, it is convenient to use Beer's law for extinction:

TABLE 1

Extinction Coefficients for Direct Beam and Sky Diffuse Radiation in Several Wavelength Bands and Several Canopies

Wavelength band	Canopy	Approximate sun elevation angle ∿10°	∿30°	∿50°	∿70°	Overcast	Ref.
Solar	Barley	—	0.41	—	—	0.34	5
		—	—	0.37	—	0.37	5
	Barley	4.0	1.0	0.55	0.45	—	15
	Corn	2.5	0.95	0.72	0.63	—	16
	Sunflower	—	—	—	—	0.54	17
						0.90	18
	Soybean	2.0	0.9	0.8	0.8	—	19
	Rice	3.0	1.0	0.6	0.35—0.65	—	20,21
	Grass	1.0	0.65	0.5	0.4	—	22
Visible	Corn	1.5	1.0	0.67	0.56	—	23
	Spruce	—	0.6	—	—	0.45	24
	Pine	—	—	—	—	0.4	25
	Sunflower	—	—	—	0.94	—	26
	Sumac	—	—	1.01	0.82	—	26
Near-IR	Spruce	—	0.45	—	—	0.38	24
	Sunflower	—	—	—	0.42	—	26
	Sumac	—	—	0.44	0.37	—	26

$$T = \exp(-KF) \quad (1)$$

where T is the canopy transmittance, F is the leaf area index accumulated from the top of the canopy downward to the height of interest, and K is the extinction coefficient. Table 1 contains a few extinction coefficients that have been measured by various researchers for different incident radiation conditions in several canopies. With the variability apparent in Table 1 and the uncertainties associated with the precise conditions of measurement, clearly an empirical extinction coefficient is of qualitative, or at best, semiquantitative value only. However, the concept of simple exponential extinction appears to be widely applicable for canopies of full cover. Thus, considerable effort has been expended to relate the extinction coefficient to detailed descriptions of the canopy structure.

The first attempt to relate light extinction to canopy geometry mathematically for canopies of full cover approximating random leaf positioning was made by Monsi and Saeki.[27] Their basic equations have been refined and discussed extensively in the literature.[28-31] The approach of Monsi and Saeki was generalized by deWit,[32,33] Ross and Nilson,[11] and Duncan et al.[34] to accommodate nearly any canopy geometry that involved random leaf positioning. Thus, any leaf angle distribution could be used with incident direct beam and sky diffuse radiation to specify the distribution of radiation over variously oriented leaves in visible and near-IR wavelength bands with considerable rigor. Some simplified analytical solutions have been discussed by Cowan.[35] The work of deWit has also been extended to thermal wavelengths.[1,33,36] Some applications of these general equations are available in the literature.[1,8,37]

When leaves are distributed nonrandomly in space, such as conifer needles on branches,[24] plants in widely spaced rows,[38] or widely spaced individual plants,[9] the radiation incident on individual leaves is much more difficult to calculate. The model of Welles[9] appears to be quite general because it predicts the distribution of radiation (visible, near-IR, or thermal) over foliage contained within ellipsoids of any shape and arranged in any array. A brief review of models representing nonrandomly distributed arranged in any array. A brief review of models representing nonrandomly distributed foliage is contained in Norman.[3]

III. COMBINATION OF RADIATION MODELS WITH OTHER LEAF MODELS

A. Radiation Interception Examples

The mathematics describing the radiation incident on individual leaf surfaces with various orientations is moderately complex even for the simplest case of random leaf positioning. It is much easier to specify the average transmittance over the horizontal plane, and scores of models are available in the literature for accomplishing this.[2] Unfortunately, such horizontal averages are not adequate for describing plant-environment interactions at the leaf level of organization because of the variety of leaf angles common to most canopies. Further, average canopy flux densities in the horizontal plane are not adequate for predicting average plant response in that plane if the relation between radiation and plant response is nonlinear. For example, if layer photosynthesis is to be predicted from the average horizontal visible flux density using leaf light-assimilation curves, errors of 50% can easily occur under clear sky conditions. (On an overcast day, these errors will be small.) Thus, dealing with the leaf level of organization may be necessary even for canopy averages.

At this point, it is worth illustrating the application of a radiation interception model to evaluate a plant response. A good example is the calculation of canopy photosynthesis from intercepted light and a leaf light-assimilation relationship. Most of the remainder of this paper will be devoted to an analysis of several methods for interfacing leaf photosynthesis information to light-interception models. This calculation can proceed by several methods which are listed in order of their decreasing complexity:

1. Determine the beam and diffuse light flux densities in layers distributed throughout the depth of the canopy, calculate the incident light and leaf photosynthesis for classes of leaves that are shaded and exposed at various angles to the direct sun, and evaluate layer photosynthesis from the sum of all these leaf classes weighted by their respective areas.
2. Determine the beam and diffuse light flux densities in layers distributed throughout the depth of the canopy, calculate the light incident on sunlit and shaded leaves separately, assuming a mean angle between the direction of the sun and the plane of sunlit leaves, and evaluate layer photosynthesis from the sum of sunlit and shaded photosynthesis rates weighted by their respective areas.
3. Determine the average light flux density in various layers distributed throughout the depth of the canopy and apply the light-assimilation relation layer-by-layer using this average flux density in the horizontal plane.

These three methods will be applied in detail under the conditions listed in Table 2. The canopy approximates a soybean canopy with exposure near mid-day under relatively clear sky conditions. The leaf photosynthesis parameters are estimated from data for soybeans presented by Singh et al.[39] (Figure 1). The spherical leaf angle distribution used in this example commonly occurs in "real" canopies and provides a good basis for comparison because it contains the full range of leaf angles possible.

1. Case 1

This is probably the most complex and complete method for calculating canopy

TABLE 2

Incident Light Conditions and Canopy Characteristics for Sample Calculations of Photosynthesis

Incident radiation
Visible flux density

Direct beam	(horizontal plane)	= 300 W/m^{-2}
Sky diffuse	(horizontal plane)	= 100 W/m^{-2}
Total	(horizontal plane)	= 400 W/m^{-2}
Direct beam	(direction of sun)	= 366 W/m^{-2}

Solar zenith angle = 35°

Canopy characteristics
Leaf area index = 2.5
Number of layers = 5
Leaf angle distribution — spherical (5 leaf angle classes) (approximates soybean)

Leaf characteristics
Transmittance = 0.05
Reflectance = 0.10
Maximum net photosynthetic rate = 1000 $\mu g/m^{-2}/s$
Dark respiration rate = 139 $\mu g/m^{-2}/s$
Light compensation point = 16 W/m^{-2}
Biochemical efficiency in low light = 15.4 $\mu g/J$

photosynthetic rates from leaf photosynthetic rates because it requires dividing the canopy into N layers and M leaf angle classes in a manner similar to deWit.[32,33] In this example, only five layers and five leaf angle classes are used; in practice, ΔF, the leaf area index in each layer, usually is chosen as 0.1 or 0.2 to minimize errors in scattering estimates (especially in the near-IR) and M = 9. However, such large values of M usually are not necessary as will become evident later.

The direct beam transmittance below layer j of a canopy with a spherical leaf distribution is

$$
\begin{aligned}
T_{B,j} &= \exp\left[-(F_j + \Delta F/2) / (2 \cos\theta)\right] \\
&= \exp\left[-(F_j + 0.25) / 1.638\right]
\end{aligned}
\qquad (2)
$$

where F_j is the leaf area index at the midpoint of layer j, and θ is the zenith angle of the sun (Table 3). The product of $T_{B,j}$ and beam flux density incident on the horizontal plane above the canopy ($I_{B,O}$) provides an estimate of the horizontally averaged, beam flux density in each layer (Table 3).

$$I_{B,j} = I_{B,O} \, T_{B,j} \qquad (3)$$

It is worth noting that $I_{B,j}$ is not the flux density incident on sunlit leaves. The transmittance of diffuse sky radiation ($T_{d,j}$) can be approximated by an integral over the hemisphere of Equation 2 for isotropic radiation;[35]

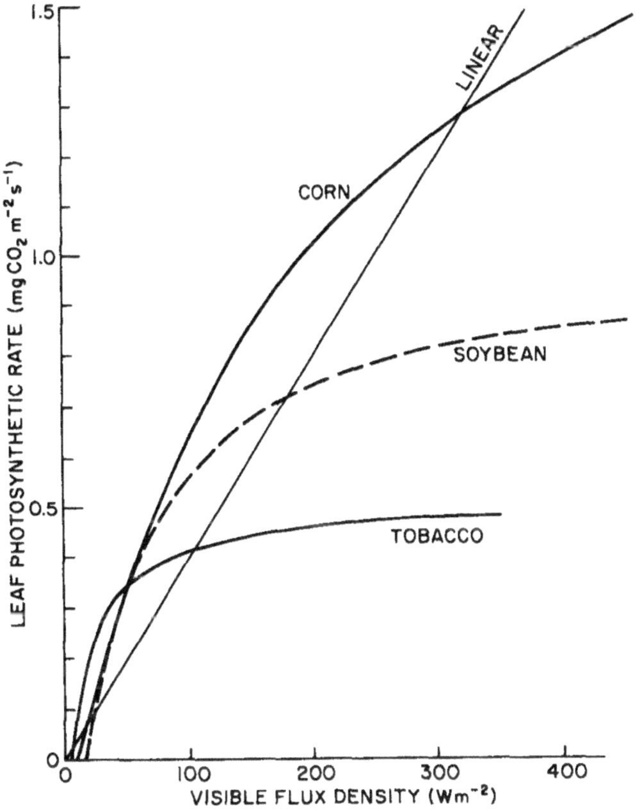

FIGURE 1. Light-assimilation curves used in sample calculations; curves are typical of corn,[40] soybeans,[39] and tobacco.[40] A straight line also is included to evaluate effects of nonlinearity.

$$T_{d,j} = (1 - F_j/2)\exp(-F_j/2) - (F_j/2)^2\, Ei(-F_j/2) \qquad (4)$$

where $Ei(-F/2)$ is the exponential integral tabulated in any handbook of mathematical tables (Table 3). The approximate sky diffuse flux density incident on leaves in a layer ($I_{d,j}$) is derived from the product of $T_{d,j}$ and the sky diffuse flux density above the canopy ($I_{d,o}$) (Table 3):

$$I_{d,j} = I_{d,O}\, T_{d,j} \qquad (5)$$

This diffuse flux density, $I_{d,j}$, is calculated at the midpoint of each layer to reduce computation errors associated with the relatively thick layers of $\Delta F = 0.5$. The direct beam and sky diffuse radiation intercepted by individual leaves is scattered radiation because the leaf transmissivity (τ) and leaf reflectivity (ρ) are not zero. The values given in Table 2 are reasonable average values for vegetation.[6] An approximate method for accounting for the scattered beam radiation is to evaluate Equation 2 at a leaf area index given by $(1 - \rho - \tau)(F_j + \Delta F/2)$ and difference this result from Equation 2 evaluated at $F_j + \Delta F/2$. The scattered sky diffuse radiation can be estimated by differ-

TABLE 3

Radiation Interception Results for Incident Radiation, Canopy and Leaf Characteristics Given in Table 2

Layer j	F_j[a]	$T_{B,j}$[a]	$I_{B,j}$[a] (W/m^{-2})	$T_{d,j}$	$I_{d,j}$ (W/m^{-2})	$I_{s,j}$ (W/m^{-2})	$I_{B,j} + I_{d,j} + I_{s,j}$ (W/m^{-2})	Sunlit LAI ΔF_j*	Shaded LAI $\Delta F - \Delta F_j$*
0	0	1.000	300	1.000	100	0	400	—	—
1	0.25	0.859	258	0.798	80	9	347	0.431	0.069
	0.50	0.737	221						
2	0.75	0.633	190	0.535	54	18	262	0.318	0.182
	1.00	0.543	163						
3	1.25	0.466	140	0.370	37	22	199	0.234	0.266
	1.50	0.400	120						
4	1.75	0.344	103	0.261	26	23	152	0.172	0.328
	2.00	0.295	89						
5	2.25	0.253	76	0.186	19	22	117	0.128	0.372
	2.50	0.217	65						
								Total 1.283	1.217

[a] The variables F_j, $T_{B,j}$, and $I_{B,j}$ are evaluated at the layer midpoints and at the bottom of each layer. All other variables in the table are evaluated at the layer midpoints.

encing the results of Equation 4 evaluated at $(1 - \varrho - \tau)\ F_j$ and evaluated at F_j. This method appears to be acceptable for visible radiation, but is unacceptable for near-IR radiation. Thus, if leaf energy balances, which require near-IR attenuation, are to be incorporated, the more rigorous model also could be used for visible wavelengths. The average visible flux density in any canopy layer from all effects is $I_{B,j} + I_{d,j} + I_{s,j}$ (Table 3).

Careful consideration of canopy photosynthesis requires separating sunlit and shaded leaves into separate groups because averaging over this broad range of flux densities produces errors when the light assimilation curve for individual leaves is non linear. For a spherical leaf distribution, the sunlit leaf area index in a layer is given by ΔF_j^*.

$$\Delta F_{j\bullet}^{*} = (T_{B,J} - T_{B,j+1})\ 2\cos\theta$$
$$= (T_{B,j} - T_{B,j+1})\ 1.638 \qquad (6)$$

where $T_{B,O} = 1.0$. The fraction of leaf area in a layer that is sunlit is given by $\Delta F_j^*/\Delta F$; the shaded leaf area index in each layer is given by $\Delta F - \Delta F_j^*$.

The sunlit leaves in the canopy must be divided into leaf classes (i) that are distinguished by various angles between the leaf normal and the direction of the sun (α_i). For a canopy with a spherical leaf angle distribution, the fraction of sunlit leaf area (f_i) exposed at various angles to the sun is independent of solar zenith angle and given by

$$f_i = \sin\ \alpha_i\ d\alpha \qquad (7)$$

where α_i is the midpoint angle of each leaf class (for $M = 5$, $\alpha_1 = 9°$, $\alpha_2 = 27°, \ldots \alpha_5 = 81°$ (Table 4). The spherical leaf distribution is somewhat simpler to work with than the general case because the distribution of leaf inclination angles to the horizontal is the same as the distribution of leaf-sun angles. Sometimes leaf-sun angle classes are divided into equal increments of sin α_i rather than equal increments of α_i; this is preferrable, but probably makes little difference in the results. The sunlit leaf area index in each angle class i and layer j (ΔF^*_{ji}) is given by

$$\Delta F_{ji}^{*} = f_i\ \Delta F_{j\bullet}^{*} \qquad (8)$$

The direct beam flux density appropriate to each leaf angle class is $I^*_{B,i}$ (Table 4):

$$I_{B,i}^{*} = I_{B,O}\ \cos\alpha_i/\cos\theta \qquad (9)$$

where $I_{B,O}/\cos\theta$ is the beam flux density on a plane perpendicular to the direction of the sun. The total flux density incident on each angle class of sunlit leaves is calculated with the assumption that all leaves in a layer are exposed to the same diffuse flux densities as a horizontal leaf:

$$I_{B,i}^{*} + I_{d,j} + I_{s,j} = I_{B,i}^{*} + I_{d,j}^{*} \qquad (10)$$

TABLE 4

Summary of Results from the Calculation of Leaf, Layer, and Canopy Photosynthesis for the Spherical Leaf-Distribution Example

						Sunlit leaves				Shaded leaves		
						Light flux (W/m²)		Photosynthesis (μg/m²/s)		Light flux (W/m²)	Photosynthesis (μg/m²/s)	
Layer (j)	Leaf-angle class (i)	α_i	Fraction sunlit leaf area (f_i)	Sunlit LAI (ΔF^*_{ij})	Shaded LAI ($\Delta F - \Delta F^*_{j}$)	Beam ($I^*_{B,i}$)	Total	Per unit leaf area	Per unit soil area	($I^*_{d,j}$)	Per unit leaf area	Per unit soil area
1	1	9	0.049	0.0211		362	451	870	18.4			
	2	27	0.142	0.0612		326	415	860	52.6			
	3	45	0.221	0.0953		259	348	836	79.7			
	4	63	0.279	0.1202		166	255	786	94.5			
	5	81	0.309	0.1332		57	146	667	88.8			
Layer totals			1.000	0.431	0.069				334.0	89	529	36.5
2	1	9	0.049	0.0156		362	434	866	13.5			
	2	27	0.142	0.0452		326	398	855	38.6			
	3	45	0.221	0.0703		259	331	829	58.2			
	4	63	0.279	0.0887		166	238	774	68.7			
	5	81	0.309	0.0982		57	129	635	62.4			
Layer totals			1.000	0.318	0.182				241.4	72	463	84.1
3	1	9	0.049	0.0115		362	421	862	9.9			
	2	27	0.142	0.0332		326	385	850	28.2			
	3	45	0.221	0.0517		259	318	823	42.5			
	4	63	0.279	0.0653		166	225	763	49.7			
	5	81	0.309	0.0723		57	116	606	43.8			
Layer totals			1.000	0.234	0.266				174.1	59	398	105.8

4	1	9	0.049	0.0084		362	411	859	7.2			
	2	27	0.142	0.0244		326	375	847	20.7			
	3	45	0.221	0.0380		259	308	818	31.1			
	4	63	0.279	0.0481		166	215	754	36.2			
	5	81	0.309	0.0531		57	106	581	30.8			
Layer totals			1.000	0.172	0.328				126.0	49	337	110.5
5	1	9	0.049	0.0063		362	403	856	5.4			
	2	27	0.142	0.0182		326	367	844	15.4			
	3	45	0.221	0.0282		259	300	814	23.0			
	4	63	0.279	0.0357		166	207	746	26.6			
	5	81	0.309	0.0396		57	98	558	22.1			
Layer totals			1.000	0.128	0.372				92.5	41	278	103.4
						Canopy total			968.0			440.3

Note: The grand total photosynthetic rate for the whole canopy is 1408 μg CO_2 per m^2 soil area per second.

Shaded leaves receive only diffuse light ($I_{d,j} + I_{s,j} = I^*_{d,j}$), and they can be treated as a single class of leaves (Table 4).

The photosynthetic rate per unit leaf area can be calculated from the light incident on sunlit leaves making various angles with the direction of the sun and on shaded leaves separately if the light assimilation response is known. A general equation often used to describe net carbon dioxide uptake [$P_{ji}(I)$] for leaves in layer j and inclination angle i as a function of light is

$$P_{ji}(I) = \frac{P_{max}\, b\,(I_{ji} - I_o)}{P_{max} + b\,(I_{ji} - I_o)} \quad \text{for } I \geq I_o$$

$$P_{ji}(I) = R\left(\frac{I_{ji}}{I_o} - 1\right) \quad \text{for } I < I_o \qquad (11)$$

where P_{max} is the maximum net photosynthetic rate at high light flux densities, b is the slope of the $P_{ji}(I)$ vs. I_{ji} curve near the light compensation point I_o, and I_{ji} is a general symbol used to denote incident light flux density in layer j and leaf angle class i. For diffuse light, the symbol $P_{j\cdot}(I^*_{d,j})$ is used to indicate that the rate is assumed independent of leaf angle and identical for all shaded leaves in a layer (Table 4). The photosynthetic rate per unit of ground area for each layer j and leaf angle class is obtained from $P_{ji}(I^*_{B,i})\,\Delta F^*_{ji}$ for sunlit leaves and $P_{j\cdot}(I^*_{d,j})(\Delta F - \Delta F^*_{ji})$ for shaded leaves in a layer (Table 4). Thus the layer photosynthetic rate is

$$\sum_{i=1}^{M} [P_{ji}(I^*_{B,i})\,\Delta F^*_{ji}] + P_{j\bullet}(I^*_{d,j})(\Delta F - \Delta F^*_{j\bullet}) \qquad (12)$$

and the canopy photosynthetic rate is

$$\sum_{j=1}^{N} \sum_{i=1}^{M} [P_{ji}(I^*_{B,i})\,\Delta F^*_{ji}] + \sum_{j=1}^{N} P_{j\bullet}(I^*_{d,j})(\Delta F - \Delta F^*_{j\bullet}) \qquad (13)$$

Several interesting observations can be made from the results in Table 4. About one third of the total canopy photosynthesis is the result of diffuse light on a clear day near noon. Under any other conditions, it is likely to be even greater. This is the result of light saturation at the high direct-beam flux densities. Although leaves oriented nearly perpendicular to the solar beam have the highest photosynthetic rate per unit leaf area, they have the lowest rate per unit soil area because relatively few leaves are so oriented in a canopy with a spherical leaf distribution. About one half of the total canopy photosynthesis results from the top two layers or the top-unit leaf area index. After the light has penetrated to below a leaf area index of 2, shaded leaves are contributing more to the layer photosynthesis than sunlit leaves.

2. Case 2

This case is very much simpler than the previous case because it is not necessary to deal with leaf inclination classes, although it is necessary to work with sunlit and shaded leaves separately. Using this approach, measurements of light extinction, including sunfleck area estimates, could be used to obtain reasonable canopy photosynthesis values with some knowledge of the mean angle between the foliage and the sun. This is a tremendous simplification over Case 1 where the leaf angle distribution is required.

A simple numerical characteristic of foliage inclination is discussed by Ross[6] and is called the inclination index of the foliage χ_L; it has the following properties:

$\chi_L = 1$	Horizontal leaves
$0 < \chi_L < 1$	Predominently horizontal leaves
$\chi_L = 0$	Equal leaf area fractions at all inclination angles (uniform leaf angle distribution)
$-1 < \chi_L < 0$	Predominently vertical leaves
$\chi_L = -1$	Vertical leaves

Values of χ_L are tabulated for numerous canopies in Ross.[6] For a canopy with a spherical leaf distribution, $\chi_L = -0.22$. This simple canopy inclination index can be related to the mean angle between the foliage and the sun, using the so called "G-function" discussed in Ross.[6] Using approximations in Ross,[6] this mean angle, α, was estimated at about 60° and $\cos \alpha = 0.50$. (A more careful integration of α, $\sin \alpha$, $d\alpha$ yields 57°.) Therefore, the direct beam contribution to all sunlit leaves becomes $366 \cos \alpha = 183$ W/m^{-2}. The results from this method are contained in Table 5, using light penetration results from Table 3. From the results in Table 5, it is clear that the nonlinearity in the light assimilation relation produces negligible errors in the comparison of Case 1 and Case 2. In fact, layer and canopy photosynthetic rates in Table 5 are within a few percent of those in Table 3. This suggests that the most important distinction is between sunlit and shaded leaves rather than sunlit leaves at various angles. This example provides evidence that this is a good method; however, it may not always work this well. If we go one step further and use this mean leaf inclination angle to calculate diffuse sky light penetration and sunlit leaf area index (rather than using the actual spherical leaf angle distribution as in Table 5), then using the mean inclination angle introduces an additional error associated with sunlit leaf area. This occurs because the extinction coefficients for the spherical leaf angle distribution are different than for a canopy of azimuthal symmetry with all leaves inclined at the mean angle for a spherical canopy (Table 6). However, the extinction coefficients for the two angular distributions have a similar behavior even for a wide range of solar zenith angles. Using this mean leaf inclination angle to obtain sunlit leaf area results in an overestimate of canopy photosynthesis of about 10%. Therefore it is desirable to attempt a better estimate of the leaf angle distribution for light extinction than is obtained from using a simple mean inclination angle. However, the results in Table 5 suggest that when sunlit foliage area is estimated reliably, then the mean leaf-sun angle can be used for estimates of photosynthesis for sunlit leaves.

3. Case 3

The simplest method for estimating canopy photosynthetic rates uses the average visible flux density in the horizontal plane to evaluate the photosynthetic rate from the light assimilation relation (Table 7). In this example, this method produces an overestimate of the canopy photosynthetic rate of 30%. In general this is not an acceptable method.

TABLE 5

Canopy and Layer Photosynthetic Rates Using Input Data From Table 2, Light Penetration Data From Table 3, and Using Only the Mean Angle Between the Leaves and the Sun

			Sunlit leaves				Shaded leaves		
			Light flux (W/m^{-2})		Photosynthesis ($\mu g/m^{-2}/s$)		Light (W/m^{-2})	Photosynthesis ($\mu g/m^{-2}/s$)	
Layer (J)	F_j	Sunlit LAI	Beam	Total	Per unit leaf area	Per unit soil area		Per unit leaf area	Per unit soil area
1	0.25	0.431	183	272	798	344	89	529	36
2	0.75	0.318	183	255	786	250	72	463	84
3	1.25	0.234	183	242	777	182	59	398	106
4	1.75	0.172	183	232	769	132	49	337	110
5	2.25	0.128	183	224	762	98	41	278	103
				Canopy total		1006			439
				Grand total		1445	$\mu g\ CO_2/m^{-2}/s$		

TABLE 6

Extinction Coefficients for Two Canopy Leaf Angle Distributions: (1) Spherical ($K_{spherical}$) and (2) Azimuthal Symmetry and All Leaves Inclined at 60° to the Horizontal ($K_{60°}$)

Solar zenith angle (°)	$K_{spherical}$	$K_{60°}$[a]
5	0.50	0.50
15	0.52	0.50
25	0.55	0.50
35	0.61	0.53 (example used)
45	0.71	0.65
55	0.87	0.85
65	1.18	1.23
75	1.93	2.08
85	5.74	6.31

[a] See Ross,[6] page 30 for equation.

4. Case Summary

The three methods illustrated above cover a wide range of complexity, and the differences between them depend on many factors, such as the canopy leaf angle distribution, sun angle, fraction of the radiation above the canopy that is diffuse, and the shape of the light-assimilation relationship. In general these effects are difficult to isolate; however, if we consider canopies with horizontal leaves, then the differences in estimated photosynthesis arise mainly from differences in the light assimilation curves. Table 8 contains estimates of canopy photosynthesis by the three methods discussed for a canopy of horizontal leaves with the four different light-assimilation relations shown in Figure 1. All methods give the same results for a linear light-photosynthesis curve. As the nonlinearity increases, Case 3 overestimates by an increasing amount (28% overestimate for corn and 52% overestimate for soybeans). However, with tobacco, the most nonlinear of the light-assimilation curves, the overestimate of

TABLE 7

Canopy and Layer Photosynthetic Rates Calculated from the Direct Application of Average Horizontal Light Flux Density in the Leaf Photosynthesis Equation

Layer (j)	F_j	Visible flux density (W/m^{-2})	Photosynthetic rates ($\mu g/m^{-2}/s$) Per unit leaf area	Per unit soil area
1	0.25	347	836	418
2	0.75	262	791	396
3	1.25	199	738	369
4	1.75	152	677	338
5	2.25	117	609	305
			Grand total	1826

TABLE 8

Canopy Photosynthesis Estimates in Milligrams CO_2 per m^2 (soil area) per s for Three Light-Assimilation Relations (Figure 1) Calculated by Three Methods

Light-assimilation curve	Leaf angle distribution	Canopy photosynthesis method of calculation: Case 1	Case 2	Case 3
Linear	Horizontal	1.60	1.60	1.60
Corn		1.68	1.68	2.03
Soybeans		1.18	1.18	1.56
Tobacco		0.93	0.93	1.06
Linear	Spherical	1.57	1.57	2.15
Corn		1.81	1.92	2.55
Soybeans		1.41	1.44	1.83
Tobacco		1.01	1.03	1.13

Case 3 is only 27%; this arises because diffuse light on shaded leaves becomes increasingly important when light saturation values are low, and the treatment of diffuse light is the same for Cases 1, 2, and 3. It is interesting to note that a canopy with a spherical leaf angle distribution can produce more photosynthate (other conditions being equal) than one with a horizontal leaf habit (Table 8). Even though more of the incident radiation on the horizontal plane above the canopy is intercepted by a horizontal leaf canopy, the beam radiation intercepted by the spherical leaf distribution canopy is distributed over more leaves (horizontal sunlit LAI = 0.92; spherical sunlit LAI = 1.28) at a lesser flux density with a resultant increase in potential productivity. Again, this is particularly true for the more nonlinear light-assimilation curves (if the direct beam radiation dominates). The canopy photosynthesis estimates for Case 1 and Case 2 differ insignificantly for both horizontal and spherical leaf canopies. The discrepancy between Case 1 and Case 3 for the spherical leaf distribution canopy arises from two sources: (1) nonlinearity in the light-assimilation relation and (2) the failure of the assumption that beam radiation is uniformly distributed over all leaves. The failure of this latter assumption is particularly evident for the linear light-assimilation relation. Unfortunately, many of the results in Table 7 will change with solar zenith angle;

therefore, these examples provide some interesting insights, but a comprehensive analysis is quite difficult.

One final conclusion that is evident from these examples is the following: although Case 1 is the most comprehensive method for calculating canopy assimilation from leaf assimilation, Case 2 can provide good estimates with much less effort, and Case 3 is unacceptable except for diffuse light.

The three methods illustrated above are all limited to canopies that approximate random leaf positioning. Estimates of mean light extinction are possible for highly nonrandom canopies, whether they are clumped or regular,[41,42] and the methods discussed in this section can be modified to provide approximate photosynthesis estimates mainly through adjustments in sunlit leaf area.

Another factor not considered in the above three methods is penumbral effects that result from the finite (½°) size of the sun. The effect of penumbra is to remove energy from bright sunflecks and distribute it at lesser flux densities over larger areas of leaf. Although it is difficult to calculate penumbral effects precisely,[43,44] a relatively simple method can be used to estimate an upper limit of its effect. Using the method in Case 2, if we assume the beam flux density to be uniformly distributed over all the leaves in a layer at an intensity level given by

$$I_{B,j} \cos \bar{\alpha} / \cos \theta \qquad (14)$$

then the total flux density incident on all leaves in the layer is given by

$$I_{B,j} \cos \bar{\alpha}/\cos \theta + I^*_{d,j} \qquad (15)$$

This produces an upper estimate of penumbra effects that can never be realized by any canopy growing in the open, but it may be approximated by a very tall canopy with very small leaves. From Table 9, it appears that penumbral effects would be greatest in horizontal leaf canopies. However, in a more typical canopy, such as one with a spherical leaf angle distribution, maximum enhancement due to penumbral effects is about 15%, and in an actual canopy, the effects are likely to be much less than this; perhaps a few percent. Thus, the effect of penumbra on photosynthesis estimates is likely to be small, even in most tall canopies. It is only likely to be important for plants with horizontal leaves growing beneath other tall canopies.

B. General Model Approaches

The previous section dealt with interfacing several radiation interception models with a leaf photosynthesis relation. Although this probably is the most common use of radiation interception models, such models can be useful in a wide range of applications. For example, the radiation incident on individual leaves not only affects carbon dioxide fixation, but also the leaf energy budget. Sunlit leaves exposed to high irradiances tend to have higher leaf temperatures, lower stomatal resistances, more negative leaf water potentials, and higher photosynthetic rates than shaded leaves. The relations between leaf photosynthetic rates or water use and various environmental factors, such as air temperature, soil water tension, vapor pressure deficit, incident radiation, wind speed, etc. all are nonlinear. Therefore, a comprehensive model that is to accommodate the numerous environmental factors in a realistic manner must be

TABLE 9

Comparison of Canopy Photosynthesis in Milligrams CO_2 per m^2 (soil area) per s with Maximum Penumbral Effects and With No Such Effects for Three Light-Assimilation Relations with Horizontal and Spherical Leaf-Angle Distributions Calculated by the Method of Case 2

Light assimilation curve	Leaf angle distribution	Canopy photosynthesis: Without penumbra	Canopy photosynthesis: With maximum penumbra	Increase from maximum penumbra (%)
Corn	Horizontal	1.68	2.03	37
Soybeans		1.18	1.55	31
Tobacco		0.93	1.06	14
Corn	Spherical	1.81	2.11	17
Soybeans		1.41	1.64	16
Tobacco		1.01	1.09	8

structured to handle such nonlinearities. For example, the leaf energy budget might be evaluated separately for each leaf angle class listed in Table 4. Then the characteristic of interest (evaporation, sensible heat flux, or leaf temperature) is evaluated for each leaf angle class in each layer and accumulated to provide layer or canopy estimates in a manner directly analogous to Case 1 in the previous section. This method has been used[1,8,37] and probably provides the best estimates of canopy fluxes that presently are available. Simplifications analogous to Case 2 and Case 3 in the previous section also are possible and certainly are more commonly encountered in the literature.

IV. SYMBOLS

b	Slope of light-photosynthesis relation near light compensation point (μg/J)
f_i	The fraction of sunlit leaf area in leaf angle class i
F	General symbol for leaf area index accumulated from top of canopy downward
F_j	Cumulative leaf area index from top of canopy to midpoint of layer j
ΔF	Leaf area index in each layer j
ΔF^*_{ji}	Sunlit leaf area index in layer j and leaf angle class i
$\Delta F_{j\cdot}^*$	Sunlit leaf area index in layer j including all leaf angle classes i
i	Arbitrary leaf angle class
I_o	Visible flux density at which leaf net photosynthesis is zero
$I_{B\div o}$	Incident direct visible radiation above canopy on horizontal plane (W/m^{-2})
$I_{d\div o}$	Sky diffuse radiation above canopy on a horizontal plane (W/m^{-2})
$I_{B\div j}$	Direct beam radiation averaged over horizontal plane below layer j (W/m^{-2})
$I_{d\div j}$	Sky diffuse radiation at midpoint of layer j (W/m^{-2})
$I_{B,o}$	Incident direct visible radiation above canopy on horizontal plane (W/m^{-2})
$I_{d,o}$	Sky diffuse radiation above canopy on a horizontal plane (W/m^{-2})
$I_{B,j}$	Direct beam radiation averaged over horizontal plane below layer j (W/m^{-2})

$I^*_{d,j}$	Total diffuse radiation incident on sunlit (and shaded) leaves in layer j
$I^*_{B,i}$	Direct-beam flux density incident on sunlit leaves in leaf angle class i (same for all layers) (W/m^{-2})
j	Arbitrary layer (j = 1, top layer and j = N, bottom layer)
K	General symbol for radiation extinction coefficient
LAI	Leaf area index
M	Number of leaf angle classes, i
N	Number of canopy layers, j
P_{max}	Maximum photosynthetic rate at light saturation ($\mu g/m^{-2}/s$)
$P_{j,i}(I)$	Net photosynthetic rate as a function of visible flux density in layer j and leaf angle class i ($\mu g/m^{-2}/s$)
T	General symbol for canopy transmittance
$T_{B,j}$	Direct beam transmittance to below layer j ($T_{B,O} = 1.0$)
$T_{d,j}$	Transmittance of diffuse sky radiation to the midpoint of layer j
α_i	Angle between normal to plane of leaf and direction of sun at midpoint of angle class i
α	Mean angle between leaf normal and direction of the sun for canopy
θ	Solar zenith angle
χ_L	Inclination index of the foliage
ϱ	Leaf reflectivity
τ	Leaf transmissivity

REFERENCES

1. **Norman, J. M.,** Modeling the complete crop canopy, in *Modification of The Aerial Environment of Plants,* Barfield, B. J. and Gerber, J. F., Eds., ASAE Monograph, American Society of Agricultural Engineers, St. Joseph, Mich., 1978, 249.
2. **Lemeur, R. and Blad, B. L.,** A critical review of light models for estimating the shortwave radiation regime of plant canopies, *Agric. Meteorol.,* 14, 255, 1974.
3. **Norman, J. M.,** Radiative transfer in vegetation, in *Heat and Mass Transfer in The Biosphere,* de-Vries, D. A. and Afgan, N. H., Eds., Scripta Book Co., Washington, D.C., 1975, 187.
4. **Monteith, J. L.,** Light interception and radiative exchange in crop stands, in *Physiological Aspects of Crop Yield,* Eastin, J. D., Haskins, F. A., Sullivan, C. Y., and van Bavel, C. H. M., Eds., American Society of Agronomy, Madison, Wisconsin, 1969, 89.
5. **Monteith, J. L.,** *Principles of Environmental Physics,* American Elsevier, New York, 1973, 241.
6. **Ross, J.,** Radiative transfer in plant communities, in *Vegetation and The Atmosphere,* Vol. 1, Monteith, J. L., Ed., Academic Press, New York, 1975, 13.
7. **Monteith, J. L., Ed.,,** *Vegetation and The Atmosphere,* Vol. 2, Academic Press, New York, 1976, 439.
8. **Goudriaan, J.,** *Crop Micrometeorology: A Simulation Study,* PUDOC, Centre for Agric. Publ. and Doc., Wageningen, The Netherlands, 1977, 249.
9. **Welles, J. M.,** A Model of Foliage Temperatures for a Heated Orchard, Master's thesis, The Pennsylvania State University, University Park, 1978, 88.
10. **Norman, J. M. and Welles, J. M.,** Radiative transfer in an array of canopies, *Agron. J.,* submitted.
11. **Ross, J. K. and Nilson, T. A.,** A mathematical model of radiation regime in the plant cover, in *Actinometry and Atmospheric Optics,* Pyldmaa, U. K., Ed., 1966, Translated from Russian by Israel Program for Scientific Translations, Cat. No. 5835, 253, 1971.
12. **Kondratyev, K. Ya.,** *Radiation in The Atmosphere,* Academic Press, New York, 1969.
13. **Davies, J. A. and Idso, S. B.,** Estimating the surface radiation balance and its components, in *Modification of The Aerial Environment of Plants,* Barfield, B. J. and Gerber, J. F., Eds., ASAE Monograph, American Society of Agricultural Engineers, St. Joseph, Mich., 1978, 183.
14. **List, R. J.,** *Smithsonian Meteorological Tables,* Vol. 11, Smithsonian Miscellaneous Collections, Smithsonian Institution, 1966.

15. **Udagawa, T. and Uchijima, Z.**, Studies of energy and gas exchange within crop canopies. V. Geometrical structure of barley canopies and the penetration of direct solar radiation into the canopy, *Proc. Crop Sci. Soc. Jpn.*, 38, 364, 1969.
16. **Uchijima, Z., Udagawa, T., Hovie, T., and Kobayashi, K.**, Studies of energy and gas exchange within crop canopies. IV. The penetration of direct solar radiation into corn canopy and the intensity of direct radiation on the foliage surface, *J. Agric. Meteor. (Jpn.)*, 24, 141, 1968.
17. **Kuroiwa, S.**, Theoretical analysis of light factor and photosynthesis in plant communities. III. Total photosynthesis of a foliage under parallel light in comparison with that under isotropic light condition, *J. Agric. Meteor. (Jpn.)*, 24, 75, 1968.
18. **Saugier, B.**, Sunflower, in *Vegetation and The Atmosphere*, Vol. 2, Monteith, J. L., Ed., Academic Press, New York, 1976, 91.
19. **Ito, A. and Udagawa, T.**, Phytometrical studies of plant canopies. I. Geometrical structure of soybean canopy and light penetration, *J. Agric. Meteor. (Jpn.)*, 26, 187, 1971.
20. **Ito, A.**, Geometrical structure of rice canopy and penetration of direct solar radiation, *Proc. Crop Sci. Soc. Jpn.*, 38, 355, 1969.
21. **Udagawa, T., Ito, A., and Uchijima, Z.**, Phytometrical studies of crop canopies. III. Radiation environment in rice canopies, *Proc. Crop Sci. Soc. Jpn.*, 43, 180, 1974.
22. **Ripley, E. A. and Redmann, R. E.**, Grassland, in *Vegetation and The Atmosphere*, Vol. 2, Monteith, J. L., Ed., Academic Press, New York, 1976, 366.
23. **Norman, J. M.**, Unpublished measurements from corn taken at Guelph, Ontario, 1969.
24. **Norman, J. M. and Jarvis, P. G.**, Photosynthesis in Sitka Spruce (*Picea sitchensis* (Bong.) Carr.). V. Radiation penetration theory and a test case, *J. Appl. Ecol.*, 12, 839, 1975.
25. **Jarvis, P. G., James, G. B., and Landsberg, J. J.**, Coniferous forest, in *Vegetation and The Atmosphere*, Vol. 2, Monteith, J. L., Ed., Academic Press, New York, 1976, 178.
26. **Norman, J. M., Miller, E. E., and Tanner, C. B.**, Light intensity and sunfleck size distributions in plant canopies, *Agron. J.*, 63, 743, 1971.
27. **Monsi, M. and Saeki, T.**, Uber den lichtfaktor in den Pflanzengesselschaften und seine bedeutung fur die stoffproduktion, *Jpn. J. Bot.*, 14, 22, 1953.
28. **Warren Wilson, J.**, Stand structure and light penetration. I. Analysis by point quadrats, *J. Appl. Ecol.*, 2, 383, 1965.
29. **Warren Wilson, J.**, Stand structure and light penetration. III. Sunlit foliage area, *J. Appl. Ecol.*, 4, 159, 1967.
30. **Anderson, M. C.**, Stand structure and light penetration. II. A theoretical analysis, *J. Appl. Ecol.*, 3, 41, 1966.
31. **Anderson, M. C. and Denmead, O.**, Shortwave radiation on inclined surfaces in model plant communities, *Agron. J.*, 61, 867, 1969.
32. **deWit, C. T.**, *Photosynthesis of Leaf Canopies*, Agricultural Research Report No. 663, Institute for Biol. and Chem. Res. on Field Crops and Herbage, Wageningen, The Netherlands, 1965, 57.
33. **Idso, S. B. and deWit, C. T.**, Light relations in plant canopies, *Appl. Opt.*, 9, 177, 1970.
34. **Duncan, W. G., Loomis, R. S., Williams, W. A., and Hanau, R.**, A model for simulating photosynthesis, *Hilgardia*, 38, 181, 1967.
35. **Cowan, I.**, The interception and absorption of radiation in plant stands, *J. Appl. Ecol.*, 5, 367, 1968.
36. **Ross, J. K. and Tooming, H. G.**, Distribution of the long wave radiation fluxes and the radiation balance in plant cover, in *Actinometry and Atmospheric Optics*, Pyldmaa, U. K., Ed., 1966, Translated from Russian by Israel Program for Scientific Translations, Cat. No. 5835, 287, 1971.
37. **Stewart, D. W.**, A Simulation of Net Photosynthesis, Ph.D. thesis, Cornell University, Ithaca, New York, University Microfilms, Ann Arbor, Mich., *Diss. Abstr.*, No. 70-17101), 141, 1970.
38. **Allen, L. H., Jr.**, A model of light penetration in a wide-row crop, *Agron. J.*, 66, 41, 1974.
39. **Singh, M., Ogren, W. L., and Widholm, J. M.**, Photosynthetic characteristics of several C_3 and C_4 plant species grown under different light intensities, *Crop Sci.*, 14, 563, 1974.
40. **Hesketh, J. D. and Moss, D. N.**, Variation in the response of photosynthesis to light, *Crop Sci.*, 3, 107, 1963.
41. **Acock, B., Thornley, J. H. M., and Warren Wilson, J.**, Spatial variation of light in the canopy, in *Prediction and Measurement of Photosynthetic Productivity*, Setlik, I., Ed., PUDOC, Wageningen, The Netherlands, 1969, 91.
42. **Nilson, T.**, A theoretical analysis of the frequency of gaps in plant stands, *Agric. Meteorol.*, 8, 25, 1971.
43. **Miller, E. E. and Norman, J. M.**, A sunfleck theory for plant canopies. I. Length of sunlit segments along a transect, *Agron. J.*, 63, 735, 1971.
44. **Miller, E. E. and Norman, J. M.**, A sunfleck theory for plant canopies. II. Intensity distributions along sunfleck segments, *Agron. J.*, 63, 739, 1971.

Chapter 4

PREDICTING DARK RESPIRATION IN THE SOIL-PLANT SYSTEM

J. D. Hesketh, R. S. Alberte, and J. W. Jones

TABLE OF CONTENTS

I. INTRODUCTION

Duncan et al.[1] had difficulty in predicting respiration in their canopy photosynthesis model based upon leaf behavior and light interception into the canopy. When leaf respiration was held constant for all leaf layers in their model, net photosynthesis decreased above an optimum leaf area index because of respiration of shaded leaves. Such leaves behaved as parasites on the plant, importing photosynthate for respiration needs.

An optimum leaf area index for net photosynthesis was not borne out experimentally in earlier growth analysis studies with maize, where neither the crop growth rate nor dry matter accumulation per unit ground area decreased at high leaf area index values.[2,3] To correct their model, Duncan et al.[1] defined leaf respiration as a linear function of leaf photosynthesis, as had been reported earlier.[4-6] At about the same time, others studied this problem using gas exchange methods and confirmed that there was no optimum leaf area index for net photosynthesis and that leaf respiration decreased as leaves were shaded.[7,8] Duncan was concerned at the time[8a] about the lack of quantitative information from which to predict respiration. Since then there has been an explosion of new data and theory. However, respiration still creates considerable problems in the prediction of gross photosynthesis.

II. EXPERIMENTAL ESTIMATE OF MAINTENANCE AND GROWTH RESPIRATION

A. Theory and Methods

Several methods have evolved for experimentally estimating maintenance and growth respiration. Superimposed upon the various methods are at least two sets of notation. The differences in technique and theory make it difficult to compare results. We will try to present here methods for comparing coefficients obtained using different experimental techniques and analyses of data.

McCree[9] from earlier results[8] set maintenance respiration to that after 24 hr of darkness, i.e., the y-intercept of plots of respiration vs. photosynthesis for a 24-hr period. The slope of the respiration to photosynthesis plot was taken to be the growth respiration. This analysis was soon replaced with differential equations[10,11] first used by the microbiologist Duclaux.[12]

Thornley[13] presented the theory supporting differential equations involved. We have already introduced many of the equations in our earlier discussion of the carbon balance. The basic equations are given below:[14-16]

$$d\overline{W}^{*}/dt = P_g - Q\overline{W} \tag{1}$$

where $\overline{W}$ is the CH_2O equivalent of the oven dry weight obtained a conversion factor: (% carbon of the oven dry weight/40) (oven dry weight, $\overline{W}$) (g/plant), P_g is the gross photosynthetic rate (g CH_2O equivalent/plant/time), Q is the respiration rate per g $\overline{W}$ (g CH_2O equivalent/g CH_2O equivalent/time) and $d(\cdot)/dt$ is the time derivative or change in $(\cdot)$ with time. With

* As in Chapter 4, Volume I, we are expressing all weights as CH_2O equivalents based upon the carbon content involved, hence W as we define it here. Substituting W for $\overline{W}$ in these equations can greatly alter estimates of the various Q respiration coefficients in the equations discussed.

$$\bar{W} = \bar{W}_G + \bar{W}_C \quad (2)$$

$$d\bar{W} = d\bar{W}_G + d\bar{W}_C \quad (3)$$

$$Q\bar{W} = Q_o\bar{W}_G + Q_W d\bar{W}_G/dt + Q_C d\bar{W}_C/dt \quad (4)$$

where Q_o is the maintenance respiration rate (g CH_2O equivalent/g $\bar{\mathbf{W}}_G$/time, assuming the reserve requires no maintenance), $\bar{\mathbf{W}}_G$ is the dry weight (g CH_2O equivalent) excluding the reserves $\bar{\mathbf{W}}_C$ (g CH_2O equivalent), and Q_W and Q_C are the respiration values (g CH_2O equivalent/g CH_2O equivalent) associated with an increment of $\bar{\mathbf{W}}_G$ or $\bar{\mathbf{W}}_C$. The $\bar{\mathbf{W}}_G$ and $\bar{\mathbf{W}}_C$ increments must always be equal to or greater than zero. Combining Equations 1 and 4 gives the basic carbon balance equation discussed earlier.

Ignoring the reserves $\mathbf{W}_C$, and setting $\mathbf{W} = \mathbf{W}_G$, Thornley[13] defined a "conversion efficiency" Y_G as follows:

$$Y_G = d\bar{W}_G/(d\bar{W}_G + Q_W d\bar{W}_G) = (1 + Q_W)^{-1} \quad (5)$$

Y_G then is the fraction of CH_2O available for growth (excluding that used for maintenance) that ends up as plant dry weight and $(1-Y_G)$ is the fraction of CH_2O respired to supply energy for growth. Setting $P_g - Q_o\bar{\mathbf{W}}_G$ as CH_2O available for growth and growth respiration, from Equation 4,

$$Q\bar{W}_G = (1-Y_G)(P_g - Q_o\bar{W}_G) + Q_o\bar{W}_G \quad (6)$$

$$= (1-Y_G)P_g + Y_G Q_o\bar{W}_G \quad (7)$$

To obtain $(1-Y_G)$ from plots of $Q\bar{\mathbf{W}}_G$ vs. P_g, $\bar{\mathbf{W}}_G$ must be constant. Likewise, where $d\bar{\mathbf{W}}_G/dt = P_g - Q\bar{\mathbf{W}}_G = P_g - (1-Y_G)P_g - Y_G Q_o\bar{\mathbf{W}}_G = Y_G P_g - Y_G Q_o\bar{\mathbf{W}}_G$, $\bar{\mathbf{W}}_G$ must also be constant to obtain Y_G from plots of $d\bar{\mathbf{W}}_G/dt$ vs. P_g. Dividing by $\bar{\mathbf{W}}_G$ in both cases resolves this problem, with Q plotted vs. $P_g/\bar{\mathbf{W}}_G$ (see Equation 14 below).

Where $P_g = d\bar{\mathbf{W}}_G/dt + Q\bar{\mathbf{W}}$ from Equation 1 and $Q\bar{\mathbf{W}}_G = Q_o\bar{\mathbf{W}}_G + Q_W d\bar{\mathbf{W}}_G/dt$ from Equation 4 (ignoring the reserve term), and with $Y_G = (1 + Q_W)^{-1}$ from Equation 5,

$$(1-Y_G)P_g = (1-Y_G)(d\bar{W}_G/dt + Q_W d\bar{W}_G/dt + Q_o\bar{W}_G) \quad (8)$$

$$= [(1-Y_G)/Y_G]\, d\bar{W}_G/dt + Q_o\bar{W}_G - Y_G Q_o\bar{W}_G \quad (9)$$

and Equation 7 now reads

$$Q\bar{W}_G = [(1-Y_G)/Y_G]\, d\bar{W}_G/dt + Q_o\bar{W}_G \quad (10)$$

which is the same as Equation 4, excluding the reserve term. Assuming no change in the reserve term, we can set $\overline{W}_G$ equal to $\overline{W}$. We make the distinction here between $\overline{W}_G$ and $\overline{W}$ to stress the importance of the assumption about a constant reserve level. The above manipulations show how the two techniques and theories can be compared. Factors for converting oven dry weights (W) to their CH_2O equivalents ($\overline{W}$) can be estimated from a biochemical model (see Section IV). Thornley[10,13] presented these same derivations from another perspective.

Equation 7 represents the McCree plot of respiration as a linear function of gross photosynthesis, whereas Equation 10 represents respiration as a linear function of net photosynthesis for a 24-hr period. From regression slopes and axis intercepts of such plots, we can determine the various parameters in Equations 1 to 9. Q must be measured for the 24-hr period; usually daytime Q values are set equal to night values.

McCree[17] related sorghum night respiration (Q_N) to daily net photosynthesis (D) and plant dry weight (W), as follows:

$$Q_N = 0.14\ D + 0.0054\ W \qquad (11)$$

Setting $dW_G/dt = D - Q_N$,

$$Q_N = 0.163\ d\overline{W}_G/dt + 0.017\ W \qquad (12)$$

and assuming equal respiration during the daylight 12-hr period ($Q = 2Q_N$), $Q_W = 0.33$ and $Q_o = 0.034$. Values from Hodges and Kanemasu[18] have been similarly corrected to give $Q_W = 0.44$ and $Q_o = 0.072$. Q_N, D, and **W** must always be in the same units, either weight of CO_2 or the CH_2O equivalent.

Earlier, McCree[9] had found the ratio between day plus night respiration and gross photosynthesis to be 0.25. From Equation 6, $(1-Y_G)$ then equaled 0.25. Where $Y_G = (1 + Q_W)^{-1}$, Q_W calculates to be 0.33.

From estimates of dry matter accumulation rates (dW/dt) and respiration, one can plot Q vs. (1/W) (dW/dt), where

$$Q = Q_o + (Q_W/W)(dW/dt) \qquad (13)$$

and estimate Q_o and Q_W from axis intercepts and response slopes.[19,20] If (1/W) (dW/dt) is constant, Q must be constant, and Q_o and Q_W cannot be estimated.

Winzeler et al.[21] used

$$Q = (1-Y_G)P_g/W_G + Y_G Q_o \qquad (14)$$

to transform their photosynthesis and respiration data. The transformation did not work well for the life cycle of a barley plant (Figure 1). Changes in dead, nonrespiring W_G may have caused some of the problem. The reserve term in Equation 4 which we ignored in subsequent equations could have invalidated the transformation. Also, Q_W will vary with the molecular composition of the substrate respired. These points will be discussed in Section IV, below.

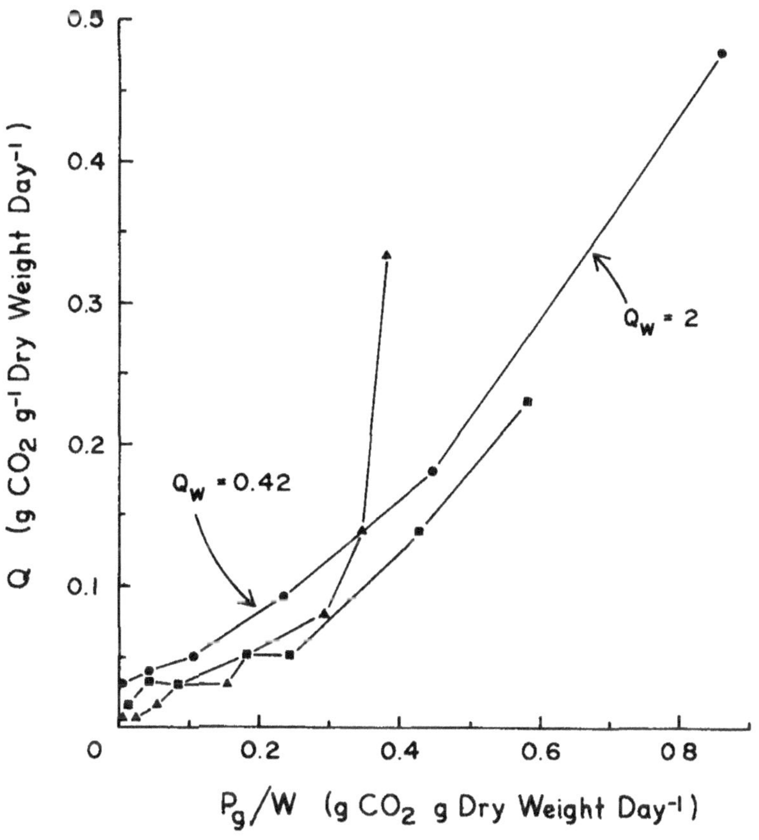

FIGURE 1. Ratio of total respiration per day (Q) to gross photosynthesis divided by the plant dry weight (P_g/W). (From Winzeler, H., Hunt, L. A., and Mahon, J. D., *Crop Sci.*, 16, 786, 1976. With permission.)

Frequently, dW/dt is constant during a large fraction of the growing period, allowing Equation 4 to be solved, with Q measured for two W values on the growth curve.

Estimates of the various parameters in the equations presented, which are based upon dry weight changes, will include a component taking into account the molecular composition of the product synthesized or losses of H and O and gains of mineral nutrients. Estimates based upon gas exchange data alone will not contain such a factor. This point will be discussed in a later section.

B. Maintenance and Growth Coefficients

Literature values[8,9,11,17,18,22-34] for the various growth coefficients are given in Table 1. Again it must be pointed out that values based upon gas exchange determinations of respiration and photosynthesis give g glucose g^{-1} glucose values, whereas those based upon gas exchange respiration values and oven dry weights give g glucose g^{-1} dry weight values, including the extra glucose needed to provide the carbon molecules found in a g dry weight. By knowing the carbon to dry weight ratio, one can make the various values comparable. Hanson and Jensen[27] analyzed this problem when attempting to compare their results with literature values.

III. EXPERIMENTAL ESTIMATE OF THE N FIXATION COEFFICIENT

Mahon[35,36] introduced a $Q_N dN/dt$ term to the respiration equation for a nodule,

TABLE I

Literature Maintenance and Growth Respiration Coefficients

Crop	Q_w Growth coefficient, g CH_2O per g CH_2O	Q_w Growth coefficient, g CH_2O per g W^a	Q_o Maintenance coefficient, mg CH_2O per day per g CH_2O	Q_o Maintenance coefficient, mg CH_2O per day per g W^a	Y_G, growth conversion coefficient (g/g)	Y_t, overall conversion coefficient (g/g)	Ref.
Clover (*Trifolium repens* L.)	0.33		30		0.75		8,9
	0.33		34		0.75		17
Sorghum (*Sorghum bicolor* L.)	0.33		13		0.75		17
Maize (*Zea mays* L.)	0.27	0.34	8—22		0.75—0.8		22,23
	0.33		15—45		0.75		
	0.31		18		0.76	0.65	24
Bean (*Phaseolus vulgaris* L.)	0.13		12				22
	0.33		28		0.75		26
Italian ryegrass (*Lolium multiflorum* L.)		0.22		46	0.82		27
Tops		0.09		37			
Roots		0.67		91			
Cotton (*Gossypium hirsutum* L.)		0.38		3	0.72	0.66	11
Bolls							
		0.33		6	0.74		
		0.39		7—42	0.72		28
Sunflower (*Helianthus annuus* L.)	0.26		34—45				23
Chrysanthemum (*Chrysanthemum morifolium* L.)		0.56		17	0.62	0.5—0.57	29
Perennial ryegrass (*Lolium perenne* L.)		0.25		14	0.8	0.65	30
Barley (*Hordeum vulgare* L.)	0.34		15				31

TABLE 1 (continued)

Literature Maintenance and Growth Respiration Coefficients

Crop	Q_w Growth coefficient, g CH_2O per		Q_o Maintenance coefficient, mg CH_2O per day per		Y_G, growth conversion coefficient (g/g)	Y_T, overall conversion coefficient (g/g)	Ref.
	g CH_2O	g W[a]	g CH_2O	g W[a]			
Uniculm barley	0.42		14		0.7		21
	0.31		30				25
(*Kindred Uniculm* 97)	0.29		20		0.78	0.65	24
Tomato (*Lycopersicon esculentum* L.)		0.19		12	0.84	0.68	20
Subterranium clover (*Trifolium subterraneum* L.)							
Nodulated	0.56		24		0.64	0.65	32
Non-nodulated	0.38		26		0.72	0.74	
Pitch pine (*Pinus rigida* Mill.)							
Shoot	0.32		19				33
Roots	0.13		5				
Wheat (*Triticum aestivum* L.)	0.44		72				18
Unicellular algae						0.73	34
Higher plants						0.76	34

[a] Oven dry weight. This creates problems because of the variable carbon content of biomass; see text and Chapter 4, Volume 1.

where Q_N is the carbohydrate need for the fixation of an increment of N. From measurements of CO_2 evolved and C_2H_2 reduced, he found that 17 g of carbohydrate were consumed per gram N fixed. He also reported growth and maintenance coefficients for the nodule.

The maintenance and growth respiration approach has indicated[32-36] that considerably more carbohydrate is needed for plants fixing N than for plants growing on NO_3^-, whereas earlier carbon balances of similar systems[37,38] had suggested similar costs for synthesizing protein from the two sources of N.

IV. MAINTENANCE AND GROWTH RESPIRATION FROM BIOCHEMICAL ANALYSES

Penning de Vries[12,22,23] developed a biochemical model which can be used to generate Q_w values. Such construction values depend upon the chemical composition of the plant material synthesized, e.g., nitrogenous compounds, carbohydrates, lipids, lignin, and organic acids (Table 2, Chapter 4, Volume I). Knowing the weight of each of these as gram dry weight for plant or organ, we can use the yield values (grams glucose per grams product) from Table 2 in Chapter 4, Volume I to calculate the gram glucose needed for synthesis of a gram dry matter (Table 2). The total glucose values account for losses of O and H and gains in N and S and other nutrients and represent coefficients that should be used in carbon balance models. The glucose values needed for respiration are representative of growth coefficients determined from respiration, % carbon, and dry matter measurements (grams glucose per grams product) or photosynthetic measurements (grams glucose per g glucose). For a whole plant, Penning de Vries increased the glucose needed by 1.05 to account for translocation.

The basis for the Penning de Vries model can be found in a *Journal of Theoretical Biology* paper[12] where some 90 synthesis reactions involving glucose, pyruvic acid, acetyl coenzyme-A, serine, aspartic acid, glutamic acid, NH_3^+, CO_2, O_2, $NADPH_2$, and ATP were tabulated. The source of much of this information was Dagley and Nicholson.[39] In the model, pathways for lignin, deoxyribose, fructosan, and chitin synthesis could not be found and were estimated. The reduction of NO_3^- was considered to require four $NADPH_2$ molecules. Three ATP molecules were required to polymerize amino acids into protein. One more ATP was required for maintenance of the biosynthetic enzymes. No ATP was assumed for biomass synthesis from polymers.

For transfer of 1 mol of cation across a membrane, 0.3 ATP was required; 70% of the ions were considered to be located in the vacuole, crossing two membranes and requiring 0.6 ATP, 20% were located in the cell walls, and the remaining 20% were assumed to be in the cytoplasm. The uptake of one carbohydrate molecule was assumed to require 1 ATP.

The merits of the above assumptions are discussed by Penning de Vries and coworkers.[12] Much of this information is based upon in vitro biochemistry, radioactive tracer studies, and thermodynamic considerations.

Other assumptions and conditions used in the model were as follows:

1. Chemical composition of the biomass as given in Table 3 for a corn plant was specified.
2. The reactions occurred under fully aerobic conditions.
3. Intra-cellular translocation occurred spontaneously, except for ATP required for one transfer of organic and inorganic ions across membranes.
4. Glycolysis, tricarboxylic acid, and oxidative phosphorylation pathways were used for $NADPH_2$ and ATP production.

TABLE 2

Glucose Requirements for Biosynthesis of Dry Matter in Various Plant Tissues

Organ	Composition (%)[a]	Total glucose[b] needed for synthesis (g/g product)		Glucose needed for respiration (g/g product)		Glucose needed for respiration (g/g glucose)	
		With NH_3^+	With NO_3^-	With NH_3^+	With NO_3^-	With NH_3^+	With NO_3^-
Leaves	25, 66.5, 2.5, 4, 2	1.37	1.60	0.17	0.67	0.12	0.42
Nonwoody stem	12.5, 74, 2.5, 8, 2	1.21	1.32	0.13	0.24	0.11	0.18
Woody stem	5, 45, 4, 50, 5	1.64	1.68	0.29	0.32	0.18	0.19
Bean seeds	35, 55, 5, 2, 3	1.43	1.73	0.31	0.74	0.22	0.43
Rice seeds	5, 90, 2, 1, 2	1.25	1.30	0.11	0.16	0.09	0.12
Peanut seeds	20, 21, 50, 6, 3	2.22	2.39	0.65	0.82	0.29	0.34
Bacteria	60, 25, 5, 2, 8	1.47	1.98	0.25	0.76	0.17	0.38
Cotton boll	13.6, 74, 3.6, 2, 6.5	1.27	1.39	0.15	0.26	0.12	0.19

[a] Nitrogenous compounds, carbohydrates, lipids, lignin, and minerals, listed in sequence.

[b] See Table 2, Chapter 4, Volume I.

Adapted from Penning de Vries, F. W. T., in *Photosynthesis and Productivity in Different Environment*, Cooper, J. P., Ed., Cambridge University Press, Cambridge, 1976, chap. 20. With permission.

TABLE 3

The Chemical Composition of a Young Maize Plant with Fractions Expressed on a Percent Dry Weight Basis

Main fraction (%)	Subfractions (%)
Nitrogenous compounds (23)	Amino acids (10)
	Proteins (87)
	Nucleic acids (3)
Carbohydrates (56.5)	Ribose (1)
	Glucose (5)
	Fructose (2)
	Annose (1)
	Galactose (1)
	Sucrose (5)
	Cellulose (40)
	Hemicellulose (40)
	Pectin (5)
Lipids (2.5)	Glyceroltripalmitate (12)
	Glyceroltristearate (3)
	Glyceroltrioleate (47)
	Glyceroltrilinolate (33)
	Glyceroltrilinoleate (5)
Lignin (nitrogen-free) (8)	
Organic acids (5)	Oxalic acid (5)
	Glyoxalic acid (5)
	Oxalo-acetic acid (20)
	Malic acid (10)
	Citric cid (30)
	Aconitic acid (30)
Minerals (5)	Potassium (80)
	Chloride (20)

From Penning de Vries, F. W. T., in *Photosynthesis and Productivity in Different Environments*, Cooper, J. P., Ed., Cambridge University Press, Cambridge, 1976, chap. 20. With permission.

5. The P/O ratio for $NADH_2$ oxidation was set to 3.

Penning de Vries et al.[12] tested alternative pathways and found little difference in Q_W or Y_G values generated from the pathways tested. They also ran a sensitivity test by varying the P/O ratio and concluded that changing it from 2 to 3 generally had little (less than 10%) effect on predicted Q_W or Y_G values, particularly when the relative growth rate, (1/W) (dW/dt), was large (Figure 2). The question was whether 1 mol (P/O = 3) or 1.5 mol (P/O = 2) of glucose are used to generate 1 mol of ATP.

Mutsaers[28] applied the Penning de Vries model to the growth of a cotton boll, after compiling considerable boll composition data from the literature. The state of N-source (Table 2) can cause large differences in predicted respiration rates. The N may be reduced in leaves using photosynthate reductant independent of carbon metabolism, although apparently reductant must be transferred indirectly through the chloroplast membrane into the cytoplasm for some of the reactive steps. This membrane transfer of reductant may involve the reduction and oxidation of a carbon compound.[40] N may go to the leaves and become converted to amino acids and then may be translocated to the fruit or seed, affecting respiration considerably by eliminating the need to reduce NO_3^- in such organs.

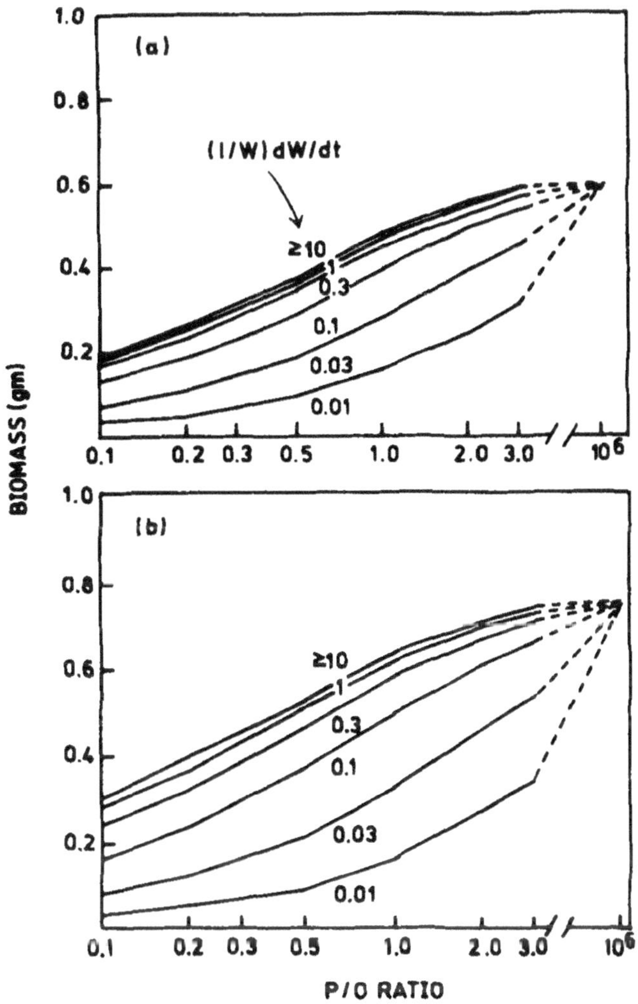

FIGURE 2. The yields of growth processes in g protein (a) or grams biomass (b) per gram glucose substrate (with ammonia, hydrogen sulfide, and minerals added) at various P/O ratios and relative growth rates (dW/dtW, g/g/day). (From Penning de Vries, F. W. T., Brunsting, A., and van Larr, H. H., *J. Theor. Biol.*, 45, 339, 1974. With permission.)

Penning de Vries[41] also speculated on the respiratory requirements for maintenance of living cells in the plant, based upon further biochemical evidence. The entire effort is a considerable innovation for predicting the respiratory needs for biochemical synthesis, but experimental data on plant composition are frequently lacking, as Penning de Vries pointed out.

V. THORNLEY'S NEW MODEL FOR RESPIRATION

Thornley[42] recently suggested a new conceptual model for respiration whereby a storage pool supplied by photosynthesis is converted into degradable and nondegradable dry weight, requiring respiration. The degradable dry weight then cycles back into the storage pool at some rate constant. Such degradation and resynthesis of plant dry weight represents maintenance respiration in the previous models. Thornley[42] was able to estimate the various rate constants in this new model from data of Ryle et al.[24]

VI. ROOT AND SOIL RESPIRATION

Constructive (Q_W) and maintenance (Q_o) respiration values were higher in shoots than in roots for pitch pine[33] but not for ryegrass[27] (Table 1). Considerable information about root respiration has accumulated from carbon and nitrogen budget analyses of nitrogen-fixing plants; however, the methodology discussed in previous sections needs to be applied to root growth in more experiments before we can confidently assign respiration coefficients to a root respiration and growth equation for many crop plants. More data are needed on the chemical composition of roots as well as on nitrate reduction that might occur in the root system.

Enclosing the root system of a potted plant for respiration measurements[44,45] is technically easy. It becomes more difficult in the field where soil and root respiration are measured by enclosing the soil surface with a chamber and determining CO_2 exchange. Such a system may be open or closed.[46-50] In a direct comparison between moving air through such a chamber with positive (10 μbars above ambient air pressure) or negative pressure (25 μbars below ambient air pressure), Kanemasu et al.[49] found a tenfold difference in CO_2 flux from the soil. Some of this difference was explained by theoretical effects of pressure on CO_2 diffusion and mass flow of air and CO_2. Therefore, the effect of pressure differences between inside such a soil chamber and the air can and needs to be corrected for.

Hodges and Kanemasu[18] reported soil respiration functions (excluding roots) of maximum and minimum air temperatures, as well as the minimum temperature for the previous day, for both day and night periods. They developed a separate function for root respiration.

Monteith et al.[48] used a closed system with soda lime as a CO_2 absorber, avoiding pressure differences, and found flux values of 8 to 20% of the estimated crop gross photosynthetic rate on a clear day. (Gross photosynthetic rates were estimated from dry matter accumulation rates, assuming $Y_T = 0.25$, or 25% of the gross photosynthate being used up in respiration, and 10% of the biomass produced going to the roots. These estimates and their use represent another example of the potential of the carbon balance approach.) The change in soil respiration (Q_{soil}) with temperature was

$$Q_{soil} = Q_{o\text{-}soils}\theta_s^{-T/10} \qquad (15)$$

where $Q_{o\text{-}soil}$ was 0.9 to 1.2 g $m^{-2}day^{-1}$ and θ_s was 3, the latter being similar to values reported by Kopf.[50] Values for CO_2 flux from cropped soils were 2 to 3 g/m²/day higher than those from bare soils, due to root and associated microbial respiration.

Biscoe et al.[31] used the same approach in the development of three comparative carbon budgets for a wheat crop (with gross photosynthesis estimated from the growth analysis, micrometeorological, and leaf-canopy light interception[51] approaches), with soil respiration contributing some 5% of the maximum net photosynthetic rate of the canopy at midday. Totsuka et al.[52] obtained soil respiration vs. temperature using the chemical absorption method in their three-way comparative carbon budget (with gross photosynthesis estimated from the growth analysis, leaf-canopy light interception, and crop enclosure approaches) for ladino clover (*Trifolium repens* L.).

Soil parameters that need to be accounted for in a model for soil CO_2 evolution would include temperature, soil moisture, oxygen, and air permeability, as well as the status of organic matter (half life) and available nitrogen. Root parameters needed would include supply of carbohydrate to the root system, root N content, root length, and metabolic activity.

Root and soil respiration should be measured when attempting to predict canopy gross photosynthesis using micrometeorological or some chamber techniques. Sealing off the soil or increasing the pressure inside a chamber[49,53] would seem to decrease CO_2 evolution enough to allow measurement of canopy photosynthesis; however, air and CO_2 leakage in or out of such a system would have to be correct.

VII. CONCLUSIONS

The literature on maintenance and growth respiration is expanding rapidly, with experimental and theoretical biochemical comparisons of the various parameters involved. Tables of grams product per gram glucose substrate for various plant components (Table 2, Chapter 4, Volume I) as derived from the Penning de Vries model apparently can be used with considerable confidence, when no experimental data are available, for predicting respiration requirements. Some information is needed on the composition of the plant being modeled. Diurnal changes in biochemical syntheses may shed some light on associated changes in respiration rate. More effort is needed to correct for dead tissue in the plant.

Respiration was singled out as a limiting factor in the early canopy photosynthesis models, and it seems that the subsequent flood of new information was inspired in part by this need for better information. As discussed in an earlier chapter, the respiration models play an important role in determining a carbon balance for a plant or plant organ, and such balances have become important in translocation and nitrogen metabolism studies, as well as in evaluating the role of soil respiration in determining canopy photosynthetic rates.[48]

More data are needed for predicting root respiration. Most information to date suggests rapid rates of respiration[27,38,43] compared with shoot respiration rates.

ADDENDUM

Since this chapter was written, Yokoi et al.[54] and Kimura et al.[55] sent us reprints with estimates of respiratory components in growing seedlings and leaves, which were based upon dry matter and respiration measurements. They present a different history of how some of this research evolved. Lambers and Steingrover[56] sent a reprint describing the application of these same techniques to roots. We mention this paper because so little has been done with root respiration.

VIII. SYMBOLS

D	Daily net photosynthetic rate, weight CH_2O equivalent per ground area or plant per day
θ_s	Factor for the increase in respiration per 10°C increase in temperature
P_g	Gross photosynthetic rate, weight CH_2O equivalent per ground area or plant per time
Q	Plant respiration rate, weight CH_2O equivalent per weight CH_2O equivalent per time
Q_c	Respiration coefficient associated with deposition or use of plant reserves, weight CH_2O per weight CH_2O
Q_N	Plant night respiration rate, same units as Q
Q_o	Maintenance respiration coefficient, same units as Q

$Q_{o\text{-}soil}$	Soil respiration rate at 0°C, weight CO_2 per ground area per time
Q_{soil}	Soil respiration rate, same units as $Q_{o\text{-}soil}$
Q_W	Growth or constructive respiration coefficient, weight CH_2O equivalent per weight CH_2O equivalent or plant dry weight
T	Temperature, °C
$\overline{W}$	Plant dry weight, oven dried at 70°C to an equilibrium weight
$\overline{W}$	Plant dry weight multiplied by its percent carbon/40, weight CH_2O equivalent on a carbon basis
$\overline{W}_C$	Reserve (starch, fructosans, sugars), dry weight, same units as **W**
$\overline{W}_G$	Plant dry weight other than reserve, same units as **W**
Y_G	Conversion efficiency coefficient, $(1 + Q_W)^{-1}$, by definition
Y_T	Total conversion efficiency coefficient, $(P_g - Q)/P_g$, by definition

REFERENCES

1. **Duncan, W. G., Loomis, R. S., Williams, W. A., and Hanau, R.,** A model for simulating photosynthesis in plant communities, *Hilgardia,* 38, 191, 1967.
2. **Williams, W. A., Loomis, R. S., and Lepley, C. R.,** Vegetative growth of corn as affected by population density. II. Components of growth, net assimilation rate and leaf-area index, *Crop Sci.,* 5, 215, 1965.
3. **Shibles, R. M. and Weber, C. R.,** Leaf area, solar radiation interception and dry matter production by soybeans, *Crop Sci.,* 5, 575, 1965.
4. **Weintraub, R. L.,** Radiation and plant respiration, *Bot. Rev.,* 10, 383, 1944.
5. **Murata, Y.,** Studies on the photosynthesis of rice plants and its culture significance, *Bull. Natl. Inst. Agric. Sci. Ser. D,* No. 9, 1961.
6. **Tsuno, Y. and Fujise, K.,** Studies on the dry matter production of sweet potato, *Bull. Natl. Inst. Agric. Sci. Ser. D,* No. 13, 1965.
7. **Ludwig, L. J., Saeki, T., and Evans, L. T.,** Photosynthesis in artificial communities of cotton plants in relation to leaf area, *Aust. J. Biol. Sci.,* 18, 1103, 1965.
8. **McCree, K. J. and Troughton, J. H.,** Non-existence of an optimum leaf area index for the production rate of white clover grown under constant conditions, *Plant Physiol.,* 41, 1615, 1966.

8a. **Duncan, W. G.,** personal communication, 1965.

9. **McCree, K. J.,** An equation for the rate of respiration of white clover plants grown under controlled conditions, in *Prediction and Measurement of Photosynthetic Productivity,* Setlik, I., Ed., PUDOC, Wageningen, 1970, 221.
10. **Thornley, J. H. M.,** Respiration, growth, and maintenance in plants, *Nature (London),* 227, 304, 1970.
11. **Hesketh, J. D., Baker, D. N., and Duncan, W. G.,** Simulation of growth and yield in cotton: respiration and the carbon balance, *Crop Sci.,* 11, 394, 1971.
12. **Penning de Vries, F. W. T., Brunsting, A. H. M., and van Laar, H. H.,** Products, requirements and efficiency of biosynthesis: a quantitative approach, *J. Theor. Biol.,* 45, 339, 1974.
13. **Thornley, J. H. M.,** *Mathematical Models in Plant Physiology,* Academic Press, New York, 1976, 123.
14. **Jones, J. W., Thompson, A. C., and Hesketh, J. D.,** Analysis of SIMCOT: nitrogen and growth, in *Proc. Beltwide Cotton Production Research Conf.,* National Cotton Council, Memphis, 1974, 111.
15. **McKinion, J. M., Jones, J. W., and Hesketh, J. D.,** A system of growth equations for the continuous simulation of plant growth, *Trans. ASAE,* 18, 975, 1975.
16. **Hesketh, J. D. et al.,** Problems in building computer models for photosynthesis and respiration, in *Environmental and Biological Control of Photosynthesis,* Marcelle, R., Ed., Dr. W. Junk, The Hague, 1975, 53.
17. **McCree, K. J.,** Equations for the rate of dark respiration of white clover and grain sorghum, as functions of dry weight, photosynthetic rate, and temperature, *Crop Sci.,* 14, 509, 1974.
18. **Hodges, T. and Kanemasu, E. T.,** Modeling daily dry matter production of winter wheat, *Agron. J.,* 69, 974, 1977.

19. **Thornley, J. H. M. and Hesketh, J. D.**, Growth and respiration in cotton bolls, *J. Appl. Ecol.*, 9, 315, 1972.
20. **Walker, A. J. and Thornley, J. H. M.**, The tomato fruit: import, growth, respiration and carbon metabolism at different fruit sizes and temperatures, *Ann. Bot. (London)*, 41, 977, 1977.
21. **Winzeler, H., Hunt, L. A., and Mahon, J. D.**, Ontogenetic changes in respiration and photosynthesis in uniculm barley, *Crop Sci.*, 16, 786, 1976.
22. **Penning de Vries, F. W. T.**, Respiration and growth, in *Crop Processes in Controlled Environments*, Rees, A. R., Cockshull, K. E., Hand, D. W., and Hurd, R. G., Eds., Academic Press, London, 1972, 327.
23. **Penning de Vries, F. W. T.**, Use of assimilates in higher plants, in *Photosynthesis and Productivity in Different Environments*, Cooper, J. P., Ed., Cambridge University Press, Cambridge, 1975, 459.
24. **Ryle, G. J. A., Cobby, J. M., and Powell, C. E.**, Synthetic and maintenance respiratory losses of $^{14}CO_2$ in uniculm barley and maize, *Ann. Bot. (London)*, 40, 571, 1976.
25. **—, Brockington, N. R., Powell, C. E., and Cross, B.**, The measurement and prediction of organ growth in a uniculm barley, *Ann. Bot. (London)*, 37, 233, 1973.
26. **Moldau, H. and Karolin, A.**, Effect of the reserve pool on the relationship between respiration and photosynthesis, *Photosynthetica*, 11, 38, 1977.
27. **Hansen, G. K. and Jensen, C. R.**, Growth and maintenance respiration in whole plants, tops, and roots of *Lolium multiflorum*, *Physiol. Plant.*, 39, 155, 1977.
28. **Mutsaers, H. J. W.**, Growth and assimilate conversion of cotton bolls (*Gossypium hirsutum* L.). I. Growth of fruits and substrate demand, *Ann. Bot. (London)*, 40, 301, 1976.
29. **Hughes, A. P.**, A comparison of the effects of light intensity and duration on *Chrysanthemum morifolium* cv. Bright Golden Anne in controlled environments. II. Ontogenetic changes in respiration, *Ann. Bot. (London)*, 37, 275, 1973.
30. **Robson, M. J.**, The growth and development of simulated swards of perennial ryegrass. II. Carbon assimilation and respiration in a seedling sward, *Ann. Bot. (London)*, 37, 501, 1973.
31. **Biscoe, P. V., Scott, R. K., and Monteith, J. L.**, Barley and its environment. III. Carbon budget of the stand, *J. Appl. Ecol.*, 12, 269, 1975.
32. **Silsbury, J. H.**, Energy requirement for symbiotic nitrogen fixation, *Nature (London)*, 267, 149, 1977.
33. **Ledig, F. T., Drew, A. P., and Clark, J. G.**, Maintenance and constructive respiration, photosynthesis and net assimilation rate in seedlings of Pitch Pine (*Pinus rigida* Mill.), *Ann. Bot. (London)*, 40, 289, 1976.
34. **Raven, J. A.**, The quantitative role of 'dark' respiratory processes in heterotrophic and photolithotrophic plant growth, *Ann. Bot. (London)*, 40, 587, 1976.
35. **Mahon, J. D.**, Root and nodule respiration in relation to acetylene reduction in intact nodulated peas, *Plant Physiol.*, 60, 812, 1977.
36. **Mahon, J. D.**, Respiration and the energy requirement for nitrogen fixation in nodulated pea roots, *Plant Physiol.*, 60, 817, 1977.
37. **Gibson, A. H.**, The carbohydrate requirements for symbiotic nitrogen fixation: a "whole-plant" growth analysis approach, *Aust. J. Biol. Sci.*, 19, 499, 1966.
38. **Minchin, F. R. and Pate, J. S.**, The carbon balance of a legume and the functional economy of its root nodules, *J. Exp. Bot.*, 24, 259, 1973.
39. **Dagley, S. and Nicholson, D. E.**, *An Introduction to Metabolic Pathways*, John Wiley & Sons, New York, 1970.
40. **Heber, U.**, Metabolite exchange between chloroplasts and cytoplasm, *Annu. Rev. Plant Physiol.*, 25, 393, 1974.
41. **Penning de Vries, F. W. T.**, The cost of maintenance processes in plant cells, *Ann. Bot. (London)*, 39, 77, 1976.
42. **Thornley, J. H. M.**, Growth, maintenance and respiration: a re-interpretation, *Ann. Bot. (London)*, 41, 1191, 1977.
43. **Herridge, D. F. and Pate, J. S.**, Utilization of net photosynthate for nitrogen fixation and protein production in an annual legume, *Plant Physiol.*, 60, 759, 1977.
44. **Mahon, J. D., Lowe, S. B., and Hunt, L. A.**, Photosynthesis and assimilate distribution in relation to yield of cassava grown in controlled environments, *Can. J. Bot.*, 54, 1322, 1976.
45. **Tsuno, Y. and Fujise, K.**, Studies on the dry matter production of the sweet potato, *Bull. Natl. Inst. Agric. Sci. Ser. D*, No. 13, 1965.
46. **Lundegardh, H.**, Carbon dioxide evolution of soil and crop growth, *Soil Sci.*, 23, 417, 1927.
47. **Moss, D. N., Musgrave, R. B., and Lemon, E. R.**, Photosynthesis under field conditions. III. Some effects on light, carbon dioxide, temperature and soil moisture on photosynthesis, *Crop Sci.*, 1, 83, 1961.

48. **Monteith, J. L., Szeicz, G., and Yabuki, K.,** Crop photosynthesis and the flux of carbon dioxide below the canopy, *J. Appl. Ecol.*, 1, 321, 1964.
49. **Kanemasu, E. T., Powers, W. L., and Sij, J. W.,** Field chamber measurements of CO_2 flux from soil surface, *Soil Sci.*, 118, 233, 1974.
50. **Koepf, H.,** Die Temperature/Zeit-Abhangigkeit der Bodenatmung, *Z. Pflanzenernaehr. Dung.*, 61, 29, 1953.
51. **Biscoe, P. V., Gallagher, J. N., Littleton, E. J., Monteith, J. L., and Scott, R. K.,** Barley and its environment. IV. Sources of assimilate for the grain, *J. Appl. Ecol.*, 12, 295, 1975.
52. **Totsuka, T., Iwaki, H., Koyama, H., Song, S. D., and Monsi, M.,** Measurements of dry matter growth by chamber method in a ladino clover field, in *Photosynthesis and Utilization of Solar Energy, Level III Experiments,* JIBP/PP, Tokyo, 1971, 33.
53. **Leafe, E. L.,** Micro-environment, carbon dioxide exchange and growth in grass swards, in *Crop Processes in Controlled Environments,* Rees, A. R. et al., Eds., Academic Press, London, 1972, 157.
54. **Yokoi, Y., Kimura, M., and Hogetsu, K.,** Quantitative relationships between growth and respiration. I. Components of respiratory loss and growth efficiencies of etiolated red bean seedlings, *Bot. Mag.*, 91, 31, 1978.
55. **Kimura, M., Yokoi, Y., and Hogetsu, K.,** Quantitative relationships between growth and respiration. II. Evaluation of constructive and maintenance respiration in growing *Helianthus tuberosus* leaves, *Bot. Mag.*, 91, 43, 1978.
56. **Lambers, H. and Steingrover, E.,** Efficiency of root respiration of a flood-tolerant and flood-intolerant *Senecio* species as affected by low oxygen tension, *Physiol. Plant.*, 42, 179, 1978.

Chapter 5

PREDICTING LEAF EXPANSION

J. W. Jones and J. D. Hesketh

TABLE OF CONTENTS

I. INTRODUCTION

The study of leaf growth has never been as fashionable as that of photosynthesis itself. Yet expansion of the photosynthetic surface is at least as important in determining photosynthate supply to a plant or plant canopy[1,2] as behavior of photosynthetic processes in the leaf.[3] To quote Milthorpe:[3] "Even a casual examination of agricultural writings reveals much confusion as to the relative importance of these two components of energy (capture) (the photosynthetic processes and extent of the photosynthetic surface), with, consequently, much misdirected effort in plant breeding, fertilizer and other agronomic programs. Thought still seems to be dominated largely by the rate and mechanisms of the photosynthetic reactions. Yet these are relatively unimportant in determining (primary productivity) compared with the effect of variations in the extent of the surface in which these reactions proceed. Further, over a fairly wide range of the environments normally experienced in agriculture, leaf growth is itself largely independent of the rate of assimilation."

Ledig[4] quantified the relative importance of differences in photosynthesis and rates of leaf growth by manipulating both in a simple photosynthesis-leaf growth model. In his model, over 140 days, a 40% difference in photosynthesis was equivalent to a 10% difference in the leaf expansion rate in determining the dry matter production rate.

It is interesting to compare wild species with their cultivated counterparts. Crop cultivars usually have much larger leaves than do their wild relatives, but their photosynthetic rates per unit leaf area have not improved and often are much less. During development of cultivars, leaf size has increased instead of the photosynthetic rate per unit leaf area. Man also has selected for larger seeds for various reasons, and this has resulted in faster leaf development during the seedling stage of growth.

The definition of plant growth has long been debated, and no single definition can encompass all levels of discussion. In this chapter we are concerned primarily with leaf area expansion and not dry weight accumulation. Cellular, leaf, and whole plant aspects of expansion will be discussed. Then an example is offered showing physiological insights one might gain from an analysis of leaf expansion and activity at the whole plant level.

II. THE CELLULAR MODEL

Levitt[5] defined growth as . . . "increase in size caused by cell enlargement when measured at constant positive cell turgor." There have been several attempts at modeling the growth of organs of plants by describing the processes of cell division and growth separately.[6] This is the classical reductionist approach of analyzing a system by studies of internal components. The understanding and modeling of the components and their interactions allows one to predict the behavior of the system better. Therefore, these models have typically considered separately the processes of meristematic cell division and non meristematic cell expansion and growth. Here we will give the development of a leaf expansion model similar to that described by Thornley.[6]

A. Cell Division

One can consider the cell division process to be analagous to population dynamics in reproducing animals. One difference is that there are two separate but linked populations of cells in a leaf: meristematic and vegetative cells. Assuming all meristematic cells, steady state growth, N_o cells at time 0 whose ages are randomly distributed, binary cell division, and a relative rate of division ($v[h^{-1}]$), the rate of change in cell numbers N_m ($dN_m[t]/dt$, number/time) is given by:

$$dN_m(t)/dt = vN_m(t) \qquad (1)$$

The solution to Equation 1 results in the familiar exponential growth equation:

$$N_m(t) = N_0 e^{vt} \qquad (2)$$

It is useful to define the population doubling time, t_d, in terms of the relative division rate. When $N_m(t)/N_0$ is equal to 2 (for a doubled population), we can solve Equation 2 for t_d, or

$$N_m(t)/N_0 = e^{vt_d} \qquad (3)$$

and

$$t_d = \ln(2)/v$$

where ln is the natural logarithm.

The above development considers each cell to be meristematic, denoted by the subscript m, and all cells are capable of division. Now suppose that a θ fraction of new cells are meristematic and the remainder are vegetative. Vegetative cells are capable of expansion and growth, but are not capable of division. It is now necessary to divide population numbers into two cell classes: meristematic ($N_m[t]$) and vegetative ($N_v[t]$).

The rate of increase of meristematic cells can now be written as:

$$dN_m(t)/dt = v'N_m(t) \qquad (4)$$

where v′ is the new relative division rate for meristematic cells and

$$N_m(t) = N_0 e^{v't} \qquad (5)$$

Remembering that t_d is the time required for the number of cells to double, but only θ fraction of the cells are dividing, we can write anexpression for v′, or

$$N_m(t)/N_O = 2\theta = e^{v't_d} \qquad (6)$$

and $v' = (1/t_d)(\ln[2] + \ln[\theta])$.

The new expression for $N_m(t)$ now becomes:

$$N_m(t) = N_O e^{[(\ln(2) + \ln(\theta))/t_d]t} \tag{7}$$

Thornley[6] defined a term α to be (for convenience): $\alpha = 1 + \ln(\theta)/\ln(2)$.

Using this definition of α, Equation 7 can be reduced to:

$$N_m(t) = N_0 e^{\alpha vt} \tag{8}$$

Therefore, in the case where only a fraction of new cells remain reproductive, the relative dividing rate of cells has been reduced from v to αv. If $\theta = \frac{1}{2}$, then $\alpha = 0$ and $N_m(t) = N_0$ for all t. If θ is less than ½, then α is negative, and the population of meristematic cells decreases with time. For leaf growth, one would expect θ to be less than ½ since cell division ceases after a period of time, about ⅓ the time before full expansion of tobacco (*Nicotiana tabacum* L.) leaves.[7]

We must also account for the population of vegetative or nondividing cells $N_v(t)$. Assume $N_v(t)$ starts at level 0 and with a production rate related to meristematic cell population by:

$$\frac{dN_v(t)}{dt} = u'N_m(t) \tag{9}$$

where u′ is the relative rate of vegetative cell production. Equation 9 may be written as

$$\frac{dN_v(t)}{dt} = u'N_0 e^{\alpha vt} \tag{10}$$

Note that N_0 is still the initial number of meristematic cells. Upon integration we obtain

$$N_v(t) = \frac{u'N_O}{\alpha v} (e^{\alpha vt} - 1) \tag{11}$$

where the (−1) arose from zero initial population of vegetative cells. Now if (1−θ) new cells are to become vegetative, then we can evaluate Equation 11 at t_d to determine the relative rate of vegetative cell production in terms of previously defined variables, or

$$\frac{N_v(t_d)}{N_0} = 2(1 - \theta) = \frac{u'N_0}{\alpha v} (e^{\alpha vt_d} - 1)$$

and

$$u' = \frac{\alpha v(2 - e^{\alpha \ln(2)})}{e^{\alpha \ln(2)} - 1} \tag{12}$$

Combining Equations 11 and 12 we can write

$$\frac{dN_v(t)}{dt} = vfN_m(t)$$

where

$$f = \frac{\alpha(2 - e^{\alpha \ln(2)})}{(e^{\alpha \ln(2)} - 1)} \tag{13}$$

The final equation for the population of vegetative cells at time t is:

$$N_v(t) = \frac{f}{\alpha} N_0 (e^{\alpha vt} - 1) \tag{14}$$

Figure 1 shows curves calculated from Equation 14 with $N_0 = 1$, $t_d = 12$ h, and for $\theta = 0.1, 0.2, 0.3$, and 0.4. Note that if θ is small, the number of vegetative cells increases rapidly at first, but after about two doubling time periods (t − 24 h), the curves for higher θ values surpass the curves for lower values of θ. High immediate production of vegetative cells implies low ultimate numbers of cells. The relative investment in meristematic cells greatly affects the ultimate population of vegetative cells.

Milthorpe and Newton[8] analyzed cell number changes in cucumber (*Cucumis sativus* L.) and suggested that the cell generation time t_d was constant but $\theta = 1/(1 + e^{\gamma t})$ where γ is a constant. Since $e^{\gamma t} \cong 1 + \gamma t$, and θ must always be less than 0.5, Dale[9] proposed that $\theta = 1/e^{\gamma t}$ where θ can approach unity at $t = 0$. Dale[9] tested these models and found in bean *(Phaseolus vulgaris* L.) that θ initialy was close to 1. The cycle time for cell division was close to 12 hr and γ was set to 0.148. The latter model fitted the data best.

B. Cell Growth

Thornley[6] discussed cell and plant growth as increases in weight. Here we will adapt his equations to leaf expansion. We will start by selecting a suitable relation for describing expansion of individual cells and will extend it to include cell populations.

The Gompertz growth equation[10] (see below) was selected because it suitably describes the declining ability of an organ to grow. The equation, which has been used to describe growth in many organisms, is as follows:

$$a_c(t) = a_{co} e^{(u(1 - e^{-st})/s)} \tag{15}$$

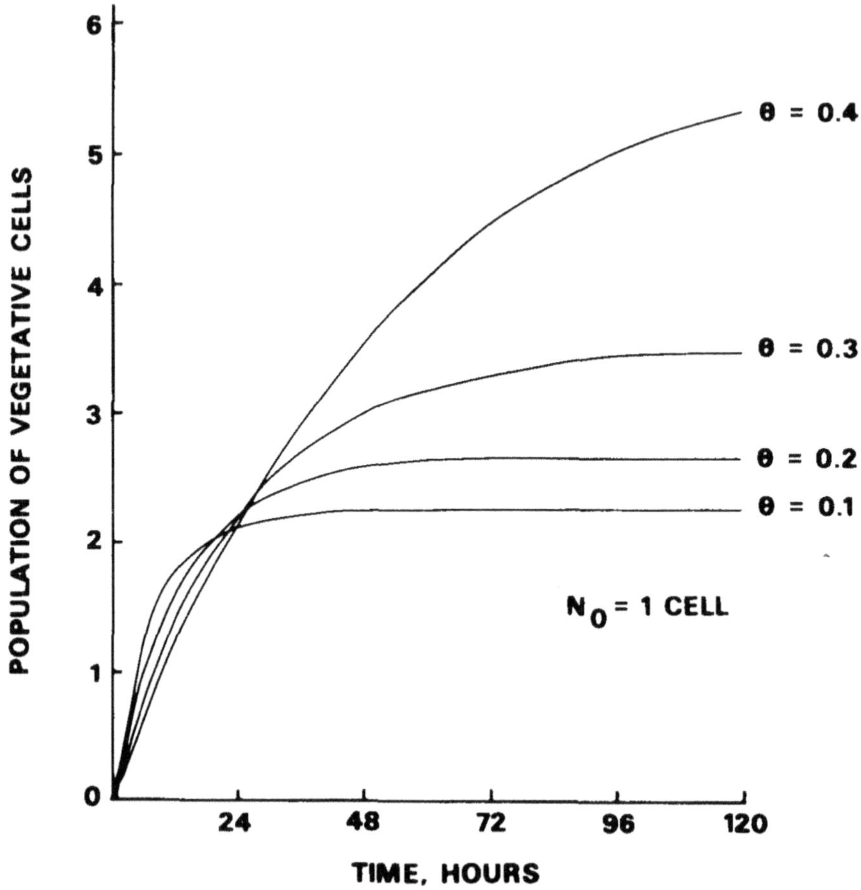

FIGURE 1. Population of vegetative cells for a single meristematic cell at time 0 (N_0 = 1), and for different values of θ.

The time rate of change of the cell cross section area is found by differentiating Equation 15,

$$\frac{da_c(t)}{dt} = ue^{-st}a_c(t) \tag{16}$$

The parameters u and s are assumed constant, and substrate for expansion is assumed to be nonlimiting for this analysis.

For an expanding leaf, we need to consider the populations of meristematic and vegetative cells in order to calculate the projected leaf area (surface of one side of the leaf) F as a function of time, F(t),

$$F(t) = F_m(t) + F_v(t) \tag{17}$$

where $F_m(t)$ and $F_v(t)$ are defined immediately below.

We consider each meristematic cell to have an average area, a_m, and the contribution of meristematic cells to leaf area (one surface only), $F_m(t)$, can be expressed as (from Equation 8):

$$F_m(t) = \frac{\bar{a}_m}{d_c} N_o e^{\alpha v t} \tag{18}$$

where d_c = number of cells in leaf cross section, assumed to be constant.

The leaf area contribution from vegetative cells is more difficult to calculate. Here, we show the initial step in a complicated integration process and refer the reader to Thornley[6] for further details. We let $a_c(t - t')$ be the area of a vegetative cell that arose from divisions at time t′. Then $F_v(t)$, the contribution of vegetative cells to leaf surface area, can be calculated by the following integral:

$$F_v(t) = \int_0^t \frac{1}{d_c} \frac{dN_v(t')}{dt} a_c(t - t')dt' \tag{19}$$

It was assumed that the relative cell division rate, v, is equal to the relative area expansion of the cell, u, in order to arrive at the following integrated expression for $F_v(t)$, or

$$F_v(t) = \frac{N_O}{d_c} \bar{a}_m f e^{us} \left(\frac{s}{u}\right)^{(\alpha u/s - 1)} e^{\alpha u t} Q \tag{20}$$

where $Q = [\Gamma(\alpha u_s, u_s) - \Gamma(\alpha u_s, u_s e^{-st})]$, and $\Gamma(r, z) = \int_0^z q^{r-1} e^{-q} dq$, the incomplete gamma function.

Figure 2 shows the results of the leaf area development model (Equations 17, 18, and 20) derived from cell division and expansion considerations. It shows the relative contribution of meristematic cells and all cells in leaf area expansion based on a single meristematic cell at time t = 0. The parameters for the results presented in the figure are

$$N_O = 1$$

$$d_c = 1$$

$$a_m = 0.1\ \mu m^2$$

$$\alpha = -0.322 \text{ (with } \theta = 0.4)$$

$$u = v = 0.0578\ (h^{-1}) (\text{or } t_d = 12\,h)$$

$$s = 0.01$$

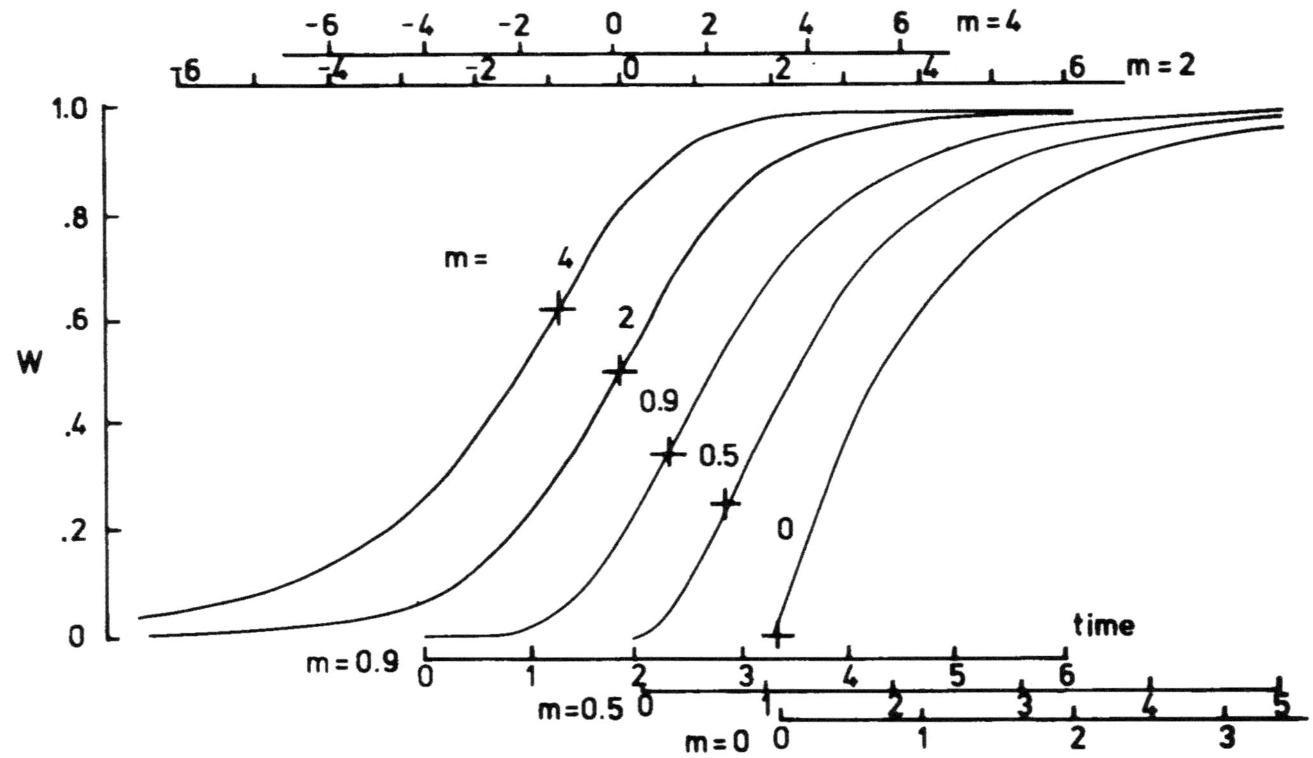

FIGURE 2. Leaf area expansion resulting from one meristematic cell at time zero. Total area (F(t)) calculated from Equation 20 and meristematic cell area ($F_m(t)$) calculated from Equation 18.

Notice that in the example the area contribution of meristematic cells approaches zero after about 300 h. Since $\theta = 0.4$, the number of dividing cells decreases. The total area, the sum of $F_m(t)$ and $F_v(t)$, is shown by the skewed S-shape curve, the skewness resulting from cell population division and subsequent expansion. The single meristematic cell produced six vegetative cells, each attaining a final size of 32 $(\mu m)^2$.

Now we have a method for calculating leaf area expansion based on six cellular parameters. The application of this model can help in our understanding of the processes affecting leaf expansion. It should be understood, however, that the scope of the model presented above is limited. It considers cell growth using an external, empirical approach and does not contain mechanistic parameters for studying cause-effect relationships in leaf cell division and expansion. For example, the substrate is always assumed to be nonlimiting, and no relations were offered for studying turgor pressure effects. Such an approach seems to be the next step — the development of a physicochemical model describing cell expansion. Such an approach would allow one to extend the leaf models (see above) for describing leaf expansion over an extended time period and varying environmental conditions.

C. Differentiation

We presently lack concepts and research results to build a comprehensive model for differentiation at the cellular level that can be generalized to predict environmental effects on growth at the field level. However, several approaches have been suggested and deserve mention here.

The first leaf cells appear as a "bump" on the side of an apical dome. Such domes can be found in the shoot or branch apices or where branches initiate near the insertion of a leaf petiole into a stem. Auxins from the main shoot apex, kinins from dividing root cells, and gibberellins from leaves interact with photosynthate and nitrogen supply to determine which buds develop into new leaves. Kinins and gibberellins neutralize auxins. Kinins enhance cell division, and gibberellins enhance cell expansion.

Plant levels of these various hormones are associated with activities of the various meristems or leaves. Given a good supply of photosynthate and nitrogen, active growth results in balances among the hormones which induce axillary bud development into branch leaves. A more detailed hypothesis requires the leaf to produce a cerain amount of photosynthate before vascular tissue develops sufficiently to supply all the water and nutrients the leaf needs for further development. This hypothesis would result in suppressed growth at branch meristems shaded from the sun.

Bonner[11] and colleagues developed a model describing how an apical meristematic cell can give rise to cell elements and differentiated tissue. In their simulation, a newly divided cell tested its environment for closeness to the organ surface. This was achieved conceptually by allowing density changes in a volatile inhibitor produced by all cells in an organ, but which diffused to surrounding air. When a cell becomes imbedded in a mass of cells where the inhibitor concentration is above a certain level, the cell stays vegetative. This logic avoids the assumption of constant θ, or ratio of meristematic to vegetative cells. All this, of course, needs to be quantified somehow from experimental data.

Bonner[11] invoked a concept for developmental pathways appropriate for cellular development into xylem and phloem elements carrying materials to and from a developing leaf. The cambium cells dividing the form zylem and phloem are usually found near the surface of the stem, and the larger the plant the larger the stem surface and potential cambium activity. In addition, Wilson[12] developed a model for cambium activity in trees.

The Bonner paper was concerned with hierarchical control programs in biological development. They started at the molecular level (molecular biology) and ended with logic for a computer program describing cell division and growth. Experimentalists interested in the potential of modeling methodology in developing and refining connepts should read the Bonner paper. Differentiation is one of the last frontiers of biology to be explored and explained, and the Bonner model has yet to be experimentally tested.

Turing[13] and Thornley[14] have developed models based upon the production, diffusion, and degradation of plant hormones (morphogens), with initiation of primordia occurring on the apical dome only where the hormone concentration or balance reaches a critical state. These models helped explain the arrangement of leaves on the plant, or phylotaxy. Thornley[14] modified the hormonal model with an additional space requirement between primordia on the dome before a new primordia can develop, to explain behavior in plants that the hormonal model could not predict.

At some stage of plant growth, flowering hormones build up to the level, or the balance among the above-listed hormones becomes such that a floral bud is initiated. Floral buds and resulting fruit compete for nutrients and also disturb the hormonal balance in the plant, resulting in slower or completely inhibited leaf growth.

III. INDIVIDUAL LEAF EXPANSION

In whole plant models, it is useful and informative to take account of plant morphogenesis or the initiation and growth of individual organs. In such models, simple relationships are needed for estimating the potential growth in biomass and size (area) of organs. Here we are concerned with leaf area expansion. Milthorpe[15] pointed out that the relative rate of leaf area expanson is largely independent of the net assimilation rate and therefore of the need to treat area expansion separately. Here we present some of the more useful empirical formulas for estimating the growth in area of individual leaves.

A. Leaf Area Expansion Formulas

Richards[10] gave several equations (the Richard's function) for estimating the growth of an organ (or plant) and the same relationships apply to leaf area, F(t). Maksymowych[16] and Williams[17] reported that early leaf expansion is approximately exponential, and the leaf growth curve is S-shaped but not symmetrical. Williams[17] recommended the Richard's function for a general formula to describe the growth of a leaf. The Richard's function is:

$$F(t) = F_f(1 - ke^{-bt})^{1/(1-m)} \tag{21}$$

for $m < 1$, and $F(t) = F_f(1 + ke^{-bt})^{1/(1-m)}$ for $m > 1$, and where k, b, and m = empirical growth constants, and F_f = final leaf size.

An expression describing the rate of leaf area expansion is frequently needed in plant growth models. The Richard's function, Equation 21, can be differentiated to obtain such an expression:

$$\frac{dF(t)}{dt} = \frac{\mp F_f k b e^{-bt}}{1-m}(1 \pm ke^{-bt})^{m/(1-m)} \tag{22}$$

The maximum dF(t)/dt occurs when t is equal to ln(b/k). Expansion rate dF(t)/dt asymptotically approaches zero as t approaches infinity. For practical purposes, the time period over which expansion occurs is finite and varies with the parameters k, b, and m. Also, expansion rate in the above formula is dependent on knowledge of the final leaf area, F_f.

Computer programs are available for fitting this function to experimental data.[18-20] Depending on the numerical value of m, the Richard's function takes on the form of one of three useful growth functions.[10] For m ranging from 0 to 1, the function shifts from the monomolecular function:

$$F(t) = F_f(1 - ke^{-bt}) \tag{23}$$

to the Gompertz equation,

$$F(t) = F_f(e^{-ke^{-bt}}) \tag{24}$$

As m approaches two, the function approaches the autocatalytic function:

$$F(t) = F_f/(1 + ke^{-bt}) \tag{25}$$

Figure 3 shows the shapes of curves generated by the Richard's function for various values of m (taken from Williams[17]). Each curve was calculated using a scale of F_f equal to 1 and curves were separated for clarity. Note that the time scale has to be shifted for application of this function for m-values of two and four. Points of inflection are noted and vary with m, pointing out that m is the important shape factor in the function.

Richards[12] used his function to fit data of Pearl et al.[21] on the growth of melon (*Cucumis melo* L.) hypocotyls grown in darkness at six different temperatures. The m-values for the different temperatures ranged from 1.0 at 30°C to 3.0 at 15°C. Richards pointed out that the k values could not be compared for curves of different form, or for different m values. Also, final leaf size is affected by temperature. These characteristics restrict the physiological interpretation of such results, and therefore the use of such functions is not suitable for detailed exploration or prediction of growth. Nevertheless, these formulas provide a useful means of summarizing leaf area expansion data for various conditions.

B. Temperature Effects

Research on the effect of temperature on the growth of leaves provides a good example of how these formulas can be used. It is convenient to express the rate of leaf area expansion as:

$$dF(t)/dt = K_T(T)[dF(t)/dt]_{T_s} = \text{constant} \tag{26}$$

where $K_T(T)$ is the temperature modifying function and $[dF(t)/dt]_{T_s}$ is the linear phase or maximum expansion rate at some temperature T_s covered by the modifying func-

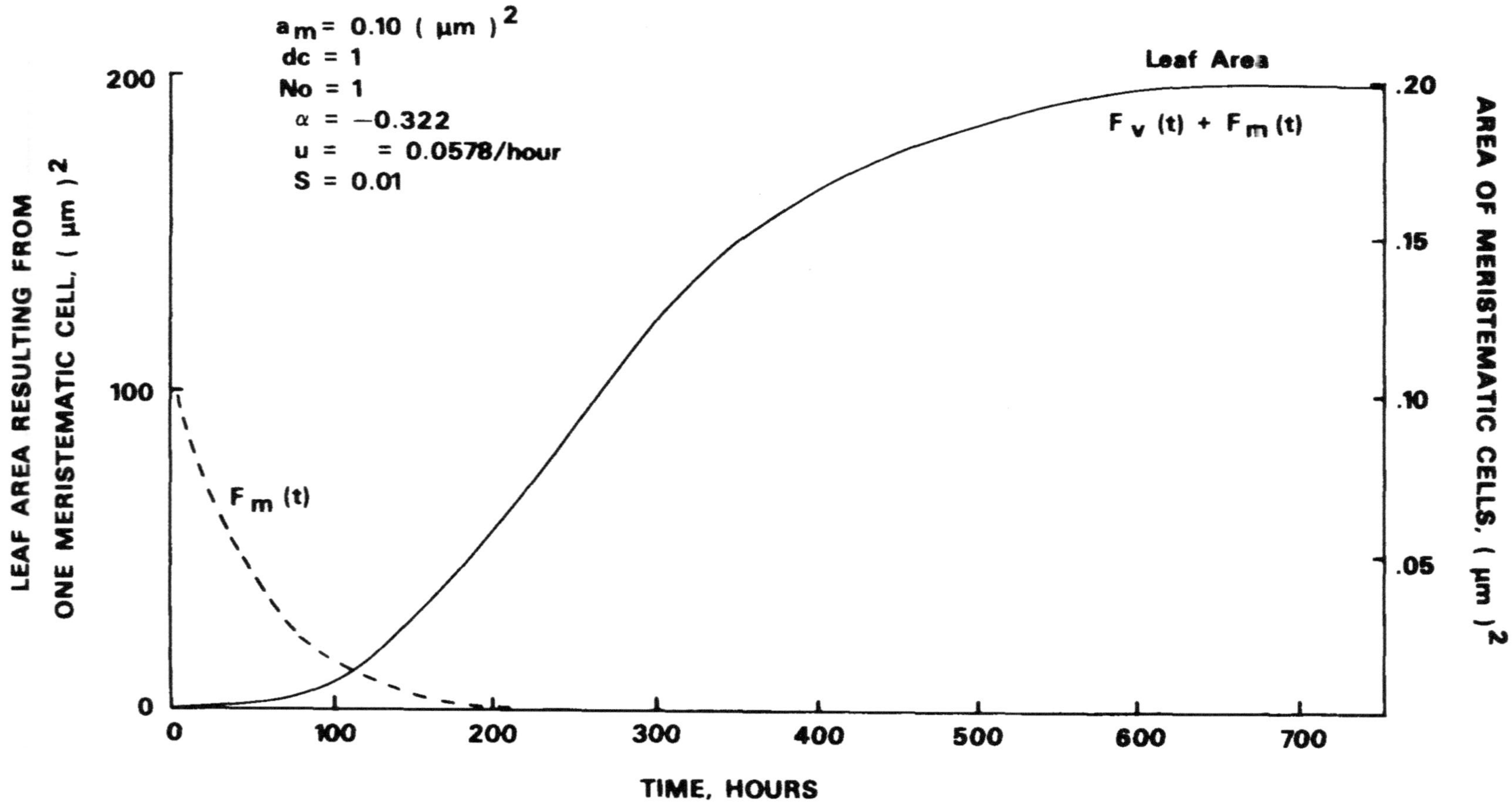

FIGURE 3. Five examples of the infinite series of curves devised by Richards[10] for fitting empirical growth data. From Williams, R. F., *The Shoot Apex and Leaf Growth*, Cambridge University Press, 1973, Chapter 2. With permission.

tion. Frequently, the temperature function is assumed to be linear with temperature over a range of temperatures, or

$$dF(t)/dt = [(T-T_0)/(T_S-T_0)]\{dF(t)/dt\}_{T_S}, \text{ for } T > T_0 \quad (27)$$

Milthorpe[15] determined the effect of temperature on the linear phase of leaf area expansion in cucumber; we have since reported similar values for cotton (*Gossypium hirsutum* L.)[22] and soybeans (*Glycine max* [L]. Merrill,[23] Figure 4. The relationship between temperature and maximum leaf growth, as occurred during the linear phase of expansion, was linear for cotton, but not for soybeans. Using values of T_0 = 12.7° T, and T_s = 22.7°C where [dF(t)/dt] = 19 cm²/day for cotton,

$$dF(t)/dt = 19\,(T-12.7)/10 \quad (28)$$

Therefore the expansion rate increases by 19 cm²/day for every 10°C rise in temperature above 12.7°C. For soybeans the effect of temperature on leaf area expansion could be expressed by Equation 26 and a nonlinear function of $K_T(t)$.

Temperature affects the time periods of leaf expansion. Therefore it is erroneous to apply the above leaf area expansion rate formulas over a constant time period for leaves exposed to different or varying temperatures. A practical method to approximate leaf expansion for such conditions is to use a degree-day concept.

A degree-day is defined as:

$$DD = \int_0^t (T-T_0)dt \quad (29)$$

where time is in days and for $T > T_0$. Over a range of normal temperatures for some crops, the number of degree-days required for expansion is about constant. The above rate equations would be applied over a constant number of degree-days.

Two commonly used methods exist for calculating degree-days required for development. Controlled temperature studies are used to determine the length of time required for leaf expansion for constant temperatures. The degree-days required for expansion at each temperature T can be calculated by multiplying $(T - T_0)$ by the measured time period of expansion. A plot of required time periods vs. $(T - T_0)^{-1}$ will be about linear, the slope being equal to the degree-days required for expansion. This approximate method may not be useful for widely fluctuating temperature conditions. There is a need to further develop and apply the thermodynamic model of poikilotherm development of Sharpe and DeMichele[59] for such conditions.

A second method to approximate the degree-days required for leaf development is to use data collected for varying temperature conditions. For time periods corresponding to periods of leaf expansion, degree-days from leaf initiation to full expansion are calculated, using Equation 29, and averaged.

Because of the empirical nature of the above analyses, the use of the equations to estimate leaf area expansion for varying environmental parameters is difficult. The fact that the formulas require the final area, F_f, clearly demonstrates some of the dif-

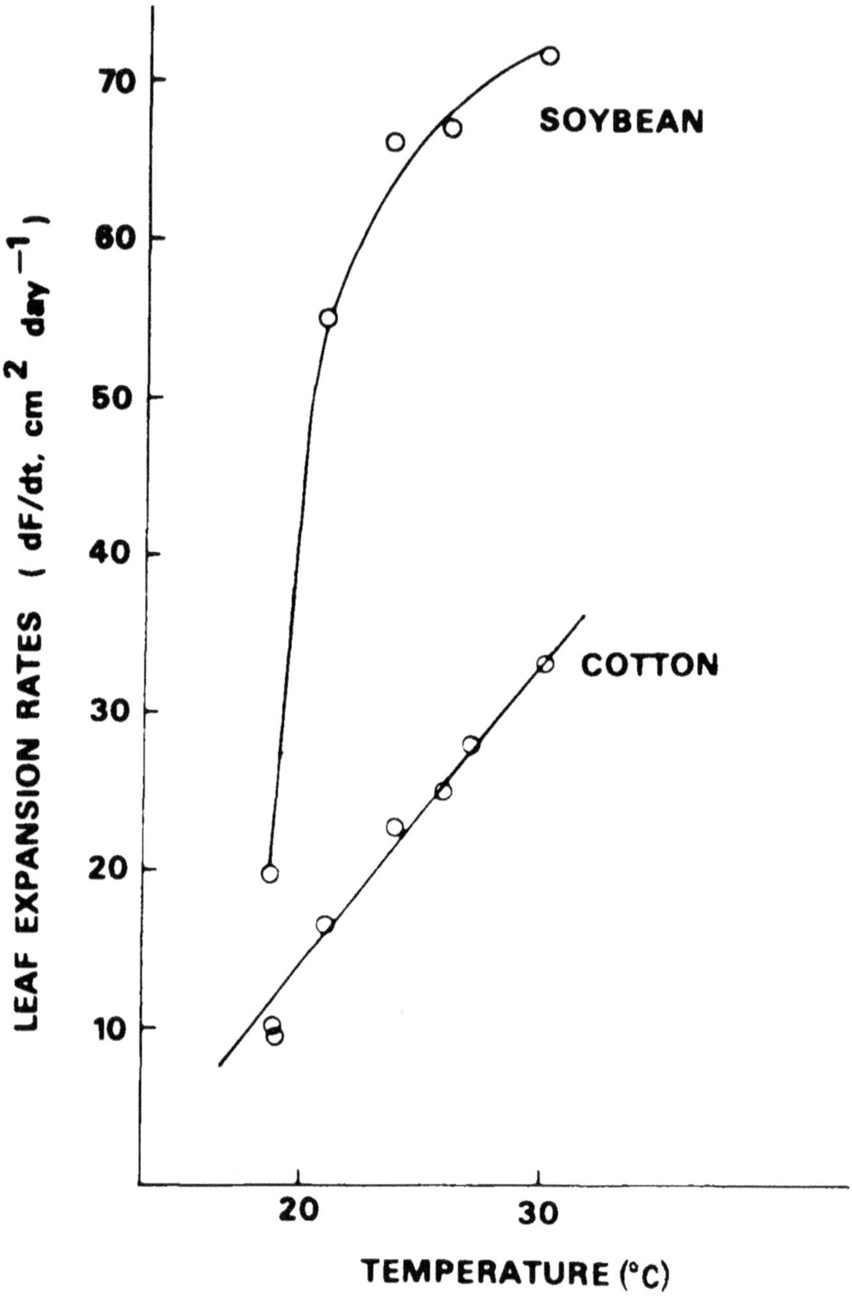

FIGURE 4. Leaf area expansion rate for single leaves (dF/dt) during the linear phase of expansion as affected by temperature regime.

ficulties encountered. Maximum dF(t)/dt values during the linear phase of growth seem to be linearly related to F_f[23]. Richards[10] fitted his function to the growth of a melon hypocotyl in seven different temperature regimes (data of Pearl et al.[21]). The final size of the hypocotyl as well as b, k, and m varied considerably with temperature. Equation 27 gives a method for altering the area expansion rate based on temperature and data from temperature studies. If T is not constant, or T = T(t), then the integration of Equation 27 can only be accomplished if we know empirically how F_f and other factors change with varying temperatures. Also, for comprehensive models of leaf area expansion, we need the nutritional requirements for growth such as those reported by Williams (his Table 5.3).[17] In practice, the measurement and generalization of F_f for various node positions on a plant has to suffice for empirical simulation of leaf surface area expansion.

C. Water Stress Effects

Water stress in plants directly affects leaf area expansion rates earlier and more severely than it affects photosynthesis and respiration rates.[24] Reduced leaf area decreases the ability of a plant to intercept light and produce photosynthate. Therefore, the effects of water stress on total crop productivity is difficult to measure and to predict.

The reduction in rates of leaf area expansion caused by reduced leaf water potential implies that a minimal turgor pressure must exist for expansion. As leaf water potential is reduced, turgor is decreased and cell enlargement is reduced and stopped. Boyer[24] found that the rates of leaf enlargement were decreased to zero when leaf water potentials were about −12, − 4, and −8 bars for soybean, sunflower, and corn, respectively (Figure 5). Photosynthesis decreased little until leaf water potential decreased below the above values and was still occurring at leaf water potentials less than −17 bars. In earlier work, Boyer[25] found that growth of sunflower leaves occurred only over a 2-bar turgor pressure change and estimated that the minimum turgor required for cell expansion was 6.5 bars. Measurements of day and night growth indicated that expansion occurred mostly at night when leaf water potential and turgor were high.

Heatherly et al.[26] found that leaf enlargement declined at leaf water potential below −4.5 bars and ceased when potentials were below −10 or −12.9 bars. Following rewatering, leaf growth rate resumed (Figure 6). Boyer[27] found that leaves may adapt to decreased leaf water potential after a period of several days, supportive of the role of osmotic adaptation in the control of turgor.

Water stress seems to affect cell enlargement more than it affects cell division, as evidenced by similar numbers of cells in plants exposed to water stress as compared with control plants. Cell size is less in stressed plants.[28] Cell enlargement, stomate control, and phloem translocation seem to be more dependent on turgor whereas biochemical processes appear to be affected more by osmotic potentials.[29] Therefore, a model of leaf area expansion from a physicochemical point of view is needed for considering osmotic and turgor components of total water potential.

For more practical applications, an empirical approach may be useful. The equations for whole leaf area expansion may be multiplied by a water stress factor empirically derived from experimental data, or

$$\frac{dF(t)}{dt} = \left(\frac{dF(t)}{dt}\right)_0 f(\Psi_l) \tag{30}$$

where $f(\Psi_l)$ is an empirical function for reducing leaf growth rate during periods of low leaf water potential, Ψ_l. The $[dF(t)/dt]_o$ represents the leaf growth expansion rate at time t for nonstressed plants. The above simple relationship causes no change in the duration of leaf expansion, but will reduce the ultimate leaf size in proportion to the integral of $f(\Psi_e)$ over time. Expansion rates are assumed to respond immediately to changes in Ψ_l and we do not account directly for osmotic adaptation.

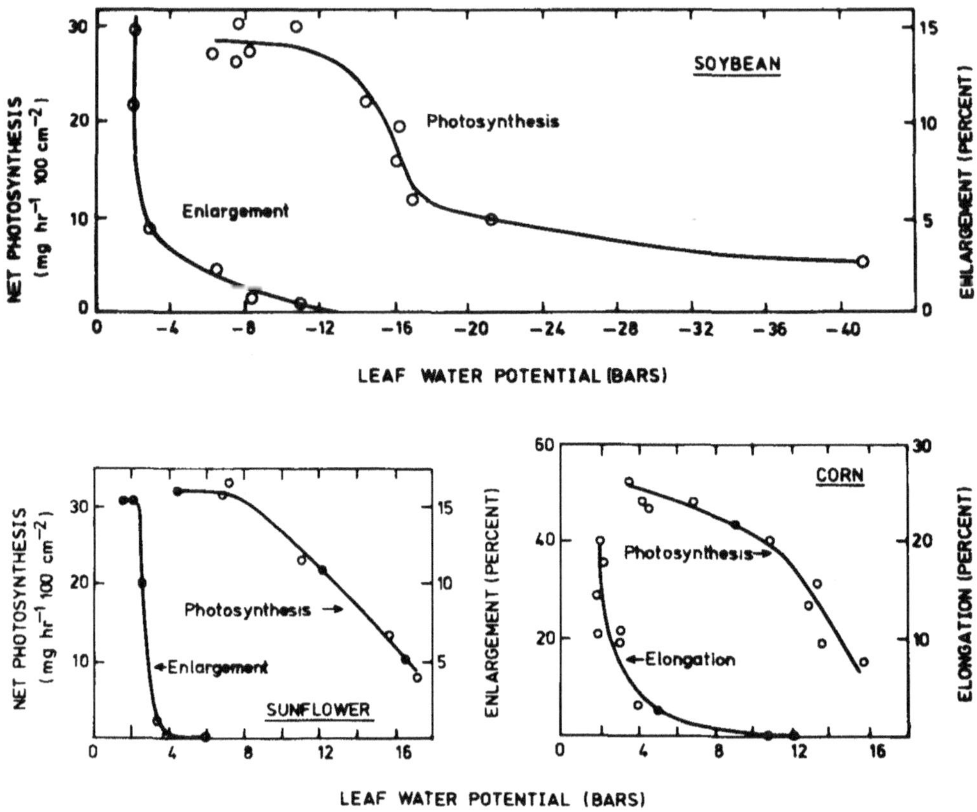

FIGURE 5. Rates of leaf enlargement and net photosynthesis in corn, soybean, and sunflower plants at various leaf water potentials. The photosynthesis data were collected from two different plants for each species. The plants were 45 to 60 cm tall. The growth data for soybean and sunflower represent enlargement of the fourth and sixth leaves from the base of the plant, the leaves having an area of about 20 and 60 cm^2, respectively, at the beginning of the 24-hr growth period. For corn, growth was determined as elongation of the sixth leaf blade. The corn leaf blades were initially 25 to 35 cm long. From Boyer, J. S., *Plant Physiol.*, 46, 233, 1970. With permission.

IV. WHOLE PLANT LEAF EXPANSION

Here we discuss the development of leaf area of whole plants from two points of view. First of all, we calculate the whole plant leaf area by summing the areas of individual leaves. This is analogous to the treatment of leaf area development based on cell division and expansion. Secondly, and using a higher order approach, we consider the rate of expansion of total leaf area of a plant.

A. The Plastochron Concept

A plastochron is the number of days between successive like organs of equivalent size on some meristematic apex. For leaves, such values are usually obtained from ln(F) plots with F values of 5 to 10 cm^2 being the size limit.[8,30,31] Such values are affected by temperature,[15,32] and one can approximate a degree-day value per plastochron. The plastochron concept is attributed to Michelini[33] as an index for describing leaf or plant age, but Zaitzev[34] reported its use for predicting morphogenic events in cotton (from Hector[35]) in 1927, with data for estimating degree-day values for a plastochron.

To determine a plastochron, length or length × width measurements are needed for

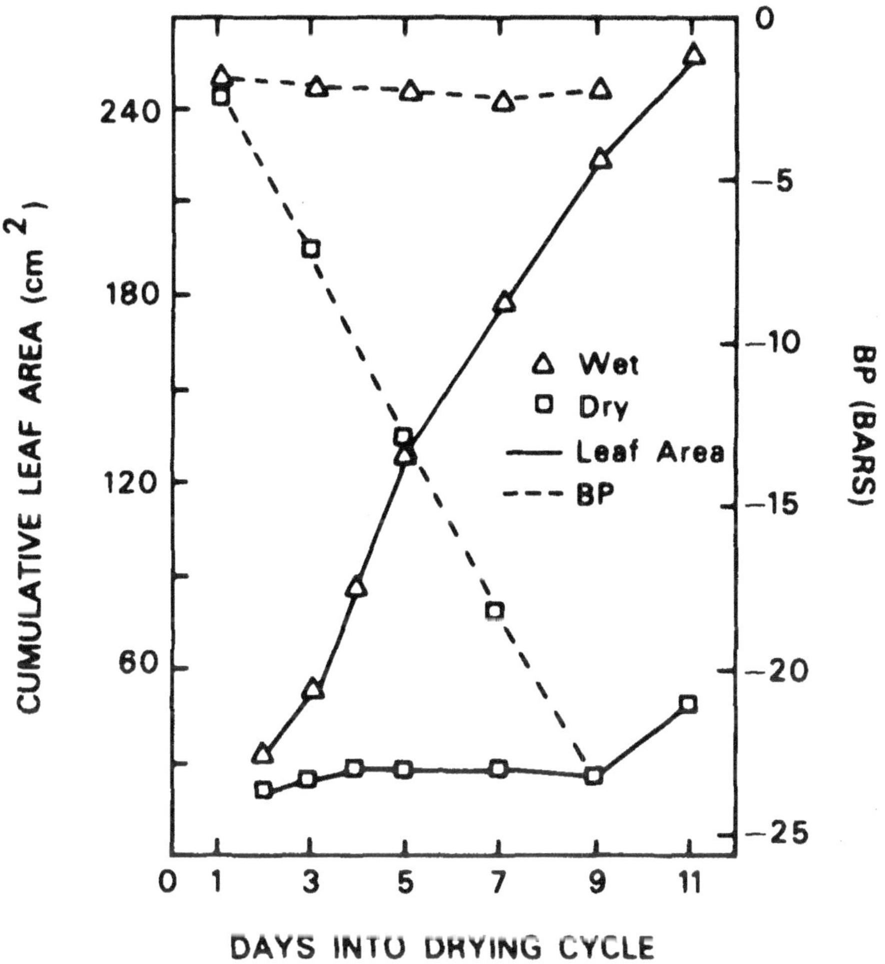

FIGURE 6. Relationship of cumulative leaf area of soybeans growing in wet and drying soil to base (presunrise) xylem pressure potential (BP). From Heatherly et al. *Crop Sci.*, 17, 381, 1977. With permission.

young expanding leaves which are larger and smaller than the dimensions chosen for the reference leaf ("like organs of equivalent size"). The Plastochron Index (PI) can then be determined from logarithmic plots of dimensions of successive leaves, using the following relationship:

$$PI = n + [\ln(F_h) - \ln(F_O)]/[\ln(F_h) - \ln(F_{h-1})] \quad (31)$$

where n is the number of leaves larger than the reference leaf, counting down the plant from the main stem apex. The latter part of Equation 31 is for the fraction of a plastochron that has taken place between the smallest leaf counted and the hypothetical reference leaf, with F_h the dimensions of the leaf larger than F_0, the reference leaf, and F_{h-1} the dimensions of the next smaller leaf. Assuming $\ln(F_i)$ is a linear function of time, the numerator and denominator of Equation 31 represents two time periods or a fraction of a plastochron. It also is assumed that the ultimate leaf size F_f is a constant.

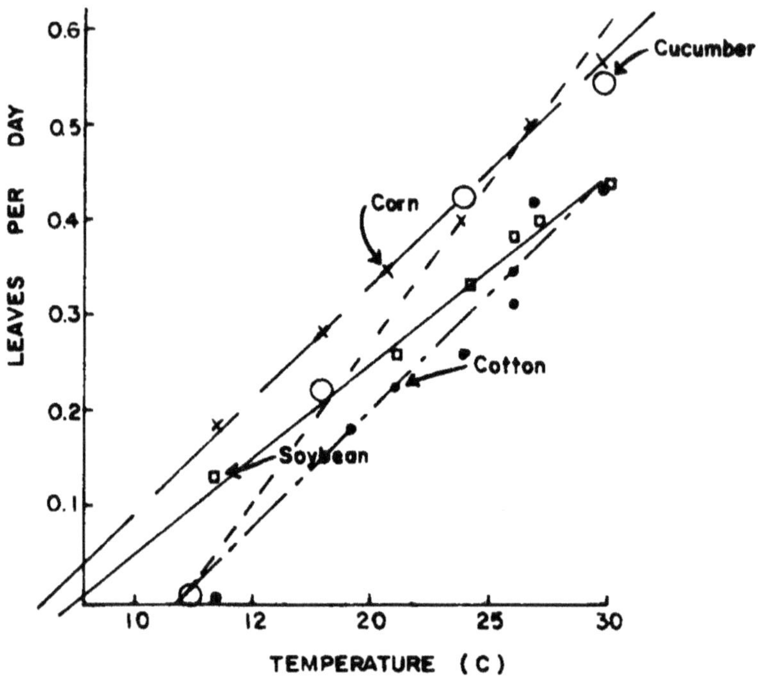

FIGURE 7. Leaf production rates vs. temperature. The reciprocal of leaves per day is the apparent plastochron. Cucumber data from Milthorpe, F. L., *J. Exp. Bot.*, 10, 233, 1959; other data from Hofstra, G., et al., *Can. J. Plant Sci.*, 57, 167, 1977.

Plastochron values for reference leaves of about 10 cm^2 in area often do not vary much in a vigorous seedling, but as a plant gets older, the plastochron gets longer.[36] For reference leaves as "bumps" of similar size on the main stem apical dome, plastochron values vary greatly, and the time between bump formation can decrease greatly just before the initiation of a floral bud.[37]

In various versions of SIMCOT and COTCROP, computer models for growth and yield in cotton, leaf production rates, the inverse of a plastochron value, were used in developing a temperature correction for time. Leaf production rates are approximately a linear function of temperature from no growth to the maximum rate[23] (Figure 7). We shall discuss this relationship further in the section on the degree-day concept.

Plastochrons at the 10 cm^2 leaf area level become longer as the plant ages, but this reduction in growth rate can be related to lack of carbohydrate or nitrogen to maintain the cell division rate necessary for constant leaf production. Similarly, plastochrons become longer when plants are stressed for water, and now this can be accounted for.[38] Some species seemingly generate a constant number of main stem nodes or leaves, and with a model for predicting the time to appearance of the first main stem leaf from planting or emergence, one can predict the time required to generate a specified number of main stem leaves, with corrections for temperature, water stress, and nutrient stress.

The rate at which plants initiate and expand leaves is an important factor in determining their ability to capture solar energy and grow, transpire, and absorb or fix nitrogen. The plastochron model can be combined with a leaf expansion model to predict whole plant leaf expansion over time. Assuming the autocatalytic form of the Richards function,

$$F(t) = F_f/(1 + be^{-kF_f t}) \qquad (32)$$

where $b = (F_f - F_O)/F_O$. The equation for the whole plant is the integral or summation of leaves initiated multiplied by their area time functions, or

$$F_T(t) = F_c + \int^{t} R(t')F(t-t')dt' \qquad (33)$$

where $F_T(t)$ is the time function for the whole plant leaf area F_T, F_c is the cotyledonary area, R(t) is the rate of leaf initiation or l/plastochron, and u is a dummy variable used for integration. Assuming the plastochron to be constant for all leaves and that temperature is constant, R(t) becomes l/p and Equation 30 may be written:

$$F_T(t) = F_c + \int_0^t [F_f/p\ (1 + be^{-kF_f(t-u)})]\ du \qquad (34)$$

$$= F_c + F_f t/p + (1/kp)\ [\ln (1 + be^{-kF_f t} - \ln (F_f/F_o)] \qquad (35)$$

Assuming $F_O = 10\ cm^2$, $F_f = 200\ cm^2$, p = 3.0 days, k = 0.002/day/cm², and $F_c = 50\ cm^2$, we obtained the plot in Figure 8, which appears to start as an exponential curve (see exponential curve for comparison) and then becomes linear. With our simplifying assumptions, as t gets large, Equation 35 reduces to $F_c + F_f t/p$.

The ultimate leaf size F_f increases over the first ten nodes or so, reaching a maximum, at which point if flowering is under way or heavy branching is occurring, it decreases in size. Auxiliary branches with leaves complicate the above problem. Water and nutrient stresses slow leaf production and expansion rates.

The plastochron can be used as a unit to define other growth periods in the plant (Table 1), providing a morphogenic-temperature model for plant growth. Such a model assumes that periods between different plant events are all affected by temperature in the same manner. Also, one must be able to predict the total number of mainstem nodes when flowering first occurs.

Plastochrons and other time units need to be determined under ample supplies of photosynthate, nitrogen, and water, and effects of diurnal amplitudes of temperature as well as different combinations of previous environments or stresses need to be accounted for. For example, the boll period is longer for plants grown continuously in cool temperatures (23/20°C day/night 8/16-hr temperature regime) than for plants moved from a warm (29/26°C regime) to the cooler regime at commencement of flowering.

Temperature-growth relationships can be made specific for the various time periods, once experimental data establish that such differences exist. Temperature effects on growth will be discussed in greater detail below.

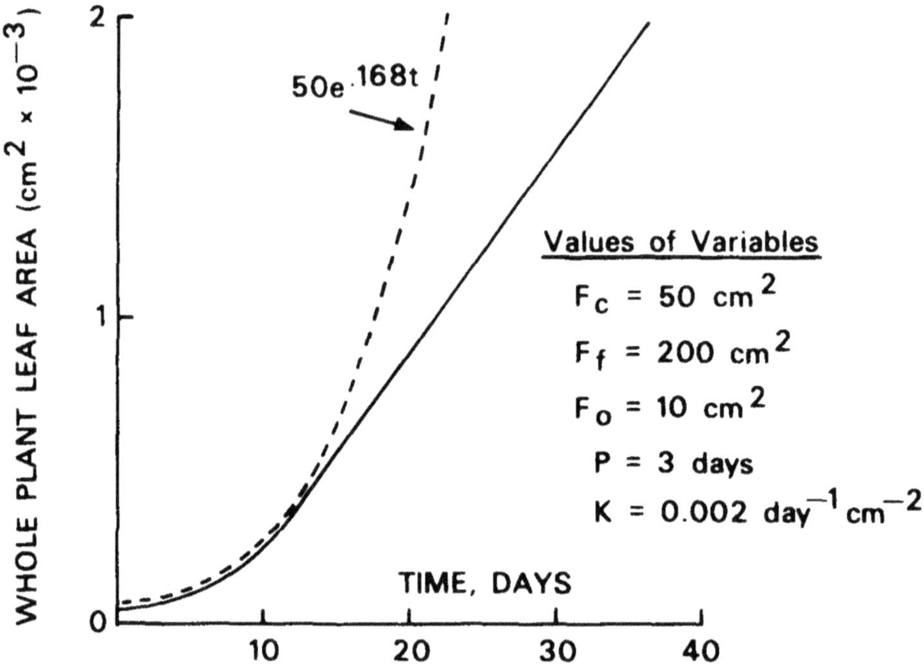

FIGURE 8. Whole plant leaf area based upon the plastochron concept and expansion of individual leaves from Equation 35.

TABLE 1

Plastochron and Degree Day Values for Cotton and Soybean Life Histories

	Cotton		Soybeans	
Life history event	Plasto-chrons	Degree days	Plasto-chrons	Degree days
Planting				
To 8 cm^2 cotyledonary area	2.3	90	2.3	100
To 8 cm^2 first true leaf	4.0	156	3.0[a]	150
Total			1.5	75
			4.5	225
8 cm^2 leaf on first vegetative branch	10.3	400	8.5	425
Plastochrons, vegetative branch leaves	2[b]	78	2[b]	100
Minimum temperature used in degree day calculations	12.7	—	7.5	—
Degree days for plastochron	39.0	—	50	—
Leaf duration	20	780	—	—
Plastochrons per flower,	1.0	39		
between sympodia (cotton)	3.0	117		
within sympodia				
Squaring period (cotton)	8	312		
Boll period (cotton)	18	700		

Note: A critical missing event is the date of first flower; one needs to know the number of mainstem nodes on the day of flowering to predict this.

[a] Cotyledon to unifoliolate, unifoliolate to first trifoliolate.

[b] Approximation — this value would approach 1.0 with ample available photosynthate.

B. Growth Analysis

1. Traditional

With F = leaf area, W = dry weight, and N = nitrogen weight, each on a per plant or ground area basis, the following parameters can be estimated for samples of weight and area as described in the Growth Analysis chapter:

$$(1/F)\ (dF/dt) = R_F \tag{36}$$

the relative rate of leaf area expansion or relative leaf growth rate (RFGR),

$$F/W = \alpha_W \tag{37}$$

the leaf area ratio (LAR),

$$dF/dW = \beta_W \tag{38}$$

the leaf area partitioning coefficient (LAP)[1,39]

$$F/N = \alpha_N \tag{39}$$

and

$$dF/dN = \beta_N \tag{40}$$

As stated above, R_F is affected by the morphogenetic potential of the genome as well as air humidity, temperature (Figure 9),[39] leaf turgor, and nutritional factors. As shown above, R_F is not constant with stage of growth. α_W or β_W can change with light intensity or photosynthate supply in some species to maintain a fairly constant R_F independent of dry matter production.

The importance of the relative leaf area expansion rate, R_F, on the growth rates of several species was shown in Figure 10.[1] Using nine species and varying temperature to obtain different relative leaf area expansion rates within each species, it was found that the relative growth rate was highly correlated with R_F, but was poorly correlated with the net assimilation rate [NAR, (l/F)(dW/dt)], the leaf area partition coefficient (LAP, β_W, Equation 38), and the leaf weight partition coefficient [LWP, dW_F/dW, the ratio of the change in leaf weight (W_F) to the total change in plant dry weight (W) or $(dW_F/dt)/(dW/dt)$]. They concluded that the competitive ability of species, regardless of photosynthesis rate, is enhanced by an increase in the leaf area partition coefficient, as predicted by Jackson.[40]

Partitioning was considered in detail in the Growth Analysis chapter, where it was pointed out that the traditional, as well as the newer carbon balance techniques, combined with summed organogenic models for the whole plant, are necessary for fundamental studies of partitioning.

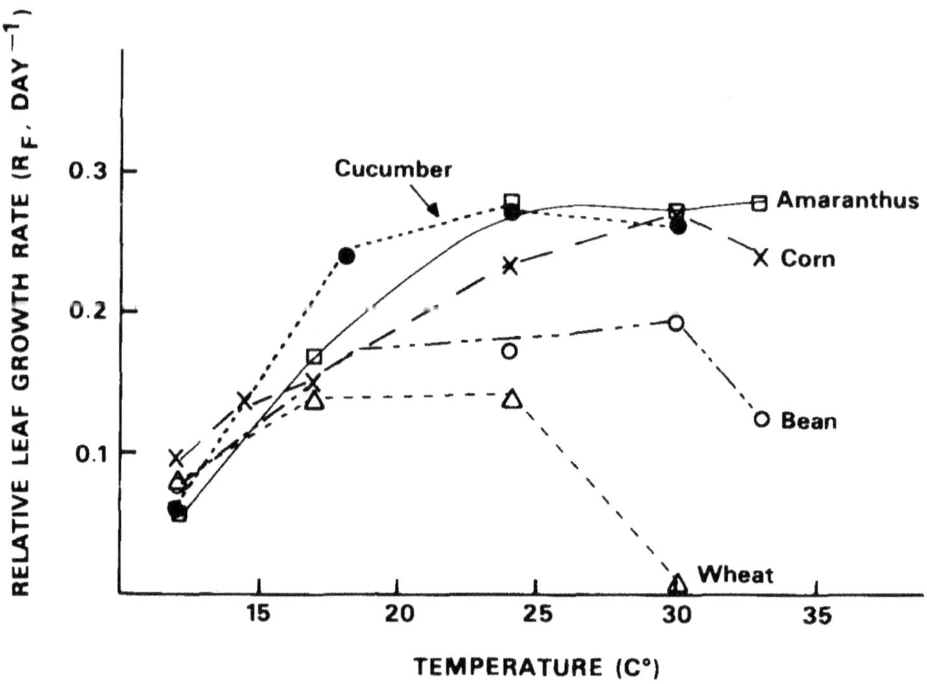

FIGURE 9. Relative leaf growth rate for whole seedlings as affected by temperature. Cucumber data from Milthorpe, F. L., *J. Exp. Bot.*, 10, 233, 1959; other data from Hesketh, J. D. and Baker, D. N., *J. Ariz. Acad. Sci.*, 5, 216, 1969. With permission.

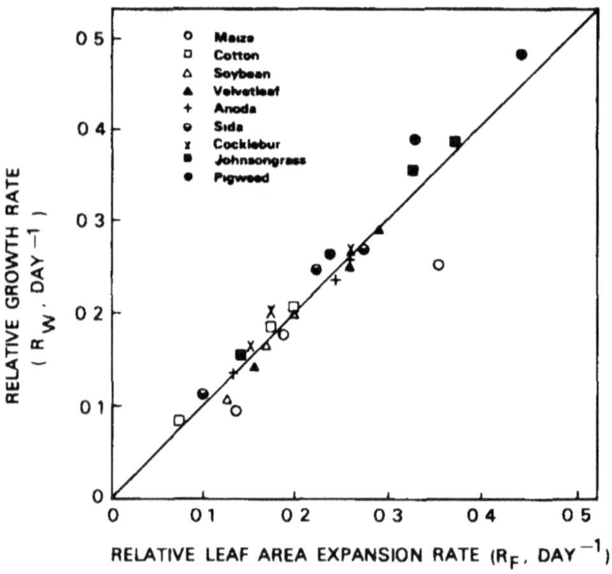

FIGURE 10. Relative growth rates (R_w) vs. relative leaf area growth rates. From Potter, J. R. and Jones, J. W., *Plant Physiol.*, 59, 10, 1977. With permission.

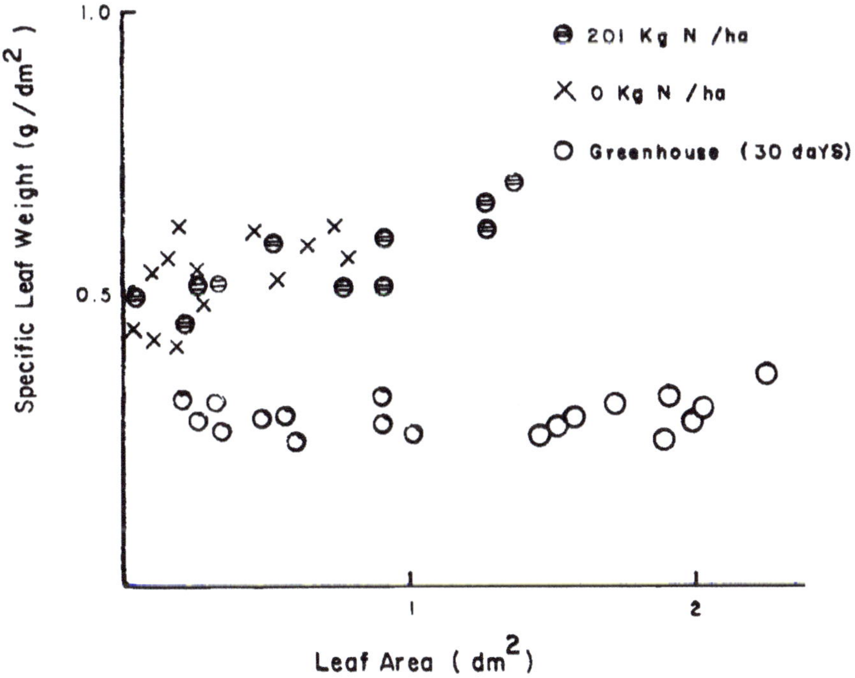

FIGURE 11. Leaf weight per unit leaf area (specific leaf weight) for field and greenhouse plants. Areas were determined on successive leaves from the main shoot apex down the main stem for plants of comparable size. Field-grown plants were from an N-fertility trial. From Hesketh, J. D., Lane, H. C., and Thompson, A. C., *Proc. Beltwide Cotton Prod. Res. Confs.*, National Cotton Council, Memphis, Tenn., 1976, 61. With permission.

2. Nutritional Requirements for Leaf Expansion

Specific leaf weights for cotton plants in an N fertility trial[41] and in a greenhouse in winter are shown in Figure 11 for expanding and nonexpanding cotton leaves. As discussed earlier, the specific leaf weight and the leaf area dry weight ratio can vary considerably with photosynthate supply. In other words, leaf expansion can proceed at the maximum rate over a wide range of photosynthate supply values.[42]

The specific leaf weight changes as starch and sugar accumulate, and this change can be almost twofold.[43] One must account for starch and sugars when determining the structural specific leaf weight, or leaf weight/leaf area.

Leaf N was monitored in an N-fertility trial[44] and the percentage N varied greatly between expanding and aging leaves depending upon N supply in the plant (Figure 12). Leaves in N-deficient plants were much smaller. We had assumed in an early model[45,46] that leaf expansion proceeded at the maximum rate above a threshold value of 3% N on a dry weight basis (i.e., below 3% N, there was no expansion). However, we stated above that leaf expansion rate is related to the ultimate leaf size attained,[23] and area per leaf was considerably less in the plots receiving no N (Figure 12). From Figure 2, Chapter 4, Volume I, the relative rate of leaf expansion was linearly related to leaf % N in one species and not the other.[47] The relative leaf expansion rate of leaves was linearly related to % N in the Japanese IBP trials, Figure 13.[48] As a consequence, we have been forced to modify the earlier assumption because of new experimental evidence to the contrary.

3. Nutritional Requirements for Root and Stem Support

The amount of photosynthate and nitrogen that can be partitioned into leaf growth

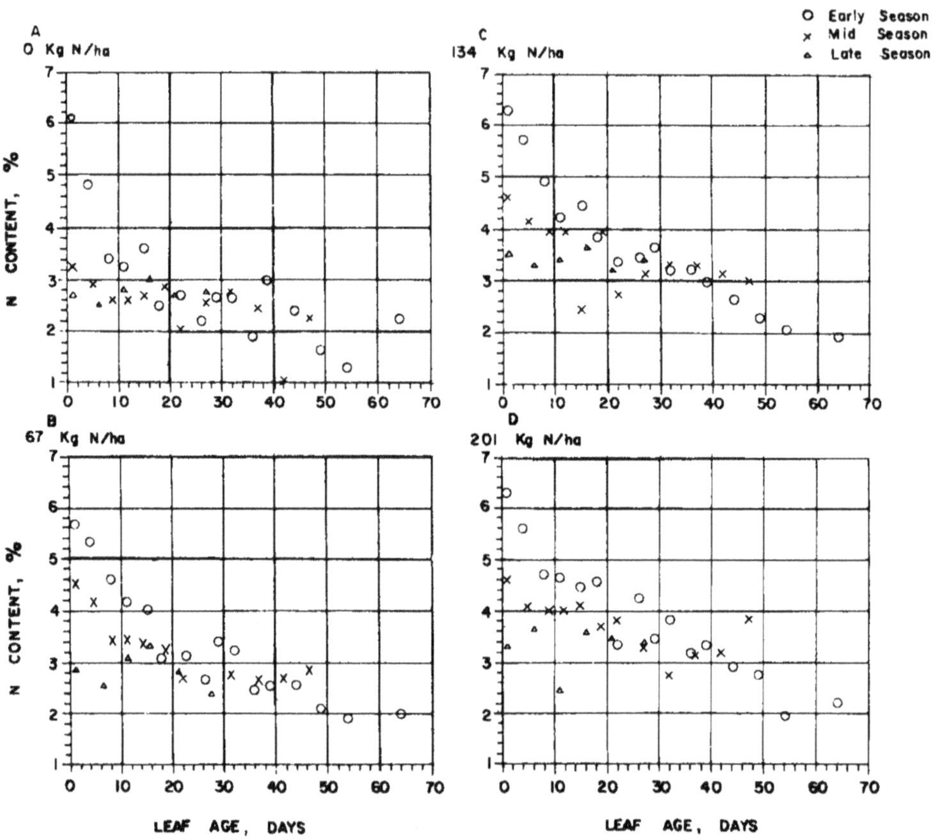

FIGURE 12. Leaf percent nitrogen vs. time for newly expanded leaves tagged on three dates and sampled subsequently. Plants were in a field N-fertility trial. From Thompson, A. C., et al., *Agron. J.*, 68, 617, 1976. With permission.

will depend upon stem and root requirements for an increment of leaf area. Bark and wood weights for cotton were determined for internodes from the main shoot apex down and were plotted vs. leaf area supported above the internode, Figure 14.[41] Such values were a linear function of supported leaf area, but they varied considerably for different plant spacings and fertility levels, as well as light levels (field vs. winter greenhouse), Figure 15. The greenhouse plants in a low light regime fell over easily and lacked the necessary root support to maintain rapid transpiration rates on clear days.

Photosynthate supply can be restricted two ways: (1) reducing the incident light level and (2) establishing more plants per unit ground area. The 4-cm spacing represents a fairly dense population, yet such plants maintained a large internode weight-leaf area ratio. Therefore, one must be careful when extrapolating results from one situation to another, even when the experimental conditions seem similar.

4. *Leaf Duration and Activity*

Maximum leaf area per unit ground area may well be controlled by leaf duration if leaves die before vegetative growth ceases. The decline in canopy area will also be controlled by leaf duration and the change from vegetative to reproductive growth. The decline in canopy photosynthetic activity will be associated with the decline in area as well as photosynthetic rates per unit leaf area as leaves age.

The effects of leaf position on the main shoot as well as temperature on leaf duration

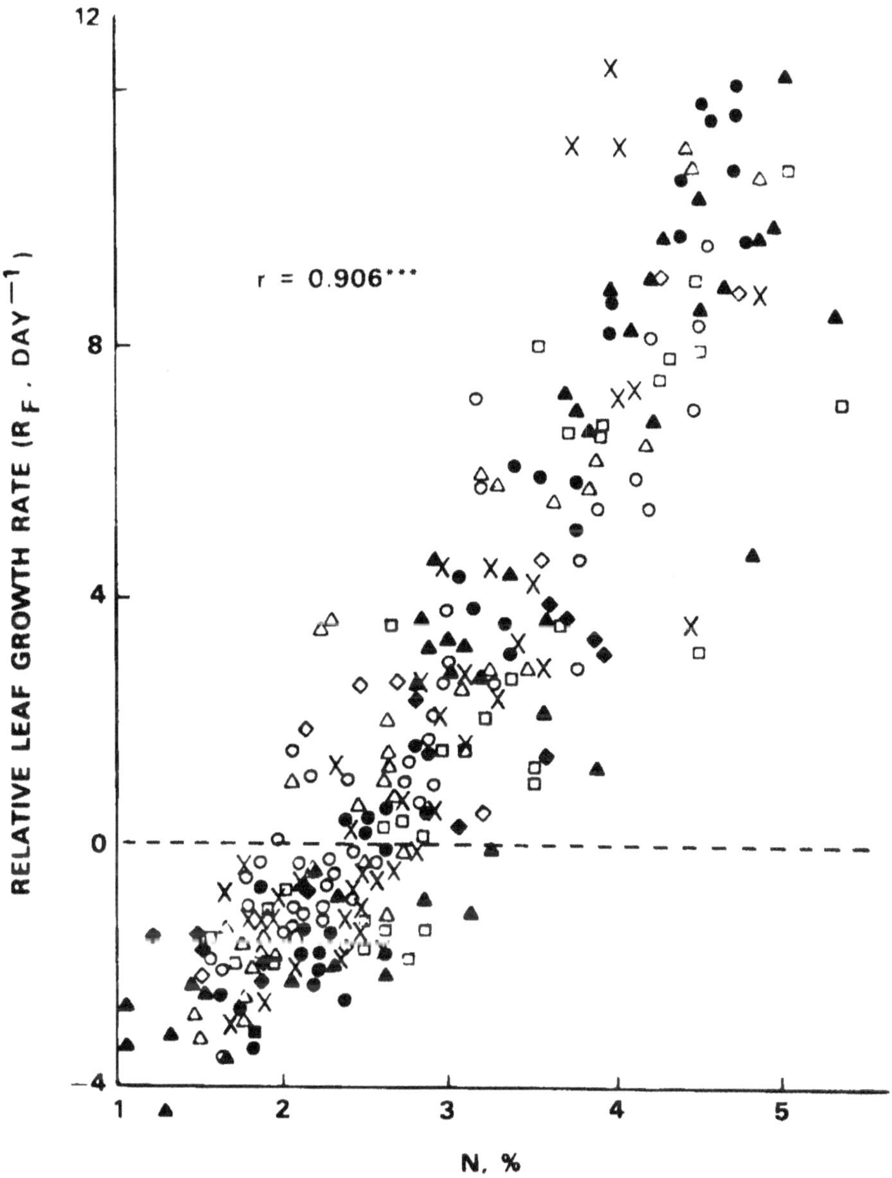

FIGURE 13. The relative leaf growth rate (R_F) vs. leaf % nitrogen for rice in the Japanese IBP effort. From Miyasaka, A., Murata, Y., and Iwata, T., *JIBP Synthesis, Crop Productivity and Solar Energy Utilization in Various Climates in Japan*, Murata, Y., Ed., Univ. Tokyo Press, 1975, 72. With permission.

are given in Figures 16 and 17 for cotton, soybeans, and sorghum.[50] The node position effect is rather dramatic and complicates any study of leaf duration. McMichael and Hesketh[38] reported effects of water stress on leaf duration, leaves shedding some 10 days sooner in stressed plants. This is one of the ways a plant can adapt to water stress,[52] as reduction in leaf area per plant due to shedding will slow transpiration and water use by the plant.

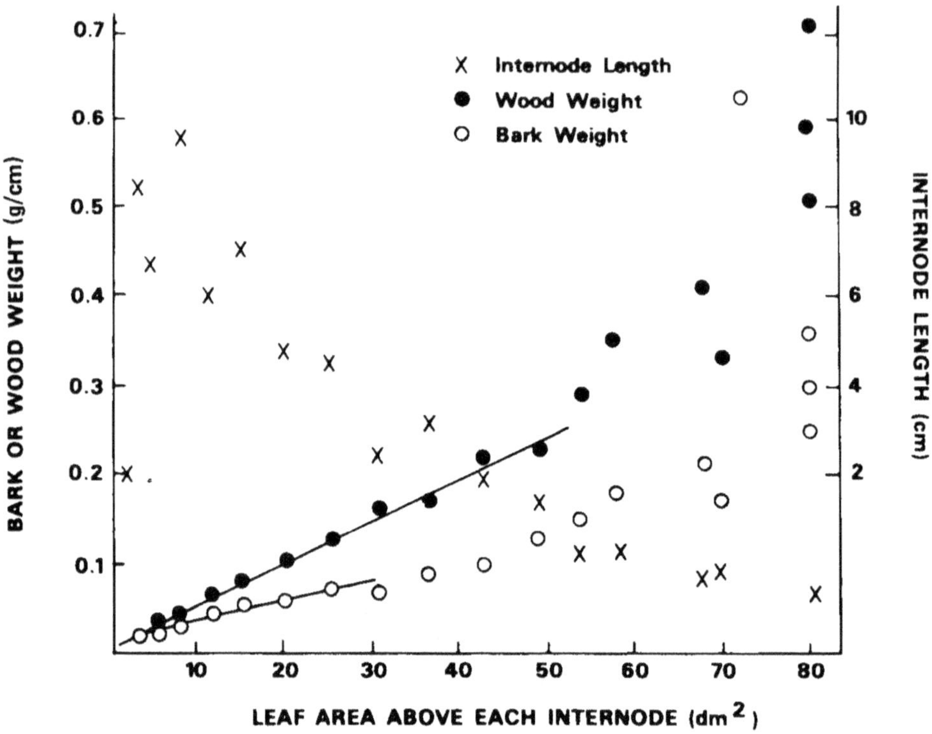

FIGURE 14. Internode bark and wood weights per unit length and lengths vs. leaf area at and above each internode or node. These data are for one plant harvested from a 201 kg/ha N plot in an N-fertility trial. From Hesketh, J. D., Lane, H. C., and Thompson, A. C., Proc. Beltwide Cotton Prod. Res. Confs., National Cotton Council, Memphis, Tenn., 1976, 61. With permission.

Abrupt removal of nitrogen from the nutrient solution for soybeans grown hydroponically induced early leaf senescence and abscission,[53] but the tenth main stem leaf endured longer in nonfertilized plots in a cotton N-trial in plants that were stunted, and consequently had few leaves to shed. The latter plants had already adapted to the treatment, whereas the soybean plants adapted to an abrupt treatment by shedding leaves.

At about 2/10 of the period from leaf emergence to death, photosynthesis reaches the maximum rate, at which point it declines almost linearly to zero rate at death.[54-57] This decline in photosynthesis is often related to a decline in leaf %N.[57] Leaf activity then depends upon the N-budget in the plant.

C. The Degree-Day Concept

We have mentioned the degree-day concept from time to time in the above discussion and other chapters. Degree-days can be substituted for time in both the plastochron and growth analysis approaches.

Wang[58] reviewed in great detail the history of the heat unit approach, as well as criticisms of the technique and improvements needed. His paper should be read by anyone interested in this problem. The main criticisms were that the approach only works over temperatures where plant responses are linear, and that plants respond differently to temperature at different stages of growth.

Other techniques are available for describing plant responses to temperatures outside the linear range. The van't Hoff-Arrhenius principle can be used to describe plant

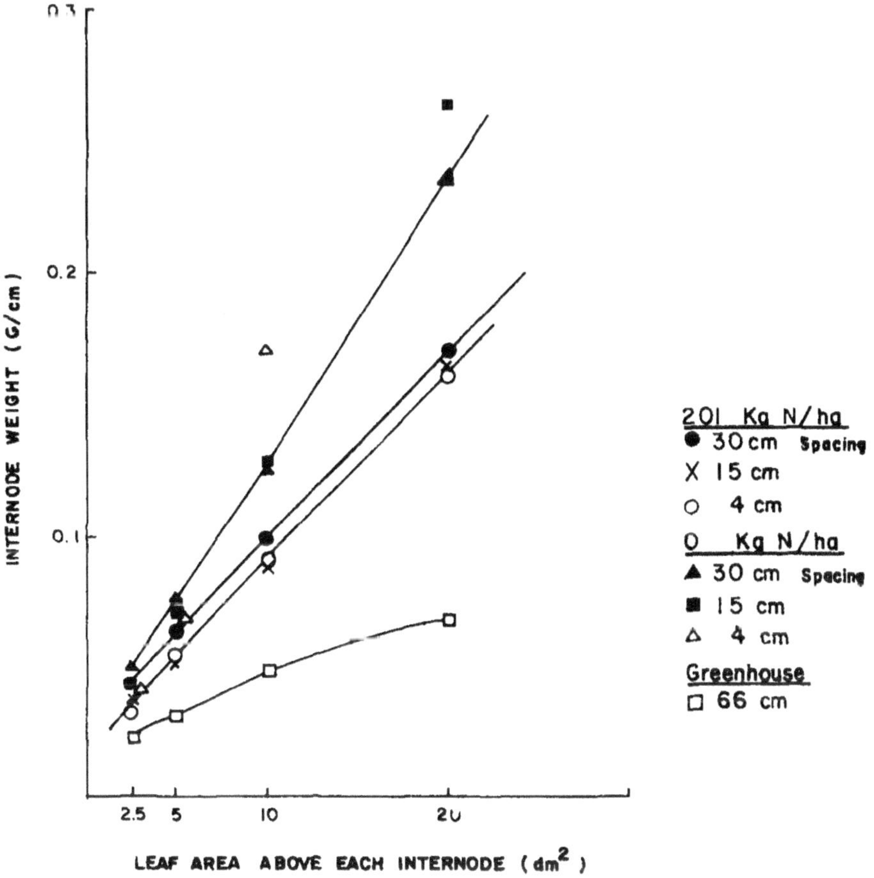

FIGURE 15. Total internode dry weights per unit length vs. leaf area above each internode of the main stem. Data taken from an N fertility trial and greenhouse experiments. From Hesketh, J. D., Lane, H. C., and Thompson, A. C., Proc. Beltwide Cotton Prod. Res. Confs., National Cotton Council, Memphis, Tenn., 1976, 61. With permisssion.

growth response to temperature below that for optimum growth. Tenhunnen et al.[59] and Sharpe and DeMichele[60] have applied the kinetic theory of Johnson, Eyring, and co-workers[61,62] to describe reaction kinetics of development, applicable over the entire temperature range.

Sharpe and DeMichele[60] assumed that development is regulated by a single control enzyme whose reaction rate determines the development rate of the organism. The control enzyme was assumed to exist in an active state, a low temperature inactivate state or a high temperature inactive state. The rates of inactivation and activation of enzymes were formulated as randomly distributed variables whose means were described by the Eyring equation. The probability of enzyme being in the active state was calculated, using a steady-state assumption. Development rate was then described by an enzyme kinetic equation expressing the conversion of substrate into a reaction product. The rate of product formation, R_D, was then derived assuming nonlimiting substrate to be

$$R_D = \frac{T' e^{(\theta - \Delta H_A/T')/R}}{1 + e^{(\Delta S_L - \Delta H_L/T')/R} + e^{(\Delta S_H - \Delta H_L/T')/R}} \qquad (41)$$

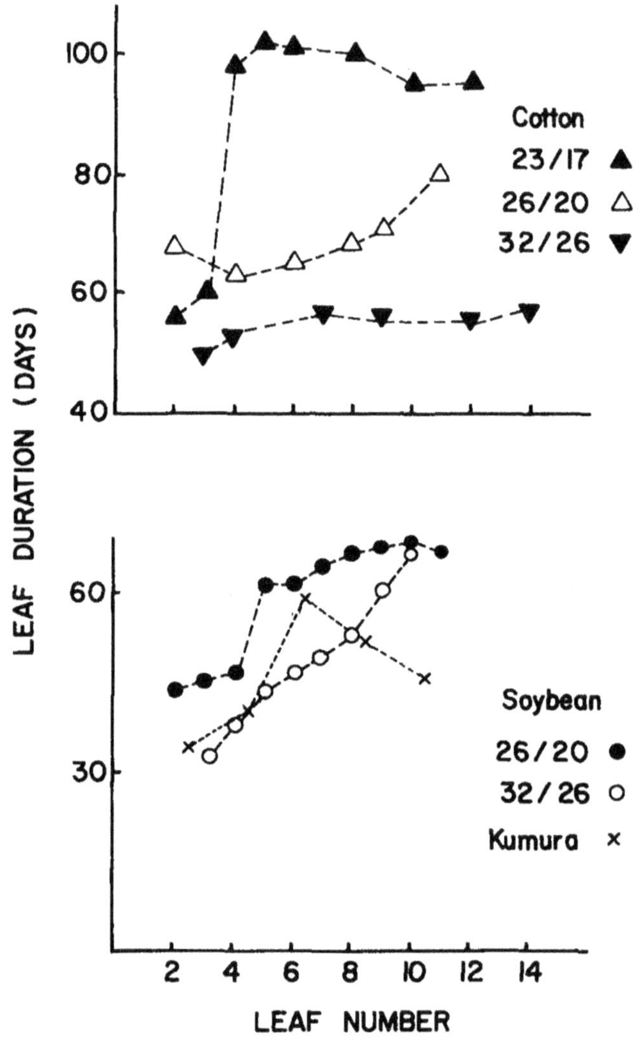

FIGURE 16. Leaf duration vs. leaf number up the main stem for cotton and soybeans. Only true leaves were considered in cotton, and only trifoliolates in soybean. Data are from plants grown in 32/26, 26/20, 23/17°C day/night 8-/16-hr temperature regimes of the Duke University phytotron, as well as from Kumura and Naniwa. From Hesketh, J.D., McMichael, B. L., and Teramura, A., Proceedings Beltwide Cotton Prod. Res. Conf., National Cotton Council, Memphis, Tenn., 1977, 59, and from Kumura, A., Studies on dry matter production in the soybean plant, v. photosynthetic system of a soybean plant population, *Proc. Crop. Sci. Soc. Jpn.*, 38, 74, 1969. With permission.

Their final equation (Equation 41) contained only one variable, absolute temperature, T. Sharpe and DeMichele[60] used nonlinear regression techniques to estimate the constants for the above equation. The general applicability of their equation was shown by applying it to development rates of several poikilothermic organisms including plants. There is potential to exend this work to include substrate limiting conditions and changes in the total concentration of enzyme due to nutritional and hormonal factors.

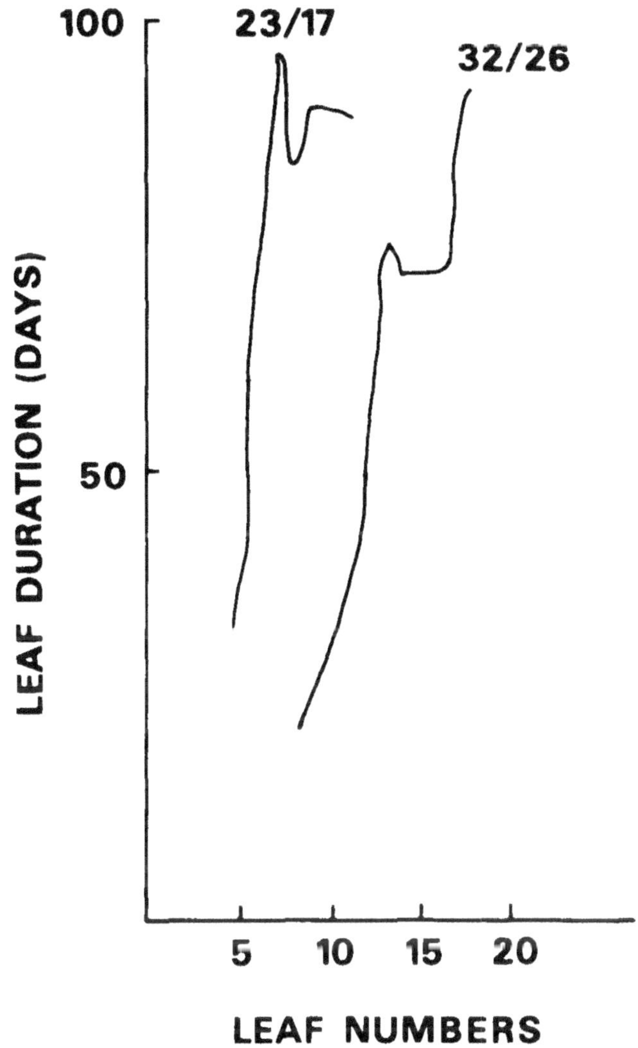

FIGURE 17. Leaf duration vs. leaf number in sorghum. Data obtained from plants in two greenhouses with 32/26 and 23/17°C day/night 8/16 hr temperature regimes, Duke University phytotron. From Hesketh, J. D., McMichael, B. L., and Teramura, A., Proc. Beltwide Cotton Prod. Res. Conf., National Cotton Council, Memphis, Tenn., 1977, 59. With permission.

All three groups derive theoretical equations which must be fitted to data describing the response of a physiological process to temperature. One assumes, using such an approach, that the better the theoretical basis for the equation used for fitting data, the better the equation will predict.

The degree-day concept requires that one set a minimum temperature for growth and this occurs when the leaf production stops (12.7°C in cotton and 7.5°C in soybeans, Figure 7, and 6.5°C in corn). The slope of the leaf production vs. temperature curve gives leaves per degree-days; the reciprocal of which gives the degree-day per plastochron (39 for cotton, 50 for soybeans, 42 for corn, and 28 for cucumber, with base temperatures of 12.7, 7.5, 6.0 and 12.0°C, respectively).

The plastochron and degree-day concepts are stress-dependent[58,63,64] in that low light or photosynthate supply; low N, P, or K supply rates; or water stress, can slow growth and invalidate predictions using these concepts. These stress effects then must be accounted for in a predictive model. Seeds usually store enough N, P, and K as well as carbohydrates to carry the seedling through about five or six plastochrons. Quantitative descriptions of the effects of stress (with stress quantitatively described) on the plastochron and the degree-day concepts, are needed. If photosynthate or nitrogen are limiting, then the plastochron can be slowed to the rate at which photosynthate and nitrogen become available. However, the plant adapts to stress by growing smaller organs, shedding organs, redistributing photosynthate reserves and N, or — in the case of water stress — maintaining turgor pressure by changes in osmotic pressure. These adaptive mechanisms are only now being worked out, and will be discussed in other chapters.

D. Hackett's Model: A Case Study

There is a good example in the literature that uses some of the concepts discussed in this chapter in a simple, straightforward manner. Hackett[7] and Hackett and Rawson[54] presented a simulation model of the growth of a tobacco plant based on organogenesis, leaf expansion, and a carbon balance. In Hackett's[7] first paper, simulation results were presented, and inferences were offered based on those results. Here we review his assumptions, his model and results, and some important aspects of that work.

Hackett[7] developed the model of a tobacco plant to study whether photosynthesis may frequently be suppressed due to the lack of demand for assimilates. In his model, individual leaves emerged based on a constant plastochron. Leaves expanded based on the autocatalytic Richard's function (see above). Using the growth curves of individual organs and growth and maintenance respiration[65,66] for the plant organs, his model calculated a demand for carbon. Then, a section of his model calculated photosynthesis rates of nonsenescent leaves. Figure 18 shows his flow chart for organogenesis and the carbon balance. An iterative technique was used to balance carbon synthesis with demand by reducing the photosynthesis of each leaf on successive iterations. Table[27] summarizes the major assumptions made by Hackett in the original model. In addition, Hackett assumed a maximum of 26 leaves with leaf size different, depending upon node position on the plant.

The agreement between Hackett's simulated results and actual data taken by Pitrie[67] was very good, especially the data on leaf dry weight which included senescence and thickening factors. Comparisons were made for number of leaves, leaf area, and leaf, stem, and root weights. According to the simulated results, there was excess photosynthate produced, up to 1.70 times that required (Figure 19). Figure 19 also shows the percentage photosynthetic active leaf area and the photosynthetic history of leaf numbers 5, 10, 15, 20, and 25 (from Hackett). Hackett made three inferences based on these results. First, crop morphogenesis is a necessary component for successful simulation of a crop. Second, he inferred that if photosynthesis was indeed suppressed due to lack of demand, there would not be a need to determine accurately respiration rates. Finally, he inferred that the understanding of leaf senescence might be furthered by considering leaves in the context of the whole plant.

Hackett and Rawson[54] reported on experiments to analyze critically the assumptions of Table 2. They found that leaf emergence rates were not constant and that the Gompertz curve fit leaf expansion better than did the autocatalytic function used earlier. Also, they found that photosynthesis rates of individual leaves were not constant over leaf ages but varied with leaf age. Peak rates of photosynthesis were associated with the time of maximum rate of expansion and also the time of cessation of cell division

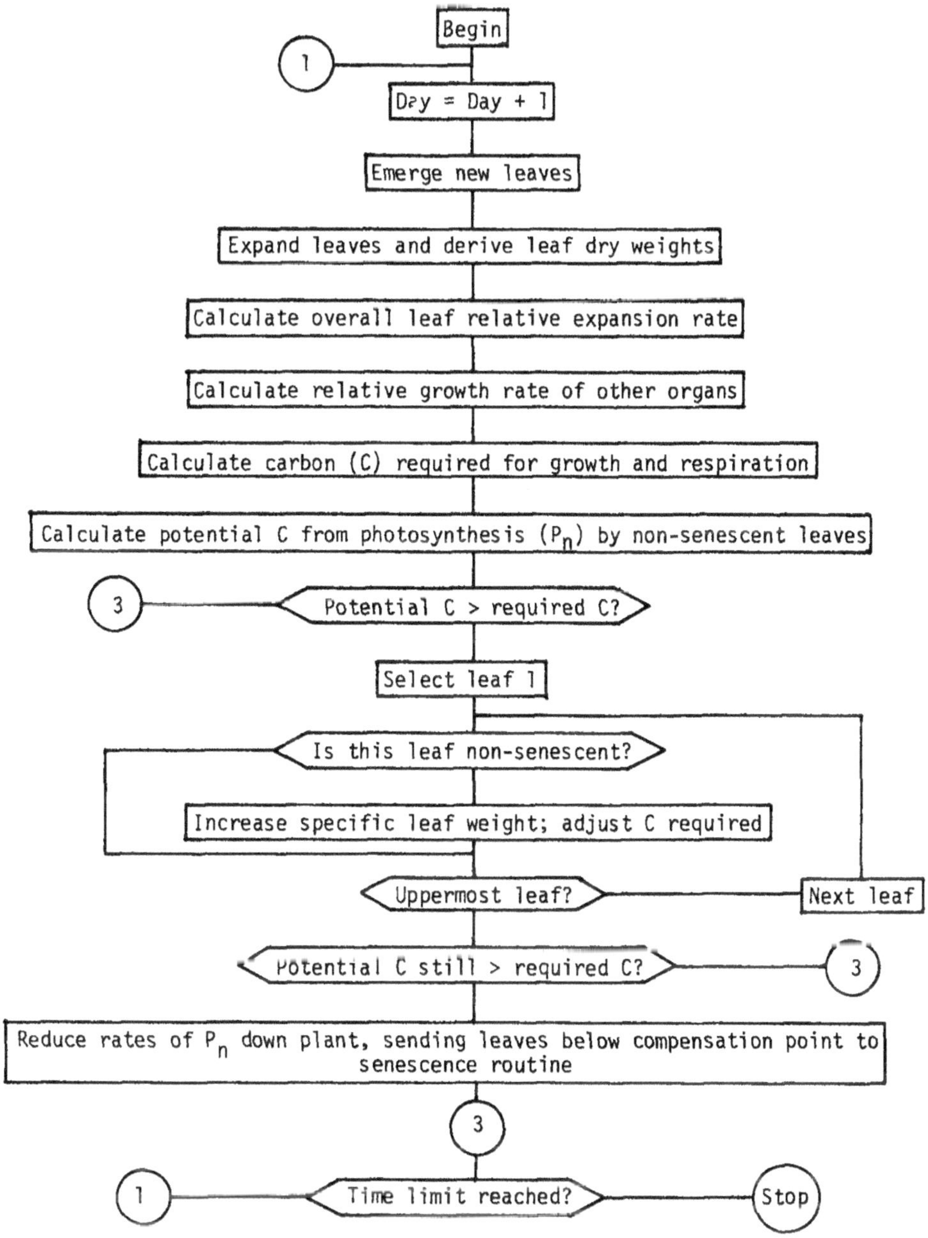

FIGURE 18. Flow chart summarizing the method of comparing the carbon required, the potential carbon availability and the logic for thickening and senescing leaves. From Hackett, C., *Aust. J. Biol. Sci.*, 26, 1057, 1973. With permission.

in tobacco leaves. Although these experiments invalidated specific results from Hackett's earlier work, they supported earlier conclusions that a model predicting organogenesis is needed to predict effectively tobacco growth. They criticized direct comparisons of P_n rates among dicotyledonous species without considering ontogenetic events, and suggested that a better measurement for comparisons might be the integrated value of (P_n) (L).

TABLE 2

Some Assumptions Made by Hackett and Rawson[53] When Calculating the Carbon Budget of the Tobacco Plants Grown by Petrie et al.[67]

Assumption:	That leaf emergence rate was approximately constant.
Comments:	Unlikely, but there were no data on how rate change with time.
Assumption:	That leaves expanded after emergence according to the autocatalytic growth formula (given in Richards[10]).
Comments:	Little firm evidence.
Assumption:	That leaves could assimilate carbon at maximum rate per unit area from the time of emergence.
Comments:	Unlikely, but there is no rule for relating P_n to time. (P_n is rate of net photosynthesis per unit leaf area, as defined in the test.)
Assumption:	That unless the P_n of a leaf had become negative, the leaf was potentially able to perform at peak P_n if demand was sufficient.
Comments:	Data of King et al.[68] and Hodgkinson et al.[69] support; data of Woledge[69] conflict.
Assumption:	That decline in leaf photosynthetic activity was due to leaf photosynthetic redundancy, and that redundancy led to senescence.
Comments:	Data of Lovell et al.[71] support; also supported by greening of upper tobacco leaves in the field following picking or topping.

From Hackett, C., and Rawson, H. M., *Mechanisms of Regulation of Plant Growth,* Biellski, R. L., Ferguson, A. R., and Cresswell, M. M., Eds., Bull. 12, The Royal Society of New Zealand, Wellington, 1974, 269. With permission.

V. A PLEA FOR MORPHOGENIC STUDIES

The expansion and maintenance of a leaf surface on a crop are important aspects determining the crop productivity. In this chapter, we have taken a hierarchical approach for predicting leaf expansion. First it was demonstrated that integration of cellular processes could provide predictions of individual leaf expansion. Then we showed that leaf expansion for whole plants could be described using growth analysis techniques and by using descriptions of individual organ growth. The example of Hackett[54] is demonstrative of the approach we suggest for estimating whole plant leaf expansion. That approach requires estimates of initiation and growth of organs of plants subjected to various stresses. In order to apply that approach for crops, more morphogenic studies are needed. In particular, more research is needed on the effects of temperature, water, and other stresses on rate of initiation, growth, and senescence of individual organs on a plant. (This approach is also discussed in Chapter 4, Volume I on partitioning.) Integration of growth processes for whole plants need, then, to be formulated into models and used to test hypotheses concerning whole plant behavior.

It is noted that many of the models presented in this chapter are empirical in nature. For the purists who strive for model development based strictly on physical and chemical formalisms, a challenge is issued to provide alternative mathematical formulas to the above. The challenge is to strive for realism and for integration into whole plant models, with considerations for temperature, water, and nutritional effects. Future progress is believed to lie in the application of the scientific method coupled with model based analysis of crop behavior.

ADDENDUM

As this chapter was about to go to press, we found a case where the relative leaf expansion rate (R_F) multiplied by the plastochton value (days per leaf) was relatively constant over a wide range of temperatures, Figure 20. The relative growth rate, R_W

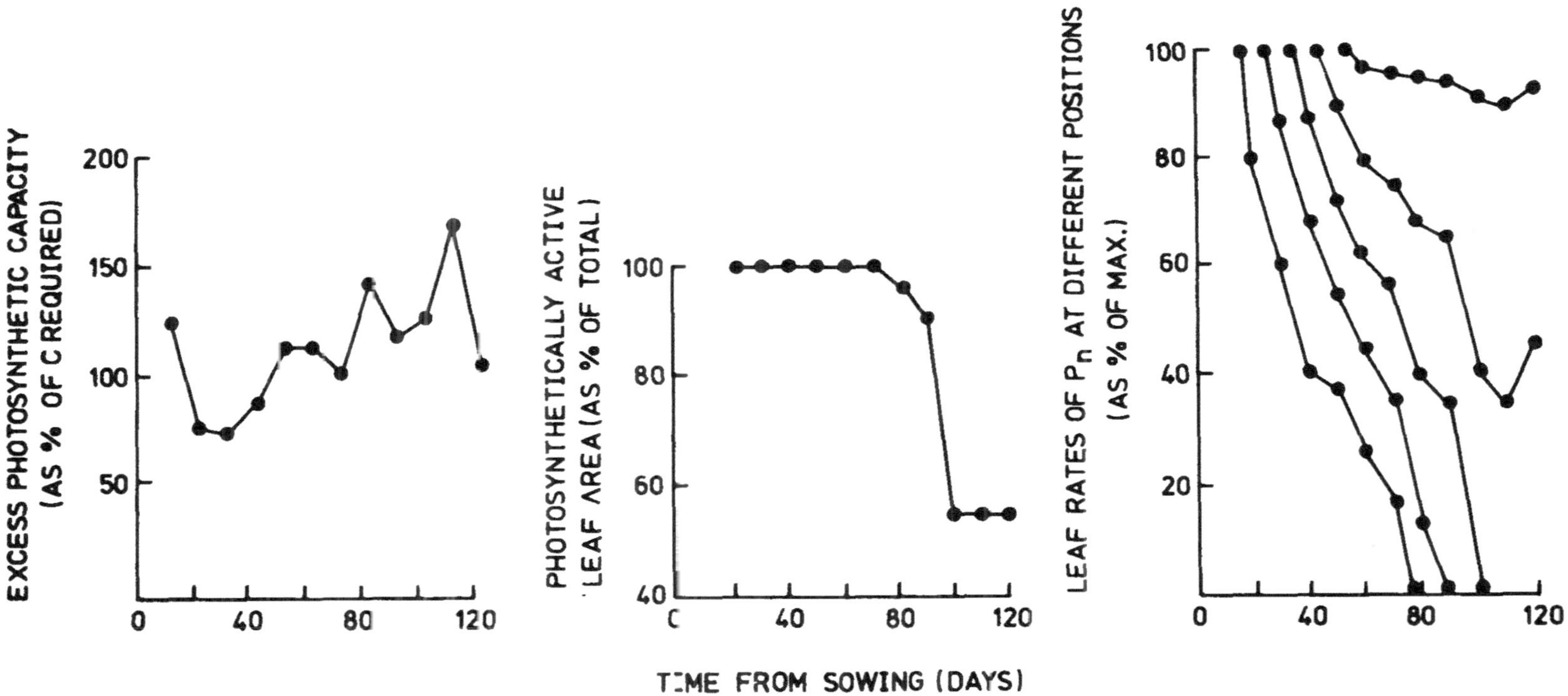

FIGURE 19. (A) Predicted excess photosynthetic capacity of tobacco plants from Hackett's simulations, (B) proportion of photosynthetically active leaf area remaining on the simulated plants, and (C) photosynthetic duration of selected leaves. From Hackett, C., An exploration of the carbon economy of the tobacco plant. I. Inferences from a simulation, *Aust. J. Biol. Sci.*, 26, 1057, 1973. With permission.

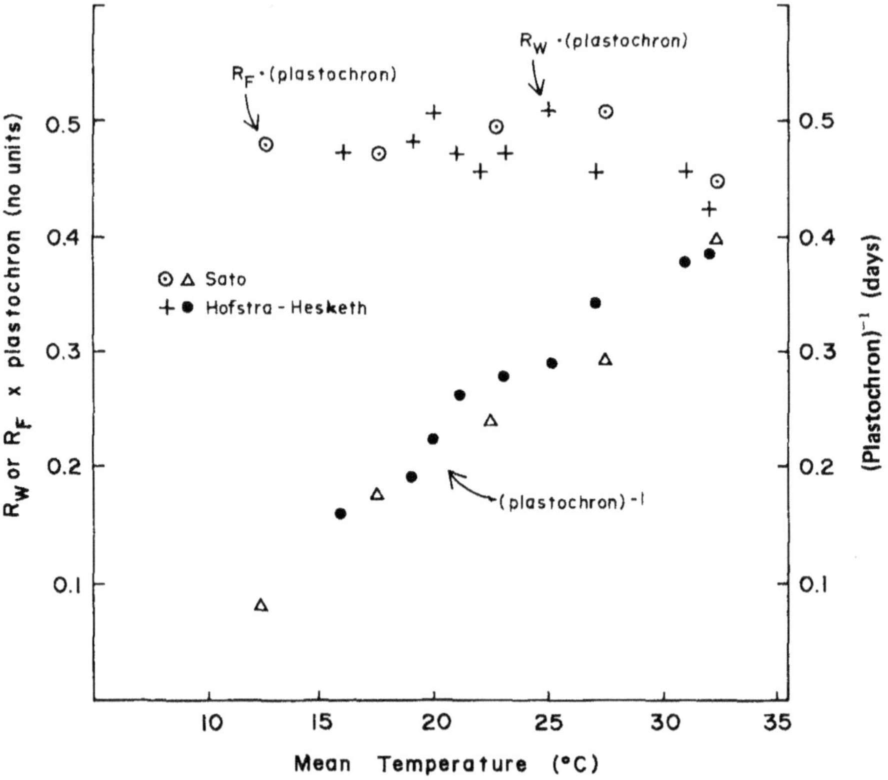

FIGURE 20. The effect of temperature on soybean plastochron values and growth rates (R_F the relative leaf expansion rate and R_W the relative rate of dry matter accumulation) multiplied by such plastochron values. Plastochron values here are from different experiments than those shown in Figure 7. Adapted from Sato, K.[72] Also from Hofstra, G. and Hesketh, J. D., unpublished data.

= (l/W)(dW/dt) behaved similarly for plants grown under comparable conditions to the same stage of growth (soybeans with 6 trifoliolates on the main stem) except for enriching the air with CO_2. Both sets of data[72,23] (unpublished, G. Hofstra and J. Hesketh) were for spaced plants and may well represent maximum growth rates for soybeans. Growth data are now needed for plants stressed for carbohydrate, either induced by shading or crowding, as well as for nitrogen. The use of plastochron values in the analysis of such data would seem to offer considerable potential for developing simple relationships that can be used to predict plant growth.

VI. SYMBOLS

$a_c(t)$	Cross sectional area of a vegetative cell at time t, μm^2
a_{co}	Initial cross section of a vegetative cell before expansion, μm^2
$\bar{a}_m$	Average area of meristematic cells at time t, μm^2
α	Ratio of relative rates of meristematic division, v′/v
α_N	Leaf area: nitrogen ratio, cm^2/g
α_W	Leaf area: weight ratio, cm^2/g
β_N	The rate of area increase divided by the rate of N increase, cm^2/g
β_W	Leaf area partition coefficient, the rate of area increase divided by the rate of weight increase, cm^2/g

d_c	Number of cells in a leaf cross section
F	Leaf area per plant, cm^2
F_c	Area of cotyledons, cm^2
F_f	Final leaf area, cm^2
F_O	Leaf area chosen for PI of zero
F(t)	Total area of leaf um^2 or mm^2
$F_m(t)$	Area of leaf due to meristematic cells at time t, cm^2
$F_T(t)$	Whole Plant leaf area, mm^2
$F_v(t)$	Area of leaf due to vegetative cells at time t, cm^2
γ	Empirical constant used to describe changes in θ with time
$\Gamma(r,z)$	Incomplete gamma function for parameters r,z; $\Gamma(r,z) = \int_O^z q^{r-1}e^{-1}dq$
h	Node position for the leaf just larger than F_0
k,b,m	Empirical growth constants for Richard's equation
$K_T(t)$	Temperature modifying function, $0 < K_T(t) < 1$
LAP	Leaf Area Partition coefficient, same as β_W
LAR	Leaf Area Ratio, the ratio of plant leaf area F to the dry weight **W**, of F/**W**, cm^2/g
LWP	Leaf Weight Partition coefficient, leaf dry weight ($\mathbf{W}_F$) to total plant dry weight (**W**) ratio, of $\mathbf{W}_F/\mathbf{W}$
NAR	Net Assimilation Rate of dry matter (**W**) per unit leaf area (F) per unit time (t), (1/F)(d**W**/dt), $g/cm^2/day$
N_0	Number of meristematic cells at time t = 0
$N_m(t)$	Number of meristematic cells at time t
$N_v(t)$	Number of vegetative cells at time t
n	The serial number counting from the leaf larger than F_0 down the main stem
PI	Plastochron age of leaf counting down from the apex, days
p	Plastochron
R	The gas constant
R_D	Rate of development in thermodynamic model for poikilothermic development
R_F	Relative leaf area expansion rate, t^{-1}
R_W	The relative rate of dry matter production, (1/**W**)(d**W**/dt)
R(t)	Rate of leaf initiation (= 1/P), day^{-1}
s	Empirical parameter describing progressive reduction in u due to senescence
T	Temperature, °C
T′	Absolute temperature, °K
T_0	Lower temperature threshold below which growth ceases, °C
T_s	Standard temperature at which dL(t)/dt is measured
t_d	Doubling time for meristematic cells, h
t′	Dummy time variable used for integration
u	Specific cell area expansion rate, μm^2 /n
u′	Relative rate of vegetative cell production from meristematic cells
v	Relative rate of meristematic cell division, h^{-1}
v′	Relative rate of meristematic cell division when $(1-\theta)$ fraction of cells are vegetative
W	Total plant weight, g
$\mathbf{W}_F$	Leaf dry weight, g
ΔH_A	Enthalpy of activation for conversion of substrate into product
ΔH_H	Enthalpy of enzyme activation difference between the high temperature inactive state and the active state

ΔH_L	Enthalpy of enzyme activation difference between the low temperature inactive state and the active state of equilibrium
ΔS_H	Entropy of enzyme activation difference between the high temperature inactive state and the active state
ΔS_L	Entropy of enzyme activation difference between the low temperature inactive state and the active state
θ	Fraction of all cells that are meristematic
ϕ	Constant in the poikilothermic development model of Sharpe and DeMichele[53]

REFERENCES

1. **Potter, J. R. and Jones, J. W.**, Leaf area partitioning as an important factor in growth, *Plant Physiol.*, 59, 10, 1977.
2. **Heath, O. V. S. and Gregory, F. G.**, The constancy of the mean net assimilation rate and its ecological importance, *Ann. Bot. (London)*, 2, 811, 1938.
3. **Milthorpe, F. L., Preface**, in *The Growth of Leaves*, Milthorpe, F. L., Ed., Easter School in Agr. Sci., Proc., Butterworths, London, 1956, ix.
4. **Ledig, F. T.**, A growth model for tree seedlings based on the rate of photosynthesis and the distribution of photosynthate, *Photosynthetica*, 3, 263, 1969.
5. **Levitt, J.**, *Responses of Plants to Environmental Stresses*, Academic Press, New York, 1972.
6. **Thornley, J. H. M.**, *Mathematical Models in Plant Physiology*, Academic Press, New York, 1976.
7. **Hackett, C.**, An exploration of the carbon economy of the tobacco plant. I. Inferences from a simulation, *Aust. J. Biol. Sci.*, 26, 1057, 1973.
8. **Milthorpe, F. L. and Newton, P.**, Studies on the expansion of the leaf surface. III. The influence of radiation on cell division and leaf expansion, *J. Exp. Bot.*, 14, 483, 1963.
9. **Dale, J. E.**, Cell division in leaves, in *Cell Division in Higher Plants*, Yeoman, M. M., Ed., Academic Press, London, 1976, 315.
10. **Richards, F. J.**, The quantitative analysis of growth, in *Plant Physiology, A Treatise*, Vol. 5A, Steward, F. C., Ed., Academic Press, New York, 1969, 3.
11. **Bonner, J.**, Hierarchical control programs in biological development, in *Hierarchy Theory: The Challenge of Complex Systems*, Pattee, H. H., Ed., George Braziller, New York, 1973, 51.
12. **Wilson, B. F.**, Modeling developmental processes in trees, in Modeling the Growth of Trees, *Proc. Workshop Tree Growth Dynamics*, Murphy, C., Hesketh, J., and Strain, B., Eds., Oak Ridge National Lab., Tenn., EDFB-IBP-72-11, 1972, 171.
13. **Turing, A. M.**, The chemical basis for morphogenesis, *Philos. Trans. R. Soc. London Ser. B*, 237, 37, 1952.
14. **Thornley, J. H. M., Phyllotaxis. I. A** Mechanistic model, *Ann. Bot. (London)*, 39, 491, 1975.
15. **Milthorpe F. L.**, Studies on the expansion of the leaf surface. I. The influence of temperature, *J. Exp. Bot.*, 10, 233, 1959.
16. **Maksymowych, R.**, *Analysis of Leaf Development*, Cambridge University Press, London, 1973, 3.
17. **Williams, R. F.**, *The Shoot Apex and Leaf Growth*, Cambridge University Press, London, 1975, 9.
18. **Causton, D. R.**, A computer program for fitting the Richards function, *Biometrics*, 25, 401, 1969.
19. **McIntyre, G. A., Grassia, A., and Ward, M. W.**, Miscellaneous programs in the fields of analysis, linear and nonlinear regression, biological assay, association and frequency distributions, C.S.I.R.O. (Australia), Div. Math. Stat. Tech. Rep. No. 5, 1971, 21.
20. **Hunt, R. and Parsons, I. T.**, A computer program for deriving growth functions in plant growth analysis, *J. Appl. Ecol.*, 11, 297, 1974.
21. **Pearl, R., Edwards, T. I., and Miner, J. R.**, The growth of *Cucumis melo* seedlings at different temperatures, *J. Gen. Physiol.*, 17, 687, 1934.
22. **McKinion, J. M., Jones, J. W., Hesketh, J. D., Lane, H. C., and Thompson, A. C.**, Simulation of Plant Growth: Morphogenetic Control of Leaf Expansion, Proc. Beltwide Cotton Prod. Res. Confs., National Cotton Council, Memphis, Tenn., 1975, 56.
23. **Hofstra, G., Hesketh, J. D., and Myhre, D. L.**, A plastochron model for soybean leaf and stem growth, *Can. J. Plant. Sci.*, 57, 167, 1977.

24. **Boyer, J. S.**, Leaf enlargement and metabolic rates in corn, soybean, and sunflower at various leaf water potentials, *Plant Physiol.*, 46, 288, 1970.
25. **Boyer, J. S.**, Relationships of water potential to growth of leaves, *Plant Physiol.*, 43, 1056, 1968.
26. **Heatherly, L. G., Russell, W. J., and Hinkley, T. M.**, Water relations and growth of soybeans in drying soil, *Crop. Sci.*, 17, 381, 1977.
27. **Boyer, J. S.**, Recovery of photosynthesis in sunflower after a period of low leaf water potential, *Plant Physiol.*, 47, 816, 1971.
28. **Slatyer, R. O.**, *Plant Water Relationships*, Academic Press, New York, 1967, 287.
29. **Wiebe, H. H., Campbell, G. S., Gardner, W. H., Rawlins, S. L., Cary, J. W., and Brown, R. W.**, *Measurement of Plant and Soil Water Status*, Bull. 484, Utah Agricultural Experiment Station, 1971, 56.
30. **Pitrie, A. H. K.**, Physiological ontogeny in plants and its relation to nutrition. III. The effects of nitrogen supply on the drifting composition of leaves, *Aust. J. Exp. Biol. Med. Sci.*, 15, 385, 1937.
31. **Erickson, R. O.**, Modeling of plant growth, *Annu. Rev. Plant Physiol.*, 27, 407, 1976.
32. **Hesketh, J. D., Myhre, D. L., and Willey, C. R.**, Temperature control of time intervals between vegetative and reproductive events in soybeans, *Crop. Sci.*, 13, 250, 1973.
33. **Michelini, F. J.**, The plastochron index in developmental studies of *Xanthium italicum* Moretti, *Am. J. Bot.*, 45, 525, 1958.
34. **Zaitzev, G. S.**, The effect of temperature on the development of the cotton plant (trans.), Trans. Turkestan Plant Breeding Inst. Tashkent, No. 7, 1928.
35. **Hector, J. M.**, Introduction to the Botany of Field Crops. II. Noncereals, Central News Agency, Ltd., Johannesburg, South Africa, 1936, 814.
36. **Hesketh, J. D., Baker, D. N., and Duncan, W. G.**, Simulation of growth and yield in cotton. II. Environmental control of morphogenesis, *Crop Sci.*, 12, 436, 1972.
37. **Abbe, E. C. an Phinney, B. O.**, The growth of the shoot apex in maize: external features, *Am. J. Bot.*, 38, 737, 1951.
38. **McMichael, B. L. and Hesketh, J. D.**, Field investigations on the response of cotton to water deficits, Proc. Beltwide Cotton Prod. Res. Confs., National Cotton Council, Memphis, Tenn., 1977.
39. **Hesketh, J. D. and Baker, D. N.**, Relative rates of leaf expansion in seedlings of species with differing photosynthetic rates, *J. Ariz. Acad. Sci.*, 5, 216, 1969.
40. **Jackson, J. E.**, Relationship of relative leaf growth to net assimilation rate and its relevance to the physiological analysis of plant yield, *Nature, (London)*, 200, 909, 1963.
41. **Hesketh, J. D., Lane, H. C., and Thompson, A. C.**, Partitioning photosynthate in a field N-trial, with emphasis on internodal wood and bark growth, Proc. Beltwide Cotton Prod. Res. Confs., National Cotton Council, Memphis, Tenn., 1976, 61.
42. **Huxley, P. A.**, Some effects of artificial shading on the growth of upland cotton seedlings, *Emp. Cotton. Grow. Rev.*, 41, 100, 1964.
43. **Hofstra, G. and Hesketh, J. D.**, The effects of temperature and CO_2 enrichment on photosynthesis in soybean, in *Environmental and Biological Control of Photosynthesis*, Marcell, R., Ed., Dr. W. Junk, The Hague, 1975, 71.
44. **Thompson, A. C., Lane, H. C., Jones, J. W., and Hesketh, J. D.**, Nitrogen concentrations of cotton leaves, buds and bolls in relation to age and nitrogen fertilization, *Agron. J.*, 68, 617, 1976.
45. **Jones, J. W., Hesketh, J. D., Kamprath, E. J., and Bowen, H. D.**, Development of a nitrogen balance for cotton growth models: a first approximation, *Crop Sci.*, 14, 541, 1974.
46. **Jones, J. W., Thompson, A. C., and Hesketh, J. D.**, Analysis of SIMCOT: nitrogen and growth, Proc. Beltwide Cotton Prod. Res. Confs., National Cotton Council Memphis, Tenn., 1974, 111.
47. **Wilson, J. R.**, Comparative response to nitrogen deficiency of a tropical and temperate grass in the interrelation between photosynthesis, growth, and the accumulation of non-structural carbohydrate, *Neth. J. Agric. Sci.*, 23, 104, 1975.
48. **Miyasaka, A., Murata, Y., and Iwata, T.**, Leaf area development and leaf senescence in relation to climatic and other factors, in *JIBP Synthesis, Crop Productivity and Solar Energy Utilization in Various Climates in Japan*, Murata, Y., Ed., University of Tokyo Press, Tokyo, 1975, 72.
49. **Hesketh, J. D., Lane, H. C., and Thompson, A. C.**, Partitioning photosynthate in a field N-trial, with emphasis on internodal wood and bark growth, in *JIBP Synthesis, Crop Productivity and Solar Energy Utilization in Various Climates in Japan*, Murata, Y., Ed., University of Tokyo Press, 1976, 61.
50. **Hesketh, J. D., McMichael, B. L., and Teramura, A.**, Temperature studies of axillary branch growth and leaf duration, *Proc. Beltwide Cotton Prod. Res. Confs.*, National Cotton Council, Memphis, Tenn., 1977, 59.
51. **Kumura, A.**, Studies on dry matter production in the soybean plant. V. Photosynthetic system of a soybean plant population, Proc. Crop Sci. Soc. Jpn., 38, 74, 1969.

52. **McMichael, B. L. and Jordan, W.**, Abscission processes in cotton: induction by plant water deficit, *Agron. J.*, 65, 202, 1973.
53. **Egli, D. B., Leggett, J. E., and Duncan, W. G.**, Influence of nitrogen stress on leaf senescence and nitrogen redistribution in soybeans, *Agron. Abstr.*, 71, 1976.
54. **Hackett, C. and Rawson, H. M.**, Photosynthesis and the carbon economy of a dicotyledonous plant *(Nicotiana tobacum* L): results from simulation and experiment, in *Mechanisms of Regulation of Plant Growth*, Bieleski, R. L., Ferguson, A. R., and Cresswell, M. M., Eds., Bull. 12, The Royal Soc. New Zealand, Wellington, 1974, 269.
55. **Rawson, H. M. and Hackett, C.**, An exploration of the carbon economy of the tobacco plant. III. Gas exchange of leaves in relation to position on the stem, ontogeny and nitrogen content, *Aust. J. Plant Physiol.*, 1, 551, 1974.
56. **Rawson, H. M. and Woodward, R. G.**, Photosynthesis and transpiration in dicotyledonous plants. I. Expanding leaves of tobacco and sunflower, *Aust. J. Plant Physiol.*, 3, 247, 1976.
57. **Woodward, R. G., and Rawson, H. M.**, Photosynthesis and transpiration in dicotyledonous plants. II. Expanding and senescing leaves of soybean, *Aust. J. Plant Physiol.*, 3, 257, 1976.
58. **Wang, J. Y.**, A critique of the heat unit approach to plant response studies, *Ecology*, 41, 785, 1960.
59. **Tenhunnen, J., Yocum, C., and Gates, D.**, Development of a photosynthesis model with an emphasis on ecological applications. I., Theory, *Oecologia, (Berlin)*, 26, 89, 1976.
60. **Sharpe, P. H. H. and DeMichele, D. W.**, Reaction kinetics of poikilotherm development, *J. Theor. Biol.*, 64, 649, 1977.
61. **Johnson, F., Eyring, H., and Williams, R.**, The nature of enzyme inhibitions in bacterial luminescence: sulfanilamide, urethane, temperature and pressure, *J. Cell. Comp. Physiol.*, 20, 247, 1942.
62. **Johnson, F., Eyring, H., and Polissar, M.**, *The Kinetic Basis of Molecular Biology*, Wiley Press, New York, 1954.
63. **Whyte, R. O.**, *Crop Production and Environment*, Wm. Clower and Sons, London, 1960, 317.
64. **McKinion, J. M., Jones, J. W., and Hesketh, J. D.**, Analysis of SIMCOT: photosynthesis and growth, Proc. Beltwide Cotton Prod. Res. Confs., National Cotton Council, Memphis, Tenn., 1974, 118.
65. **McCree, K. J.**, An equation for the rate of respiration of white clover plants grown under controlled conditions, in Proc. IBP/PP Technical Meeting: productivity of Photosynthetic Systems, Model and Methods, Setlik, I., Ed., PUDOC, Wageningen, 1970, 221.
66. **Hesketh, J. D., Baker, D. N., and Duncan, W. G.**, Simulation of growth and yield; respiration and the carbon balance, *Crop Sci.*, 11, 394.
67. **Pitrie, A. H. K., Watson, R., and Ward, E. D.**, Physiological ontogeny in the tobacco plant. I. The drifts in dry weight and leaf area in relation to phosphorus supply and topping, *Aust. J. Exp. Biol. Med. Sci.*, 17, 93, 1939.
68. **King, R. W., Wardlaw, I. F., and Evans, L. T.**, Effect of assimilate utilization on photosynthetic rate in wheat, *Planta, (Berlin)*, 77, 261, 1967.
69. **Hodgkinson, K. C., Smith, N. G., and Miles, G. E.**, The photosynthetic capacity of stubble leaves and their contribution to growth of the lucerne plant after high level cutting, *Aust. J. Agric.Res.*, 23, 225, 1972.
70. **Woledge, J.**,The effect of shading on the photosynthetic rate and longevity of grass leaves, *Ann. Bot. (London)*, 36, 551, 1972.
71. **Lovell, P. H., Illsley, A., and Moore, K. G.**, The effects of light intensity and sucrose on root formation, photosynthetic ability and senesence in detached cotyledons of *Sinapis alpa* L. and *Raphanus satious* L., *Ann. Bot. (London)*, 36, 123, 1972.
72. **Sato, K.**, The growth responses of the soybean plant to photoperiod and temperature, I. Responses in vegetative growth, *Proc. Crop. Sci. Soc. Jpn.*, 45, 443, 1976.

Chapter 6

PREDICTING FLOWERING AND SUBSEQUENT LEAF EXPANSION

J. D. Hesketh, C D. Elmore, and J. W. Jones

TABLE OF CONTENTS

I. INTRODUCTION

The Kallis-Tooming[1] model for the optimization of whole plant photosynthesis has as physiological variables the specific leaf weight (dry weight per unit area) and rate of leaf expansion. Photosynthesis per unit leaf area, under near optimal conditions, was made a linear function of specific leaf weight. The leaf-area-expansion rate was controlled by the supply of photosynthate (leaf area multiplied by the photosynthetic rate per unit area) and the specific leaf weight (dry weight required for newly expanded leaves). In principle, the faster the expansion rate, the smaller the specific leaf weight, and the slower the photosynthetic rate per unit leaf area. Leaf area expansion was not curbed by other factors, and conclusions about values for maximum dry matter production were biased by the development of excessive leaf-area-index values.

Obviously, some attention must be paid to the role of floral physiology in slowing or inhibiting leaf expansion. Two additional factors needed in such a model include the plant's nitrogen (N) balance and the requirement of N for leaf growth, as well as the reduction of photosynthetic rates per unit leaf area associated with light interception characteristics of the plant canopy. An N deficit alone can curtail leaf expansion, since the photosynthetic: specific leaf weight relationship requires a factor for adequate leaf N in the proper form to maintain necessary enzyme activity for maximum photosynthetic rates. As photosynthesis per unit area and the specific leaf weight increase, N per unit leaf area should also increase.

Physical support for the canopy of leaves must be adequate to withstand specific levels of wind and rain, and the root system must be adequate to supply needed water, mineral nutrients, and hormones. However, models have been developed to handle such complexity as it interacts with canopy development.

As discussed in another chapter, the canopy development rate is controlled by the leaf production rate, leaf growth rate, and leaf area duration. The flowering process can either stop leaf production completely, or result in hormonal balances and nutrient sinks that quickly limit leaf growth.

In many plants, the main stem vegetative apex is transformed into a floral bud, and leaf production ceases. In such plants, the unexpanded leaves present at floral-bud initiation develop into progressively smaller leaves. In indeterminate plants, floral buds and fruit at axillary nodes either compete with vegetative apices for nutrients, or they produce hormones that suppress pical growth, or both, and leaf production and expansion is quickly slowed. In determinate plants with apical floral buds, leaf expansion in existingleaves is similarly curtailed.

Here we will discuss the effects of temperature and photoperiod on leaf production in maize (*Zea mays* L.), a determinate plant, and in soybeans [*Glycine max* (L). Merrill], a plant that can be either determinate or indeterminate. Another purpose of this review is to point out recent research on the genetic variability of these factors. We also will list some papers dealing with the same effects on leaf production and flowering in other crop plant species. Roberts and Struckmeyer[2] were the first to report interactions of temperature and photoperiod effects on flowering in some crop plants.

II. MAIZE

A. Predicting Leaf Number

Leaf number to tassel is an indicator of potential leaf-area production in maize and is closely correlated with days or degree-days to tasseling, pollon shed, and ear maturity. Garner and Allard[3] first reported that maize was a "quantitative short day" plant, in that short days induced earlier flowering or fewer main stem leaves before tasseling. There have been at least eight groups recently reporting research on the effects of

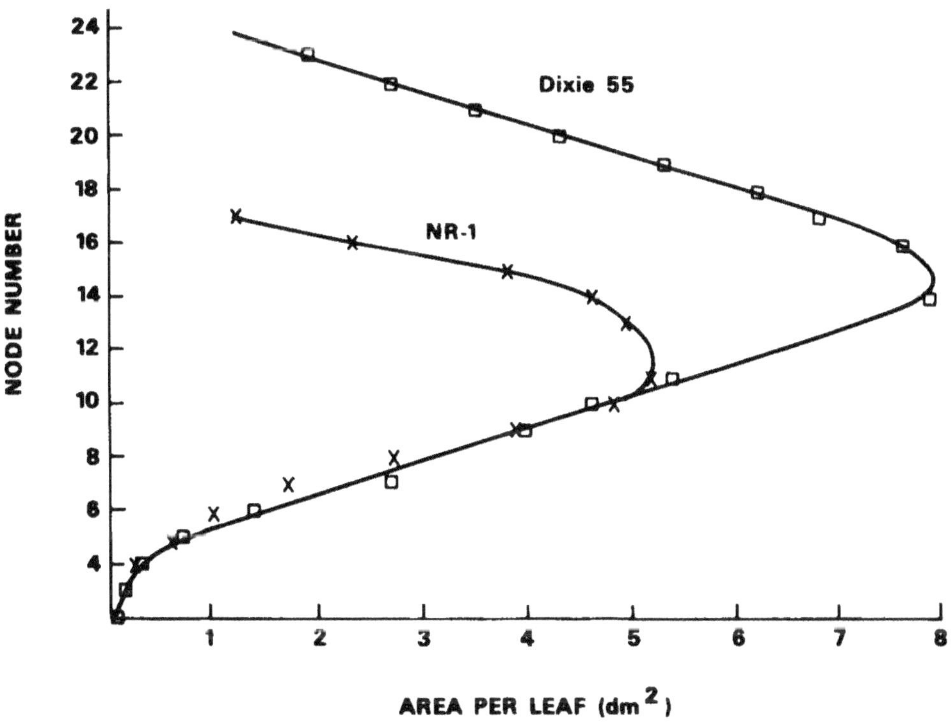

FIGURE 1. Area per leaf vs. node number in phytotron-grown maize plants.

temperature, photoperiod, genotype, and other factors on leaf number or days to tasseling.[4-13] Information so generated for specific genotypes was used to construct a model for predicting days to tasseling or total leaf number[13] which tested well against field results.

Using a degree-day model, Cross and Zuber[14] were able to predict date of pollon shed fairly accurately, for a range of maturity-types planted at three different dates in two different years at the same location. They used a base temperature of 10°C, and subtracted temperatures above 30°C to account for temperature stress.

For predictions related to research on the fundamental physiology of the processes involved, a detailed morphogenic model is needed. For field-level predictions, a modified degree-day model, taking into account photoperiod and genetic differences in responses to photoperiod and temperature, may be adequate. Research and development of the latter models must be carried on at the former, more fundamental level, with the objective of keeping the field models simple but as accurate as possible.

B. Control of Leaf Size

In maize, the tassel is initiated before many of the leaves near the apex have expanded. The ultimate size of these remaining leaves become progressively smaller. For an early- and a late maturing maize line (Figure 1), ultimate leaf sizes were similar up to Node 10, but maximum leaf size was greater in the late-maturing line. It was not clear from these data if the maximum potential leaf area was obtained in these experiments.

The increase in leaf size at the early nodes is typical of many crop plants and is conveniently related to the ability of the plant to generate photosynthate over time, where photosynthate production is approximately equal to leaf area times the leaf photosynthetic rate. Such morphogenic behavior controlling photosynthate supply, then, is fundamental to the growth analysis equations discussed in earlier chapters.

III. SOYBEANS

A. Flowering Physiology

The soybean plant has long been used in photoperiod studies,[3,4,15] and most cultivars are short-day plants. There are both determinate varieties, where stem and branch apices simultaneously change into floral buds and leaf production abruptly stops,[16] as well as indeterminate varieties, where apices remain vegetative but growth is suppressed during the pod-filling stage.

The trifoliolate node number, counting up the main stem from the unifoliolate node, is a good indicator of the plant's potential for expanded leaf area in indeterminate varieties. Total nodes produced on the main stem is a more appropriate indicator of total expanded leaf area per plant in determinate lines.

Ten maturity groups of soybean varieties have been established to identify soybean adaptation to natural photoperiods at its northernmost limit for growth (0*,00) to the latest maturing group grown in the southern continental U.S. (I to VIII), with two more groups (IX, X) including varieties adapted to natural photoperiods at lower latitudes. Garner and Allard[3,4] included several such classes of soybeans in their earliest papers on photoperiodism. Several genes have been associated with the photoperiod response,[17] as well as determinate and indeterminate growth after flowering begins.[16] Bernard[17] inserted various homozygous combinations of the three known genes for photoperiodism in their dominant or recessive state (E,e) into "Clark" and "Harosoy" indeterminate varieties. We grew these lines at different photoperiods and temperatures in greehouses and darkrooms of the Duke University Phytotron during spring and summer months when sunlight was maximum. These studies were initiated to predict flowering and maturity for a soybean growth and yield model, as well as understanding how various genotypes behave under a wide range of conditions.

Table 1 gives daylength effects on the trifoliolate node number of the first flower and the number of days to flowering for six Clark genotypes with different combinations of photoperiod genes. Days to flower and the trifoliolate node of first flower were closely correlated among lines and photoperiod treatments within a particular temperature regime (Figure 2). The slopes of the lines which might be drawn through the points in Figure 2 calculate to be 2.4 and 4.4 days leaf^{-1} for the 33/29 and 32/27°C regimes, respectively. Such values are an estimate of the leaf plastochron, or days between appearance of new leaves on the main shoot apex, and are a little longer than values reported earlier[18,19] for the first five trifoliolates in seedlings. Since such plastochron values typically become longer as more leaves are produced, the values calculated from Figure 2 are consistent with other results and offer hope for a plastochron model for predicting flowering maturity.

Table 1 reports results from one temperature treatment of an experiment run at the same time as 32/29, 32/17, and 23/20°C 8-/16 day/night temperature regimes. Mean values across daylengths and genotypes for the trifoliolate node number of the first flower were 6.9 and 4.4 nodes (n = 144) for the 32/29 (30°C mean) and 23/20 (21°C mean) treatments, respectively. In the same comparison, mean values for days to first flower were 36 and 42. Thus temperature can add to the plant size (2.5 extra nodes in our comparison over 9°C) and to the time required for planting to flowering.

In a dynamic morphogenic model, node production is associated with leaf area ex-

* Maturity group notation

TABLE 1

Effect of Daylength on Trifoliolate Node Number

Clark strain (Photoperiod gene notation)	Attribute measured	Daylength (hr)[a] 8	12	14	16
$e_1e_1E_2E_2E_3E_3$	A[b]	2.5	2.2	3.5	7.2
	B[c]	24.5	23.7	27.2	43.7
$E_1E_1E_2E_2E_3E_3$	A	2.7	3.0	5.0	17
	B	26.3	27.0	34.2	76
$e_1e_1e_2e_2E_3E_3$	A	2.5	2.0	2.7	5.0
	B	24.0	23.5	26.3	36.4
$E_1E_1e_2e_2E_3E_3$	A	2.7	2.7	4.2	10.2
	B	26.7	26.2	30.3	49.0
$e_1e_1E_2E_2e_3e_3$	A	2.0	2.0	4.2	5.2
	B	23.7	23.5	28.0	36.7
$e_1e_1e_2e_2e_3e_3$	A	2.7	2.5	3.0	4.0
	B	25.8	24.7	26.2	32.8

[a] Plants were grown in 4-liter pots in a 32/23°C, 8-/16-hr day/night temperature regime, in greenhouses of the Duke University phytotron. Daylengths were generated with photoperiod lights (incandescent) and darkrooms. N = 6.

[b] Trifoliolate node of first flower, s_x = 0.61.

[c] Days to first flower, s_x = 0.25.

pansion and the production of new floral sites. Further studies of plastochrons for branch and main stem node production during flowering are needed for development of a model. The effects of temperature, nutrition, and stress on plastochron values need to be defined.[24] Such detailed morphological analyses of growth represent a fundamental method for studying, understanding, and predicting how long it takes for a plant to flower and mature a crop of beans.

In the soybean experiments reported here, a new node was counted when the trifoliolate had expanded to 8 cm² in area. For the experiment reported in Table 1, there were 3.4 such main stem nodes above that of the first flower when the first flower appeared on plants in the 8- to 14-hr photoperiods. In the 16-hr treatment, the mean value was 4.9 nodes, excluding the $E_1e_2E_3$ genotype which had 2.8 nodes. During flowering, 6.8 additional main stem trifoliolate nodes were produced in the 8- to 14-hr photoperiod treatments. Again the 16-hr photoperiod produced different results, with a range of 6.4 additional nodes to uncurtailed node production in some genotypes. The latter plants matured a crop of pods as flowering and vegetative node production continued on the main stem and branch apices. Branch nodes produced during the floral period (or subsequently) were also affected by photoperiod, temperature, and genotype.

The large number of possible heterozygous combinations of the soybean photoperiod genes and the responses of such genotypes to environment further complicates the results. Presumably, these responses would be intermediate between those encountered with combinations of homozygous dominant and recessive genes. Since the genetic makeup of most cultivars is unknown with respect to these genes, comparisons with standard lines carrying known combinations of photoperiod genes at different latitudes, elevations, and planting dates should help in developing logic for a degree-day model predicting vegetative growth and flowering behavior[20-23] at different latitudes. Even then, specific locations may have to be "calibrated" by comparing flowering dates with those obtained at other test locations.

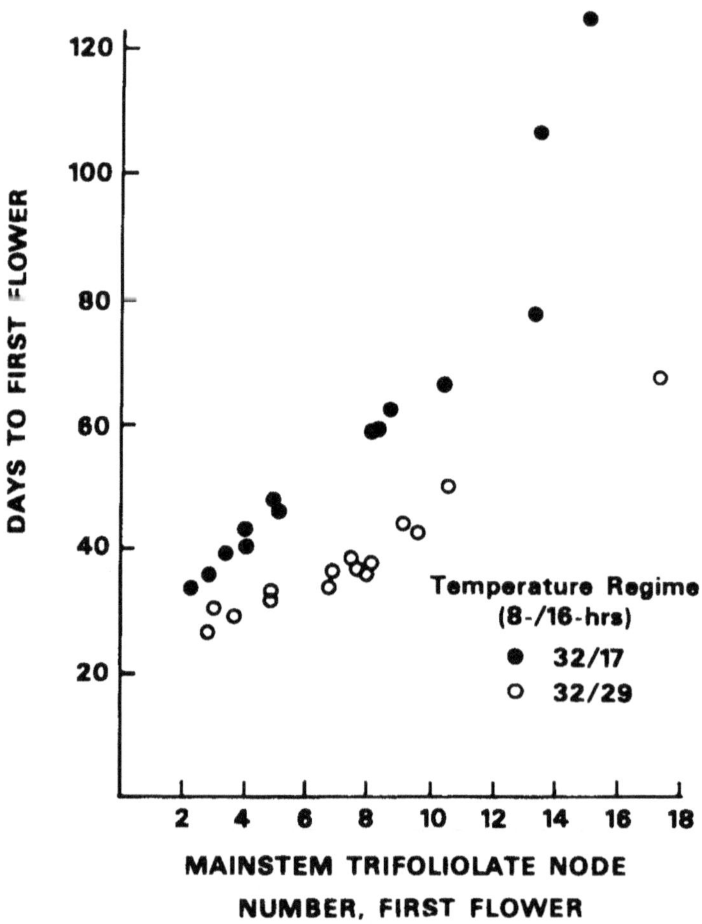

FIGURE 2. Days to first flower vs. mainstem trifoliolate node number of the first flower, in Clark soybean strains homozygous for various combinations of photoperiod genes and growing under various photoperiods and two temperature regimes.

The situation is even further complicated in the southern U.S. since some determinate soybean cultivars planted early in the growing season can be induced to flower, returning to the vegetative state as days get longer, flowering again sometime later than June 21. This phenomnon needs further study.

B. Curtailment of Leaf Size and Expansion Rate

Hofstra et al.[24] reported leaf area vs. node position for soybean plants with trifoliolate areas varying from 200 to 400 cm.[2] A major part of the expansion curve was linear, and such linear rates were also linearly related to maximum leaf area attained. Hence, leaf production, expansion rates, and ultimate area attained are all curtailed as the plant sets fruit.

IV. OTHER SPECIES

Since there are no known reviews to direct modelers to relevant literature, we cite here similar work for sorghum [*Sorghum bicolor* (L) Moench,][25-27] tobacco (*Nicotiana tabacum* L.),[28-31] wheat (*Triticum aestivum* L.),[32-34] cotton (*Gossypium hirsutum* L.),[35] alfalfa (*Medicago sativa* L.),[36-38] peanuts (*Arachis hypogaea*),[39,40] orchardgrass (*Dactylis glomerata* L.),[41a,41b] Aitken's book[43] also contains useful information. These responses of flowering to photoperiod and temperatures have to be worked out experimentally if no data are available in the literature for the crop plant being modeled, since such responses are genotypic specific.

V. PREDICTING FLORAL INITIATION AND SUBSEQUENT DEVELOPMENT OF THE FLORAL BUD TO FLOWERING

The consequences of flowering on further leaf expansion are actually predetermined at a much earlier stage of growth when initiation of a flower primordia occurs. In corn, this event can occur early in the seedling stage of growth. To predict flowering at this level, one would need a model predicting leaf primordia production rate, tassel initiation, and time for tassel growth to anthesis.

In the case of soybeans, floral buds are differentiated on axillary meristems where trifoliolate primordia and expanded leaves as well as prophylls (modified pair of leaves where a vegetative branch is attached to a stem) are attached to a stem. To predict flowering, one would need a model predicting leaf primordia production rate on the main stem as well as in vegetative branches, floral bud initiation, and the time for bud growth to flowering. When a flower is initiated, a bract and two prophylls are produced, each of which has axillary buds that can develop into flowers. The soybean model is greatly complicated by variations in time required for development of a flower from the bud, caused by genotype and environmental differences. Monographs for various crop plants such as soybeans[44] often contain chapters on morphology describing the above process. Little information is available on the dynamics of the processes involved.

The morphogenesis of almost all plants may well be described by selecting a small combination of morphogenetic principles from a slightly larger list describing possible developmental behavior near the apical domes of main stem and axillary meristems. Such a comprehensive model has yet to be built.

Associated research involves plant dissections at the microscopic level, as well as studies of organogenetic rates vs. the environment and plant status. Such an approach is also fundamental to the development of a nutritional budget for a plant.

V. CONCLUSIONS

The above discussion hopefully has impressed upon the reader the role of flowering physiology in the development of a crop canopy and in the prediction of certain events during a plant's life cycle. The latter is fundamental to the development of a degree-day model for plant growth and maturity.

Since plant morphology and morphogenic dynamics differ greatly among genera as well as species and cultivars, the crop plant to be modeled must be studied in great detail without the benefit of accumulated information for other genotypes. Such information is vital for understanding photosynthate partitioning and redistribution for the development of yield components of crop plants.

REFERENCES

1. **Kallis, A. and Tooming, H.,** Estimation of the influence of leaf photosynthetic parameters, specific leaf weight and growth functions on yield, *Photosynthetica*, 8, 91, 1974.
2. **Roberts, R. H. and Struckmeyer, B. E.,** Further studies on the effect of temperature and other environmental factors upon the photoperiodic response of plants, *J. Agric. Res.*, 59, 699, 1939.
3. **Garner, W. W. and Allard, H. A.,** Effect of length of day on plant growth, *J. Agric. Res.*, 18, 553, 1920.
4. **Garner, W. W. and Allard, H. A.,** Further studies on photoperiodism, the response of the plant to relative length of day and night, *J. Agric. Res.*, 23, 871, 1923.
5. **Chase, S. S. and Nanda, D. K.,** Number of leaves and maturity classification in *Zea mays* L., *Crop Sci.*, 7, 431, 1967.
6. **Francis, C. A., Grogan, C. O., and Sperling, D. W.,** Identification of photoperiod insensitive strains of maize, *Crop Sci.*, 9, 675, 1969.
7. **Hesketh, J. D., Chase, S. S., and Nanda, D. K.,** Environmental and genetic modification of leaf number in maize, sorghum, and Hungarian millet, *Crop Sci.*, 9, 460, 1969.
8. **Frances, C. A., Sarria, D., Harpstead, D. D., and Cassalett, D. C.,** Identification of photoperiod insensitive strains of maize (*Zea mays* L.). II. Field tests in the tropics with artificial lights, *Crop Sci.*, 10, 465, 1970.
9. **Cross, H. Z.,** Effects of photoperiod and temperature on flowering dates of maize, Ph.D. thesis, University of Missouri, Columbia, 1971.
10. **Steenson, J. C. and Goodman, M. M.,** Ecology of exotic races of maize. I. Leaf number and tillering of 16 races under four temperatures and two photoperiods, *Crop Sci.*, 12, 864, 1972.
11. **Hunter, R. B., Hunt, L. A., and Kannenberg, L. W.,** Photoperiod and temperature effects on corn, *Can. J. Plant Sci.*, 54, 71, 1974.
12. **Bonaparte, E. E. N. A.,** The effects of temperature, daylength, soil fertility, and soil moisture on leaf number and duration to tassel emergence in *Zea mays* L., *Ann. Bot.*, 39, 853, 1975.
13. **Coligado, M. C. and Brown, D. M.,** A bio-photo-thermal model to predic tassel initiation time in corn (*Zea mays* L.), *Agric. Meterol.*, 15, 11, 1975.
14. **Cross, H. Z. and Zuber, M. S.,** Prediction of flowering dates in maize based on different methods of estimating thermal units, *Agron. J.*, 64, 351, 1972.
15. **Hammner, K. C.,** *Glycine max* (L) Merrill, in *The Induction of Flowering*, Evans, L. T., Ed., Cornell University Press, Ithaca, 1969, 62.
16. **Bernard, R. L.,** Two major genes for time of flowering and maturity in soybeans, *Crop Sci.*, 11, 242, 1971.
17. **Bernard, R. L., and Weiss, M. G.,** Qualitative genetics, in *Soybeans: Improvement, Production and Uses*, Caldwell, B. E., Ed., American Society of Agronomy, Madison, Wis., 1973, 117.
18. **Hesketh, J. D., Myhre, D. L., and Willey, C. R.,** Temperature control of time intervals between vegetative and reproductive events in soybeans, *Crop Sci.*, 13, 250, 1973.
19. **Shibles, R. M., Anderson, I. C., and Gibson, A. H.,** Soybeans, in *Crop Physiology, Some Case Histories*, Evans, L. T., Ed., Cambridge University Press, Cambridge, 1975, 151.
20. **Lawn, R. J. and Byth, D. E.,** Response of soybeans to planting date in Southeastern Queensland. I. Influence of photoperiod and temperature on phasic development patterns, *Aust. J. Agric. Res.*, 24, 67, 1973.
21. **Major, D. J., Johnson, D. R., and Luedders, V. D.,** Evaluation of eleven thermal unit methods for predicting soybean development, *Crop Sci.*, 15, 172, 1975.
22. **Major, D. J., Johnson, D. R., Tanner, J. W., and Anderson, I. C.,** Effects of daylength and temperature on soybean development, *Crop Sci.*, 15, 174, 1975.
23. **Sierra, E. M.,** Energetic-photo-thermal developmental model for medium late and late soybean cultivars, *Agric. Meteorol.*, 18, 277, 1977.
24. **Hofstra, G., Hesketh, J. D., and Myhre, D. L.,** A plastochron model for soybean leaf and stem growth, *Can. J. Plant Sci.*, 57, 167, 1957.
25. **Quinby, J. R.,** The genetic control of flowering and growth in sorghum, *Adv. Agron.*, 25, 125, 1973.
26. **Quimby, J. R.,** Leaf and panicle size of sorghum parents and hybrids, *Crop Sci.*, 10, 251, 1970.
27. **Miller, F. R., Quinby, J. R., and Cruzado, H. J.,** Expression of known maturity genes of sorghum in temperate and tropical environments, *Crop Sci.*, 8, 675, 1968.
28. **Muraoka, Y. and Ohori, K.,** Developmental response of tobacco plant to temperature in relation to photoperiodic condition, *Proc. Crop Sci. Soc. Jpn.*, 24, 271, 1955.
29. **Muraoka, Y. and Ohori, K.,** Developmental responses of tobacco plants under various temperatures and photoperiods, *Proc. Crop Sci. Soc. Jpn.*, 25, 104, 1956.
30. **Muraoka, Y. and Ohori, K.,** Varietal differences of developmental responses of tobacco plants under controlled thermal and photoperiodic conditions, *Proc. Crop Sci. Soc. Jpn.*, 25, 151, 1957.

31. **Shinohara, T.**, Effects of temperature on the photoperiodic responses of tobacco, *Proc. Crop Sci. Soc. Jpn.*, 40, 267, 1971.
32. **Syme, J. R.**, Quantitative control of flowering time in wheat cultivars by vernalization and photoperiod sensitivities, *Aust. J. Agric. Res.*, 24, 657, 1973.
33. **Halloran, G. M.**, Genotype differences in photoperiodic sensitivity and vernalization response in wheat, *Ann. Bot., (London)*, 39, 845, 1975.
34. **Pirasteh, B. and Welsh, J. R.**, Monosomic analysis of photoperiod response in wheat, *Crop Sci.*, 15, 503, 1975.
35. **Hesketh, J. D., Lane, H. C., Alberte, R. S., and Fox, S.**, Earliness factors in cotton: new comparisons among genotypes, *Cotton Grow. Rev.*, 52, 126, 1975.
36. **Sato, K.**, Growth and development of the alfalfa plant under controlled environments. I. The effects of daylength and temperature on the growth and chemical composition, *Proc. Crop Sci Soc. Jpn.*, 40, 120, 1971.
37. **Pearson, C. J. and Hunt, L. A.**, Effects of temperature on primary growth in alfalfa, *Can. J. Plant Sci.*, 52, 1007, 1972.
38. **Pearson, C. J. and Hunt, L. A.**, Effects of temperature on primary growth and regrowths in alfalfa, *Can. J. Plant Sci.*, 52, 1017, 1972.
39. **Ono, Y., Ozaki, K., and Nakayama, K.**, Effects of air temperature on flowering of peanut plants, *Proc. Crop Sci. Soc. Jpn.*, 43, 237, 1974.
40. **Ono, Y., Ozaki, K., and Nakayama, K.**, Effects of air temperature on pod development and yield of peanut plants, *Proc. Crop Sci. Soc. Jpn.*, 43, 242, 1974.
41. **Broue, P.**, Flowering in *Dactylis glomerata*. I. Photoperiodic requirement, *Aust. J. Agric. Res.*, 24, 677, 1973.
41a. **Broue, P.**, Flowering in *Dactyles glomerata*, II. Interaction of temperature and photoperiod, *Aust. J. Agric. Res.*, 25, 242, 1974.
42. **Clarkson, N. M. and Russell, J. S.**, Flowering responses to vernalization and photoperiod in annual medics (*Medicago* spp.), *Aust. J. Agric. Res.*, 26, 831, 1975.
43. **Aitken, Y.**, *Flowering Time, Climate, and Genotype*, Melbourne University Press, 1974.
44. **Carlson, J. B.**, Morphology, in *Soybeans: Improvement, Production and Uses*, Caldwell, B. E., Ed., ASA Monograph No. 16, American Society of Agronomy, Madison, Wis., 1973, 17.

Chapter 7

LEAF GROWTH AND SENESCENCE

H. R. Leffler

TABLE OF CONTENTS

I. LEAF INITIATION AND DEVELOPMENT

A series of cell divisions just below the surface of the lower reaches of the apical dome initiates the development of a leaf.[1] Cells derived from these divisions create a small enlargement on the surface of the dome; this enlargement is generally considered to be the earliest visible indicator of leaf initiation. Although formation of the actual stimulus triggering these mitoses may truly represent active leaf initiation, the visually recognizable promordium is commonly regarded as the stage from which leaf development progresses.

Cells forming the leaf buttress divide and enlarge, eventually producing the axis of a juvenile leaf. Meristematic activity in the axis is predominantly apical during early stages of development, but it becomes more generalized later. As the leaf axis grows above the buttress, procambium differentiates from the central region. As elongation continues, the leaf lamina is initiated in the meristematic regions along the margins of the axis.

Although both cell division and cell enlargement contribute to leaf growth, the predominance of their contributions occurs at separate stages of leaf development. Early leaf development is associated with a relatively high mitotic index (percentage of cells undergoing active cell division), while later leaf growth predominantly reflects cell enlargement. It should be emphasized that the relatively high mitotic indices of the juvenile leaf tissue are not indicative that most cells in a zone are actively dividing. Actually, measurements of cell division in meristematic regions of young leaves frequently show from 4 to 6% of the cells to be actively dividing. Consequently, nearly 95% of the cell population is between cell divisions, even though the mitotic index may be relatively high. Supporting this are estimates of the cell doubling time: in *Lupinus* leaf primordia, the time required to double the cell number has been estimated to be between 3 and 5 days.[2]

Within the leaf primordium, the rate of mitosis varies according to position: it is higher just behind the tip than it is at the tip itself.[3] With additional leaf blade growth, the rates of cell division continue to vary according to position within the leaf. Final cell number is first established in the leaf tip, next in the middle zone, and finally in the base of the leaf. Leaf growth in each of these regions occurs as a result of cell enlargement after establishment of the final cell number.

Leaf growth studies frequently use the plastochron concept to define growth stages.[3,4] Therefore, a short discussion of the plastochron concept is appropriate. The term *plastochron* is defined as a unit measure of time; the unit of measurement is the time interval between any two successive, periodically repeated events. When employed in studies of leaf development, the plastochron is keyed to a stage of development that can be easily identified, such as leaf length.

The plastochron index (PI) derived in this manner is linear with time, but its scale incorporates the expression of plant development adjusted to a stage of leaf development. The PI of a plant is defined by the following equation

$$PI = n + (\mathrm{Log}\, L_n - \mathrm{Log}\, R) / (\mathrm{Log}\, L_n - \mathrm{Log}\, L_{n+1})$$

where the terms of the equation are listed below.

PI: Age of the plant, in plastochrons

n: Number of leaves that have passed the reference stage of development (length)

Log L_n: Logarithm of the length of leaf n, the last leaf to have passed the reference stage

Log R: Logarithm of the reference stage (length)

Log L_{n+1}: Logarithm of length of leaf n + 1 (the next leaf up the plant from leaf n)

Calculation of the PI of a plant involves four steps:

1. Establishment of a reference stage (R)
2. Counting the number of leaves that have passed the reference stage (n)
3. Measurement of two successive leaves (L_n and L_{n+1}), and
4. Actual calculation of the PI according to the formula above.

For example, if the reference stage of development is set at a length of 20 mm, L_n is 22 mm, and L_{n+1} is 16 mm, the fractional portion of the PI is (1.3424−1.3010)/(1.3424−1.2041), or 0.2993. If 14 leaves on this plant had passed the reference stage, n would be 14, and the PI would be 14.2993 or 14.30.

A related term, the leaf plastochron index (LPI), is derived from the PI and is used to index the development of a specific leaf. The LPI is calculated from the PI by adjustment of the value n. The formula becomes LPI = PI − n′, where n′ is the serial number of the leaf of interest. It follows then that LPI can be either positive or negative. For example, if development of leaf 4 were being studied on the hypothetical plant measured earlier (PI = 14.30), the LPI of leaf 4 would be 14.30 − 4 = 10.30. This would indicate that about 10.3 plastochrons had elapsed since leaf 4 was 20 mm long. If, on the other hand, the 15th leaf were being studied on that same plant, its LPI would be 14.30 − 15 = −0.70. This would indicate that leaf 15 had not reached 20 mm, and would not reach that length for another 0.7 plastochrons. When leaf 15 reached a length of 20 mm, the PI would be 15.00, and the LPI of leaf 15 would be zero.

Both the PI and the LPI, as calculated, provide precise developmental scales which, in turn, facilitate the expression of plant and leaf growth as well as the interpretation of physiological events occurring throughout the development. Many of these timing estimates would be difficult, if not impossible, without the indices. However, applicability of the PI and LPI requires that successive leaves be formed at nearly equal intervals, their growth must be exponential, and their growth curves should be parallel (Figure 1). These qualifying criteria are not always met.[6]

Expansion of the leaf blade initially requires translocation of photosynthate from leaves below.[7] Although the developing leaf initially becomes capable of some CO_2 fixation when still quite small,[8] sustained growth results from continued import of nutrients. Consequently, sink activity of the expanding leaf blade changes with the ontogeny of the leaf: import of photosynthate increases with time, but then falls to approximately zero by the time the leaf reaches about half-maximal area.[7-9] The maximum leaf expansion rate occurs relatively early in development (when slightly over one third expanded) and although final leaf sizes may very significantly, periods of expansion are similar.[10]

The balance between sink and source activities changes not only with the ontogeny of the entire leaf, but also with the relative stages of development of areas within the leaf. In very young leaves, the leaf tips import most of the photosynthate, with relatively less being transported to the more basal sections.[11] As the leaf grows, the transport of photosynthate gradually shifts to favor the base of the leaf.[9,11] Continual expansion of the leaf, then, is accompanied by the transition from import to export of photosynthate from the leaf tip to the leaf base.

The leaf tip is the first portion of the leaf to acquire the capability to fix CO_2;

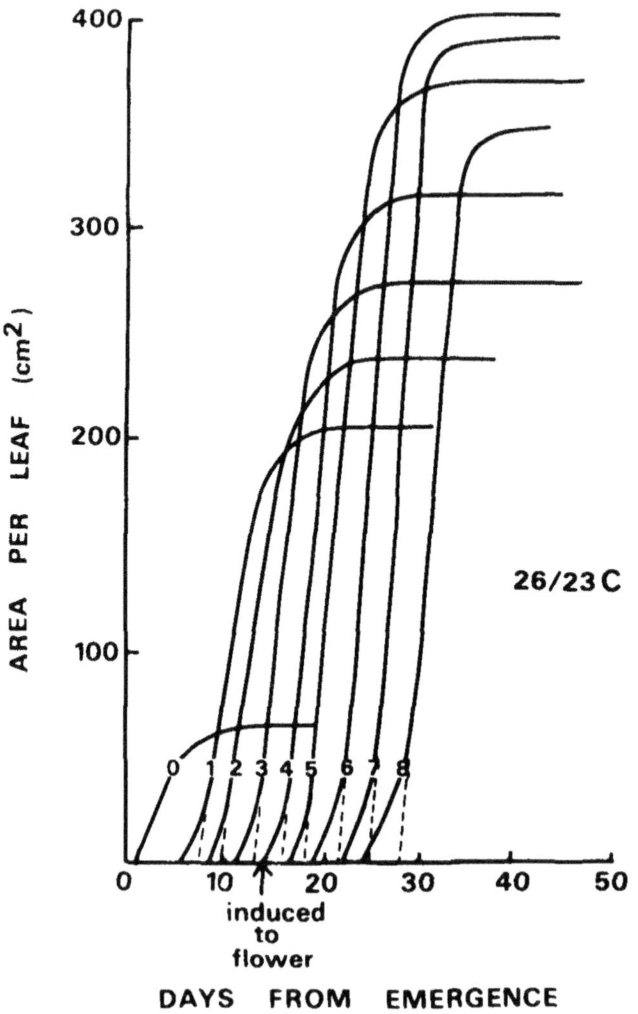

FIGURE 1. Leaf expansion of unifoliolates (0) and trifoliolates 1 to 8 for Wayne soybeans growing in the Duke University Phytotron. (From Hofstra, G., Hesketh, J. D., and Myhre, D. L., *Can. J. Plant Sci.*, 57, 167, 1977. With permission.)

subsequent development of CO_2-fixing capacity by the remaining areas of the leaf blade immediately follows and parallels the loss of sink activity.[8,9,11-13] Initial phloem loading was observed in the tip of sugar beet leaves when about 22% expanded; it was essentially complete by the time the leaf reached about 50% of maximum length.[13] Similarly, export from these leaves was first observed when they were at 35% of full length, and it increased rapidly until leaves were half-maximum length.

The relative size at which these transitions occur appears to be dependent upon the species being examined. For example, in cottonwood leaves of LPI 4 (half-maximal size), only the leaf tips exported photosynthate to the petiole; two plastochrons later (LPI 6), however, all parts of the leaf were exporting photosynthate.[14] Consequently, export capacity of cottonwood leaves did not plateau at half-maximum size, but continued to increase for some time thereafter. These temporal relationships should therefore be established for each species being studied, because extrapolating from one species to another may introduce considerable error into a modeling analysis.

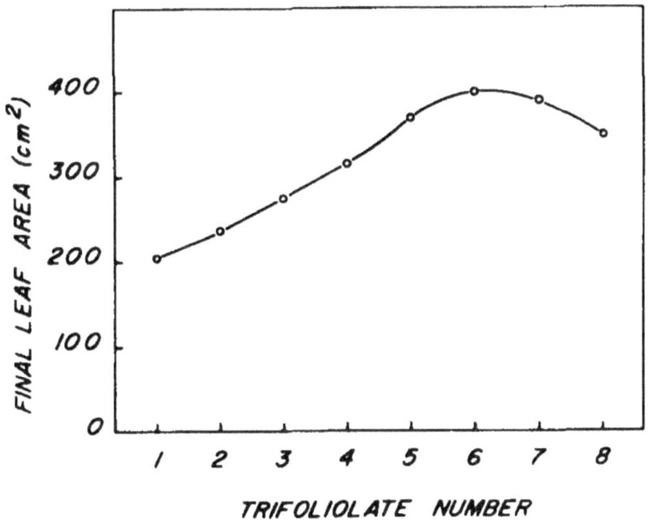

FIGURE 2. Final leaf area of trifoliolates 1 to 8 of Wayne soybeans growing in the Duke University phytotron. (Data taken from Hofstra, G., Hesketh, J. D., and Myhre, D. L., *Can. J. Plant Sci.*, 57, 167, 1977.)

The early rapid enlargement of leaf area[10] is a major contributor to the increase in leaf photosynthesis and respiration.[15-17] Photosynthesis per unit of leaf weight also increases during early leaf enlargement because of increasing activities of photosynthetic enzymes; respiration, when expressed in a similar fashion, continually declines during this period.[15] However, since respiration is so small relative to photosynthesis, these differences in developmental patterns do not significantly alter the pattern of development of net photosynthesis.[16] The overall effect of respiration may, however, be quite significant; Wilson[19] found that growth of low respiring genotypes of *Lolium* was consistently higher than that of fast-respiring genotypes.

Attainment of maximum leaf photosynthetic rate is approximately coincidental with the establishment of maximum leaf area.[8,17,19,20] Leaf photosynthetic rates then decline slowly as the aging process continues.[20]

Leaf size and maximum photosynthetic capacity are affected by environmental conditions such as nutrition, position within the canopy (see Figure 2) and temperature. Although size and activity of the leaves are affected by these factors, the enlargement period and the life span of the leaves apparently are not.[17] Growth temperatures outside the range of adaptation of the plant being studied will reduce photosynthesis.[21] Additionally, photorespiration in C_3 plants extracts a relatively larger penalty in warm conditions than it does in cooler conditions.[22] Mineral nutrition also affects leaf photosynthetic activity.[23] The peak photosynthetic activity can be maintained longer if the mineral nutrient supply to the roots can be sustained at a high level.[16] The nitrogen (N) content of the leaf is related to the photosynthetic capacity of the leaf.[16] Significantly, the N contents of leaves fall with ageing — leaf N falls as a leaf ages, and leaf N also decreases in new leaves as the canopy ages.[24] Photosynthesis may also be restricted by deficiency of minerals required by photosynthetic enzymes.[25]

II. LEAF SENESCENCE

Simon[26] described five categories of leaf senescence, three of which are of varying degrees of relevance to the subject of this section. These three types were identified as sequential senescence, sychronous senescence, and overall senescence. Each is concerned either with senescence of leaves from a plant canopy or with the senescence of the entire canopy. The other two types of leaf senescence concern deterioration of detached leaves maintained either in the light or in the dark (or very dim light).

Sequential (or progressive) senescence is the term used to describe the senescence of individual leaves within the canopy, with lower leaves senescing first; more leaves initiate senescence as additional leaves expand at the apex. If the growth rate should equal the senescence rate, the number of leaves in the canopy would remain constant. Synchronous senescence is characterized by the autumnal leaf fall of deciduous trees and shrubs, in which all leaves senesce together. Overall senescence, on the other hand, is characterized by the death of the entire plant after flowering (if the plant is monocarpic) or following senescence of the apex. The requirement for plant death can be considered to be a distinguishing criterion between synchronous and overall leaf senescence.

Although there are distinguishing characteristics among these three categories of leaf senescence, each is a normal event that initiates the termination of the life cycle of a leaf. Since each is a normally occurring process (as contrasted with senesence of detached leaves), it may be that there are degrees of similarity among the physiologies of these senescences.

Foliar senescence can be triggered by reproductive development. Lindoo and Nooden[27] described the interrelationship between fruit development and foliar senescence in soybeans. In a later report,[28] they described a series of experiments in which, by various manipulations of the reproductive tissues, they could control the degree of leaf senescence. Studies with a male-sterile line of soybeans demonstrate the consistency with which foliar senescence and reproductive growth are associated.[29] Foliar senescence occurs when reproductive tissues reach a critical number or a particular stage of development. If the sink demands of the reproductive tissues remain low enough, no leaf senescence is initiated. Similarly, once the demands of the reproductive growth reach a critical level, senescence of the leaf canopy is triggered. The depodding data of Lindoo and Nooden[27,28] and the male-sterile data of Wilson et al.[29] mutually demonstrate the influence that reproductive growth can exert upon leaf senescence.

Following the onset of senescence, the leaf protein content beings to decline.[26] This reduction in leaf protein has been demonstrated to result from the activity of proteolytic enzymes.[30-32] However, declining leaf protein is not, by itself, an indicator of the senescence phenomenon. During reproductive growth (specifically during the seed filling stage), declining leaf protein has been identified that is either independent of proteolytic activity[32] or the result of proteolytic activity not associated with the senescence phenomenon alone.[31,33]

From one fourth to one third of the nitrogen accumulated by wheat grains was estimated by Dalling et al.[33] to have come from the leaves by mobilization and translocation. Associated with the mobilization of nitrogen in the leaves was a large increase in acid proteinase activity. Additionally, the estimated loss of leaf nitrogen due to proteinase activity was highly correlated to the measured loss of leaf nitrogen. The authors suggested an important role for this acid proteinase activity in leaf senescence. However, since accumulation of nitrogen by wheat grains, mobilization of nitrogen from wheat leaves, and senescence of the wheat leaves are so closely related in time, there may remain some question that the association described by Dalling et al.[33] is strictly the result of foliar senescence. Acid proteinase activity was found to be highest during the decline of leaf protein.[34]

In corn, where these processes are somewhat more separated in time, Feller et al.[31]

reported differential activity of more than one protease in the leaves. Amino peptidase and carboxypeptidase activities of the ear leaf increased during reproductive initiation, but decreased during later grain development and canopy senescence. During the reproductive phase of development there was a slow, gradual, increase in caseolytic activity. This activity then increased rapidly with senescence; highest activity was observed during the most-rapid loss of leaf protein. These results suggest that amino peptidase and carboxypeptidase activities function during the interchanges among leaf proteins, but that the caseolytic activity in maize leaves serves a major role in the mobilization of leaf proteins during senescence.

Mobilization of leaf proteins requires protein synthesis. Although the net balance of leaf proteins declines during senescence, a select type of proteins, the proteases, increases.[35] As indicated above, several research groups have identified increases in proteolytic activities upon the onset of senescence. Martin and Thimann[35] identified a requirement for protein synthesis with this increase in proteolytic capacity. They found that they could effectively delay senescence by treating the leaves with cycloheximide at concentrations that inhibited protein synthesis. Furthermore, they observed that applications of kinetin that delayed senescence did so by suppressing protease formation.

Nuclease activities in senescing leaves also increase.[36,37] Evidence has been presented to indicate that the senescence-associated increase in ribonuclease (RNase) activity requires both nucleic acid and protein synthesis.[36] Additionally, conditions that delay senescence of detached leaf material, such as high light intensity or kinetin treatment, similarly retard this increase in RNase activity.[36]

Extrapolation from these data leads to the suggestion that during the onset of leaf senescence, complex molecules such as proteins and nucleic acids are broken down by catabolic enzyme activities. Formation of these catabolic enzymes in turn requires synthesis of protein and, probably, nucleic acid. Consequently, senescent leaves characteristically lose total protein and nucleic acid while simultaneously maintaining some synthesis of each.

During senescence, both ribosomes and polyribosomes are lost from leaves.[38] According to Eilam et al.,[39] the ribosomal RNA fraction increases during leaf growth, reaches a maximum before maximum fresh weight, and then declines. About half of the mature-leaf ribosomes are chloroplast ribosomes. Their data[39] show that the proportion of polyribosomes declines throughout leaf growth, but remains essentially constant at a low level during senescence. The loss of ribosomes is attributed to restricted synthesis. However, the remaining low level of polyribosomal material would be consistent with the apparent requirement for synthesis of some protein during senescence.

Deciduous leaves of trees are useful experimental materials with which to examine senescence-related movements of mineral nutrients (synchronous senescence). Export of foliar nutrients upon the advent of senescence has been examined in detail in oak, pine, and cottonwood.[40,41] In each of these studies, it was found that massive amounts of each of the three major nutrients moved out of the leaves as autumnal senescence approached. Although the proportion of each nutrient recycled in this fashion could be strongly influenced by the species being studied, the generality of the nutrient recycling phenomenon appears to be high. The general loss of leaf proteins during senescence (described earlier) is reinforced by these measurements of nutrient flow in forest systems.

Switzer and Nelson[42] calculated that an avarage of 80% of the nutrients required in a 20-year-old stand of loblolly pine was recycled. Similarly, Baker and Blackmon[41] measured seasonal declines of over 70% for macronutrients in a first-year stand of cottonwood. The nutrient measured greatly influences the percentage recycling: on a mass basis, the nutrient recycled to the highest degree was nitrogen; on a percentage of total nutrient, phosphorus was recycled most efficiently. Calcium recycling affected a lower proportion of content than was found for other nutrients. By incorporating a biennial developmental pattern into his study (needles of *Pinus rigida* have an 18-month life cycle), Woodwell[40] could separate the seasonal and the senescence patterns of nutrient movement. In general, export of nutrients occurred only during the season of senescence; no export of nutrients was found during the autumn season of the first growing season.

Nooden et al.[43] reported that, of the nutrients measured, only N moved from the leaf to the seed during pod fill of soybeans. Whether the other nutrients were exported from the leaf once foliar senescence had been initiated was not discussed, however. Most enzymatic activities, including photosynthesis, decline rapidly once senescence has been initiated.[44] Loss of photosynthetic capacity probably results from a loss of chloroplast integrity; Eilam et al.[39] detected a significant drop in the proportion of chloroplast ribosomes as the leaf neared senescence. Both ribulose-1,5-bisphosphate carboxylase and glucose-6-phosphate dehydrogenase activities in wheat leaves peaked shortly after anthesis and fell rapidly upon initiation of senescence.[45] Additionally, increases in proteolytic activity at senescence are associated with the disappearance of the carboxylase activity in both barley[30] and wheat[34] leaves. These temporal similarities among the loss of total leaf protein, carboxylase activity, and the increase in proteolytic activity strongly suggest that a causal association exists among them. The onset of senescence, then, stimulates the synthesis of enzymes whose function is to mobilize leaf proteins, as a type of salvage process.

III. CONCLUDING REMARKS

The crop canopy is composed of a collection of leaves of various ages. Vegetative growth predominates as the canopy is formed; increases of both leaf number and leaf size contribute to canopy growth. These individual increases combine to produce a large increase in canopy photosynthesis, since leaf photosynthesis reaches a maximum about the same time as leaf area. Therefore, photosynthesis of the canopy increases throughout the establishment period.

Within the canopy, later-formed leaves are generally larger than earlier leaves. Because the photosynthetic capacity increases as the canopy is established, later-formed leaves have relatively more photosynthate to draw upon during their sink stage of development. Consequently, they can grow more than the earlier leaves could. Toward the top of the canopy, however, leaf size again becomes smaller. Three separate phenomena contribute to this pattern:

1. Leaves may be initiated faster than the canopy can adequately provide photosynthate.
2. Leaf initiation and growth subsequently slow down, requiring a relatively ageing canopy to serve as the source of photosynthate.
3. Initiation of reproductive development creates a powerful sink that competes with and has priority over vegetative development.

During the course of its life cycle, the leaf remains a site for substantial metabolic activity. The juvenile leaf is a sink, drawing upon the reserves of the plant to develop its structure. The maturing leaf becomes photosynthetically active and provides photosynthate to other growing regions in the plant. The ageing leaf remains metabolically active, although its photosynthetic activity declines. Senescence metabolism can be considered a conservation process: internal structure and large molecules are broken down to enable the translocation of their components to other, developing, areas of the plant.

Because of the complexities of leaf development — only touched upon in this section — incorporation of developmental patterns into simulation analysis becomes a demanding process. Not only must the temporal relationships among the various ontogenetic processes be defined for each crop simulation, but the differences among them for environmental sensitivity must also be established. Once these definitions have been made, their integration into a simulation model becomes possible, albeit with some degree of difficulty.

REFERENCES

1. **Esau, K.**, Apical meristems, in *Plant Anatomy*, 2nd ed., John Wiley & Sons, New York, 1965, chap. 5.
2. **Steward, F. C.**, Cell growth and cell division. Growth forms, in *Growth and Organization in Plants*, Addison-Wesley, Reading, Mass., 1968, chap. 5.
3. **Maksymowych, R.**, Plastochron index, in *Analysis of Leaf Development*, Cambridge University Press, Cambridge, Great Britain, 1973.
4. **Erikson, R. O. and Michelini, F. J.**, The plastochron index, *Amer. J. Bot.*, 44, 297, 1957.
5. **Hofstra, G., Hesketh, J. D., and Myhre, D. L.**, A plastochron model for soybean leaf and stem growth, *Can. J. Plant Sci.*, 57, 167, 1977.
6. **Rogan, P. G. and Smith, D. L.**, Rates of leaf initiation and leaf growth in *Agropyron repens* (L.) Beauv. *J. Exp. Bot.*, 26, 70, 1975.
7. **Thrower, S. L.**, Translocation of labelled assimilates in the soybean. II. The pattern of translocation in intact and defoliated plants, *Aust. J. Biol. Sci.*, 15, 629, 1962.
8. **Turgeon, R. and Webb, J. A.**, Leaf development and phloem transport in *Cucurbita pepo:* carbon economy, *Planta*, 123, 53, 1975.
9. **Turgeon, R. and Webb, J. A.**, Leaf development and phloem transport in *Cucurbita pepo:* transition from import to export, *Planta*, 113, 179, 1973.
10. **Hackett, C. and Rawson, H. M.**, An exploration of the carbon economy of the tobacco plant. II. Patterns of leaf growth and dry matter partitioning, *Aust. J. Plant Physiol.*, 1, 271, 1974.
11. **Larson, P. R. and Dickson, R. E.**, Distribution of imported ^{14}C in developing leaves of eastern cottonwood according to phyllotaxy, *Planta*, 111, 95, 1973.
12. **Larson, P. R., Isebrands, J. G., and Dickson, R. E.**, Fixation patterns of ^{14}C within developing leaves of eastern cottonwood, *Planta*, 107, 301, 1972.
13. **Fellows, R. J., and Geiger, D. R.**, Structural and physiological changes in sugar beet leaves during sink to source conversion, *Plant Physiol.*, 54, 877, 1974.
14. **Isebrands, J. G., Dickson, R. E., and Larson, P. R.**, Translocation and incorporation of ^{14}C into the petiole from different regions within developing cottonwood leaves, *Planta*, 128, 185, 1976.
15. **Smillie, R. M.**, Photosynthetic and respiratory activities of growing pea leaves, *Plant Physiol.*, 37, 716, 1962.
16. **Osman, A. M. and Milthorpe, F. L.**, Photosynthesis of wheat leaves in relation to age, illuminance and nutrient supply. II. Results, *Photosynthetica*, 5, 61, 1971.
17. **Woodward, R. G.**, Photosynthesis and expansion of leaves of soybean grown in two environments, *Photosynthetica*, 10, 274, 1976.
18. **Wilson, D.**, Variation in leaf respiration in relation to growth and photosynthesis of *Lolium*, *Ann. Appl. Biol.*, 80, 323, 1975.
19. **Catsky, J., Ticha, I., and Solarova, J.**, Ontogenetic changes in the internal limitations to bean-leaf photosynthesis, *Photosynthetica*, 10, 394, 1976.
20. **Elmore, C. D., Hesketh, J. D., and Muramoto, H.**, A survey of rates of leaf growth, leaf aging and leaf photosynthetic rates among and within species, *J. Ariz. Acad. Sci.*, 4, 215, 1967.
21. **Bird, I. F., Cornelius, M. J., and Keys, A. J.**, Effects of temperature on photosynthesis by maize and wheat, *J. Exp. Bot.*, 28, 519, 1977.

22. **Keys, A. J., Sampio, E. V. S. B., Cornelius, M. J., and Bird, I. F.**, Effect of temperature on photosynthesis and photorespiration of wheat leaves, *J. Exp. Bot.*, 28, 525, 1977.
23. **Botrill, J. E., Possingham, J. V., and Kriedmann, P. E.**, The effect of nutrient deficiencies on photosynthesis and respiration in spinach, *Plant Soil*, 32, 424, 1970.
24. **Thompson, A. C., Lane, H. C., Jones, J. W., and Hesketh, J. D.**, Nitrogen concentrations of cotton leaves, buds and bolls in relation to age and nitrogen fertilization, *Agron. J.*, 68, 617, 1976.
25. **Ohki, K.**, Effect of zinc nutrition on photosynthesis and carbonic anhydrase activity in cotton, *Physiol. Plant*, 38, 300, 1976.
26. **Simon, E. W.**, Types of leaf senescence, *Soc. Exp. Biol. Symp.*, 21, 215, 1967.
27. **Lindoo, S. J. and Nooden, L. D.**, The interrelation of fruit development and senescence in Anoka soybeans, *Bot. Gaz.*, 137, 218, 1976.
28. **Lindoo, S. J. and Nooden, L. D.**, Studies on the senescence signal in Anoka soybeans, *Plant Physiol.*, 59, 1136, 1977.
29. **Wilson, R. F., Burton, J. W., Buck, J. A., and Brim, C. A.**, Studies on genetic male-sterile soybeans. I. Distribution of plant carbohydrate and nitrogen during development, *Plant Physiol.*, 61, 838, 1978.
30. **Peterson, L. W. and Huffaker, R. C.**, Loss of ribulose-1,5-diphosphate carboxylase and increase in proteolytic activity during senescence of detached primary barley leaves, *Plant Physiol.*, 55, 1009, 1975.
31. **Feller, U. K., Soong, T.-S. T., and Hageman, R. H.**, Leaf proteolytic activities and senescence during grain development of field-grown corn (*Zea mays* L.), *Plant Physiol.*, 59, 290, 1977.
32. **Storey, R. and Beevers, L.**, Proteolytic activity in relationship to senescence and cotyledonary development in *Pisum sativum* L., *Planta*, 137, 37, 1977.
33. **Dalling, M. J., Boland, G., and Wilson, J. H.**, Relation between acid proteinase activity and redistribution of nitrogen during grain development in wheat, *Aust. J. Plant Physiol.*, 3, 721, 1976.
34. **Peoples, M. F. and Dalling, M. J.**, Degradation of ribulose-1,5-bisphosphate carboxylase by proteolytic enzymes from crude extracts of wheat leaves, *Plant Physiol.*, 59, (suppl.), 44, 1977.
35. **Martin, C. and Thimann, K. V.**, The role of protein synthesis in the senescence of leaves. I. The formation of protease, *Plant Physiol.*, 49, 64, 1972.
36. **McHale, J. S. and Dove, L. D.**, Ribonuclease activity in tomato leaves as related to development and senescence, *New Phytol.*, 67, 505, 1968.
37. **Srivastava, B. I. S.**, Increase in chromatin associated nuclease activity of excised barley leaves during senescence and its suppression by kinetin, *Biochem. Biophys. Res. Commun.*, 32, 533, 1968.
38. **Srivastava, B. I. S. and Arglebe, C.**, Studies on ribosomes from barley leaves. Changes during senescence, *Plant Physiol.*, 42, 1497, 1967.
39. **Eilam, Y., Butler, R. D., and Simon, E. W.**, Ribosomes and polysomes in cucumber leaves during growth and senescence, *Plant Physiol.*, 47, 317, 1971.
40. **Woodwell, G. M.**, Variation in the nutrient content of leaves of *Quercus alba*, *Quercus coccinea*, and *Pinus rigida* in the Brookhaven forest from bud-break to abscission, *Am. J. Bot.*, 61, 749, 1974.
41. **Baker, J. B. and Blackmon, B. G.**, Biomass and nutrient accumulation in a cottonwood plantation — the first growing season, *Soil Sci. Soc. Am. J.*, 41, 632, 1977.
42. **Switzer, G. L. and Nelson, L. E.**, Nutrient accumulation and cycling in loblolly pine (*Pinus taeda* L.) plantation ecosystems: the first twenty years, *Soil Sci. Soc. Am. Proc.*, 36, 143, 1972.
43. **Nooden, L. D., Rupp, D. C., Derman, B. D., and Kahanak, G. M.**, Nutrient drain and diversion in relation to foliar senescence in soybeans, *Plant Physiol.*, 59 (suppl), 112, 1977.
44. **Woolhouse, H. W.**, Senescence process in the life cycle of flowering plants, *BioScience*, 28, 25, 1978.
45. **Moss, D. N. and Patterson, T. G.**, Changes in several key enzymes during senescence of field-grown wheat, *Agron. Abstr.*, 69, 90, 1977.

Chapter 8

CONCEPTS OF TRANSLOCATION WITH SPECIAL REFERENCE TO THE ASSIMILATION OF NITROGEN AND ITS MOVEMENT INTO FRUITS

John W. Radin and C. D. Elmore

TABLE OF CONTENTS

I. INTRODUCTION

In recent years there has been a pronounced trend toward dividing the plant physiological world into "whole-plant" and "cellular" subsections. Obviously, simulation modeling, as a sophisticated method for integrating the activities of each plant part, represents the ultimate in whole-plant research. The basic premise of modeling, which scarcely needs justification, is that the activities of the various plant parts are indeed coordinated, and that their relationships can be represented analytically. Such coordination is provided by the movement of hormones, signals, modifiers, substrates, inhibitors, etc., from an area of control to an area of dependence — in short, translocation. In this chapter we look first at some factors which limit or regulate translocation. In the last section we draw upon current knowledge of translocation to set up an integrated energy budget for movement of nitrogen into fruits.

Water can be induced to move in only one way, i.e., by a gradient of its chemical potential, ψ. Within this stricture, land plants have evolved two general systems of translocation. The first system carries water from soil to air via the xylem, and is dependent mostly upon radiant energy for evaporation of water in the leaves (or other evaporative surface). With respect to the plant, this system is purely passive; its functioning is physical and needs no biological intermediaries. The driving force results from the difference in energy budgets of leaf water and soil water.

The second translocation system has not been so definitively characterized. Functionally, it moves sugars and other foodstuffs from areas of supply (sources) to areas of demands (sinks) via the phloem. The most popular theory of phloem translocation requires the intervention of both metabolism and biological membranes; thus, this process is very much a biological entity. The typical complexity of biologic, as opposed to physical, systems no doubt explains the poorer understanding of phloem physiology.

II. GENERAL CONCEPTS OF TRANSLOCATION

A. The Xylem

It is obvious that plants need water and mineral nutrients, both of which are normally supplied from the soil. A specialized conducting tissue, the xylem, carries these substances in the transpiration stream from the roots to transpiring aerial organs. To a first approximation, then, one might expect a direct relationship between evaporative demand (which drives transpiration) and nutrient status of leaves. Such is not the case. It is necessary to distinguish between mineral absorption by roots and mineral translocation in the mass flow of water. Some ions (especially cations) are relatively immobile in the soil and do not follow the mass movement of water.[1-3] Rather,absorption occurs because of diffusion to the root surface, or because of interception by the growing root—exploration of previously unused soil.[4] The key concept is that nutritional sufficiency depends in large part upon physiological processes in the root — growth and absorptive capacity — rather than upon simple transpiration. Indeed, Brouwer's[5] experiments showed that transpiration rate was unchanged by root pruning. However, growth rate was curtailed, indicating that nutrient uptake or some other root function was dependent upon root size. The complexity of this subject makes it extremely difficult to predict nutrient status from simple measurements of root growth and composition of the soil solution.

There are a few documented cases in which translocation itself can limit the accumulation or redistribution of a mineral. For reasons of solubility, Ca can be delivered only in the xylem and not in the phloem.[6] Thus, nontranspiring organs are frequently marginally sufficient or even deficient in this element.[6] In an elegant study, Harris[7]

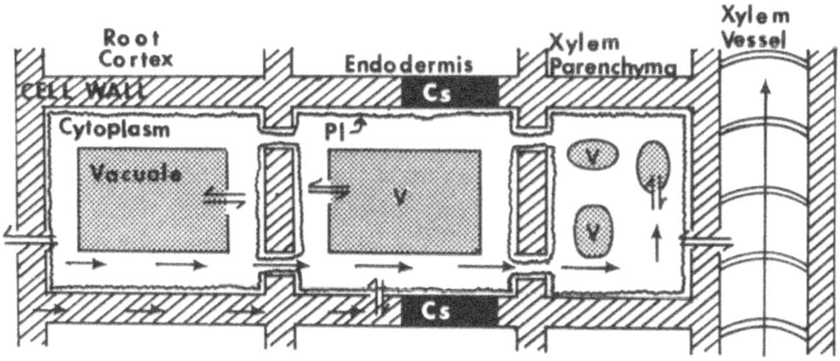

FIGURE 1. Pathways of movement of water and ions in roots. The plasmalemma (Pl) is the boundary between apoplast and symplast. The Casparian strip (Cs) is the barrier to diffusion of water and ions in the apoplast. Adapted from Lauchli, A., *Transport and Transfer Processes in Plants,* Academic Press, New York, 1976, 101.

showed that peanut gynophores (which grow in the soil) receive no Ca from the parent plant, but must absorb it independently. Because of the extreme immobility of Ca once deposited, organs which transpire heavily frequently accumulate it to excess, with no redistribution possible.[8] In a general way, then, Ca accumulation can be a measure of transpiration, and it has been so used.[9]

We turn now to the root as a site of regulation of mineral metabolism. Figure 1 is a diagram of the pathway followed by water and ions in their radial movement to the xylem. The Casparian strip is a sheath of suberin deposited in the cell walls of the endodermis and acts as an effective barrier to water or ion movement.[10] For the most part, water and solutes are thus forced to flow through living cells (but see Perry and Greenway[11]). The plasmalemma is relatively permeable to water, and little resistance to its movement is encountered. However, ions move across the cell membrane with the help of carriers. Considerable indirect evidence, based upon the kinetics and energetics of ion transport, supports the concept of carriers.[12] Although no carrier has ever been purified and characterized, a large body of literature now exists on ionstimulated ATPases in membrane fractions.[12] The assumption is that these ATPases are identical to carriers.[13]

It should be noted that roots frequently display rather large electrochemical gradients.[14] These potential differences can supply the motive force for ion uptake against a concentration gradient,[14] especially at high external ion concentrations. In such cases, transport is exergonic in spite of the concentration gradient, and is more properly termed facilitated diffusion.[15] The interpretation of uptake mechanisms with certain ions and/or at high ion concentrations remains open to question.[12] Nonetheless, the case for carriers is substantial and compelling.

Once inside a cortical root cell, ions are in a symplastic network which is continuous, through the interconnecting plasmodesmata, with the parenchyma cells of the xylem.[16] There is now good evidence that the xylem parenchyma cells actively secrete ions into the extracellular transpiration stream[17] (contrary to the classical Crafts-Broyer[18] theory of passive leakage into the xylem). In the radial transport of ions to the xylem, this is the second step which requires active transport across a membrane.

The active steps are logical sites for regulation of ion movement. The existence of such regulation has been recognized for many years. Williams[19] developed the idea that most of a plant's N supply is absorbed during early growth. The same pattern is evident with P.[19] In most, but not all, cases this decreased uptake during later stages of growth could be reversed by defruiting, suggesting that sugar availability may limit nutrient uptake.[20]

A further type of regulation has been recognized, i.e., regulation by the nutrients themselves. It has long been known that absorption proceeds more rapidly in "low-salt" than in "high-salt" roots.[21] In more recent times, some more specific aspects of regulation have been brought out. Glass[22] has shown that K absorption is inhibited by prior K nutrition in a sort of feedback loop. Similarly, NO_3^- absorption[23,24] and SO_4^{--} absorption[25,26] can be inhibited in many systems by products of their assimilation. In contrast to the negative effects exerted by nutritional sufficiency, there is a hint that P deficiency can act to promote uptake by causing increased activity of extracellular phosphatases.[27] This type of "scavenging" phenomenon has not been given the attention it deserves in higher plants.

B. The Phloem

We will not recite here a litany of the proposed mechanisms of transport in the phloem, discussions of which are available elsewhere.[28] The most popular hypothesis is the pressure-flow theory of Munch, which can be found in any general textbook of plant physiology. For the purposes of our whole-plant physiology, the mechanics of this theory will suffice, and only certain aspects of phloem translocation need be considered: loading and unloading, direction of transport, and mass flux.

The phloem is composed of living cells and is by definition part of the symplast. However, there are very few plasmodesmatal connections between the sieve element-companion cell complex and adjacent parenchyma cells.[29-31] In fact, in many species transfer cells (cells existing at the site of apoplast-symplast transfers) are located adjacent to the phloem, presumably to facilitate phloem loading and unloading.[32] It is generally agreed that both loading and unloading are active processes, requiring energy for transport.[33] Fondy and Geiger[34] consider loading to be an apoplast-symplast transfer. However, this does not mean that the only respiration associated with translocation occurs at the source or sink. Chilling a small portion of a petiole or depriving it of O_2 can effectively block movement of assimilates out of a leaf.[35,36] However, this phenomenon is probably related to a metabolic requirement to maintain phloem structure, rather than a metabolic requirement for translocation per se.[36] No firm estimates have yet appeared for the total expenditure of energy in phloem translocation.

The control of mass flux and distribution in phloem transport is not well defined. To introduce this subject, we define "source strength" and "sink strength" following Warren Wilson:[37]

1. Source strength = source size × source activity
2. Sink strength = sink size × sink activity

These definitions are obviously purely empirical, and reveal no clues as to what regulates activity. Yet they represent most of what is known about control of phloem transport. Because the definitions are purely functional, they can be applied equally well to any translocation system whatsoever. When applied to phloem, the source is usually assumed to be a photosynthesizing leaf, and the sink is any tissue which consumes assimilate on balance. When a plant is sink-limited[38] (for example, after defruiting), excess assimilate is available, a condition which can lead to depression of photosynthesis. This subject is discussed in Chapter 1 of this book. As the fruits develop, summed sink strengths increase rapidly, leading to a "source-limiting" condition.[38] Under these competitive conditions, the actual strength of a sink may not be proportional to its potential maximum strength.[39] For example, if a plant contains a source producing 10 units of photosynthate, with fruits capable of metabolizing 9 units and vegetative parts capable of metabolizing 6 units, the actual division may not be 6:4 but rather 9:1. This asymmetric division is rather strictly followed if the source is, for example, a flag leaf of wheat.[40] However, other leaves within the same plant distribute their assimilates quite differently.[40] Thus, the ability of a sink to attract assimilates under competitive pressure depends not only on potential maximum sink strength, but also upon other factors, especially proximity.[40] Vascular geometry of the plant may also play a role.[38] The others are still undefined. It is clear that the strictly operational definitions of source and sink strength given above do not solve the problems of priority, which is the crux of the matter from a whole-plant perspective.

Many nominally indeterminate crops, such as soybeans and cotton, have cultivars of differing determinacy in their growth habits. Ignoring any photoperiodic effects, this variation is most likely a function of genetic differences in sink priorities, such that the more determinate lines partition less assimilate to the vegetative growth centers. This represents almost the ideal experimental situation to investigate the control of priority, yet no substantial proposals have been forthcoming. Establishment of a high priority in a particular sink could result from successful competitive ability for any of the several kinds of solutes present in phloem sap and necessary for further development, i.e., carbohydrates, nitrogenous compounds, hormones, or mineral nutrients.[38] Without further understanding of the mechanisms of this phenomenon, modelers usually make a set of explicit assumptions about sink priorities (e.g., Fick et al.[41]), rather than to derive it from a known relationship to development.

III. NITROGEN ASSIMILATION AND TRANSLOCATION FROM AWHOLE-PLANT VIEWPOINT

It is apparent from the discussion preceding that we cannot yet simulate a crop's nutrient status solely from climatic and edaphic data. Nonetheless, given some assumptions about the amount of a nutrient removed from the soil, one can deduce a great deal about the effect of that nutrient in crop development. In this spirit we now consider the assimilation of N in the plant and its delivery to the fruit, and the energy costs of these processes. It is necessary to distinguish between plants of natural, undisturbed ecosystems (especially climax communities) and agricultural plants. The former are usually severely limited in their growth by N availability; most of the N is in organic form and is recycled rather slowly.[42] In a cultivated or otherwise disturbed soil, ammonification and nitrification increase the availability of N (as NO_3^-), with a consequent increase in productivity. The reasons for this increase in nitrification upon clearing and cultivating are in dispute, and can involve soil aeration,[42] soil pH,[43] competition between nitrifiers and rhizosphere microflora,[44] and allelopathic effects of the plants on the soil microflora.[45] Regardless of mechanism, in most cultivated soils N is predominantly available as NO_3^-, and most agricultural plants which depend upon soil N can take up NO_3^- and transport the bulk of it to the leaves for assimilation. The difference between the two types of production is well illustrated by Stewart et al.[46] They showed that in plants from a grassland habitat, nitrate reductase activity was well below what could be induced by additions of NO_3^-. In contrast, in plants from ruderal sites (characterized by soil disturbance, accumulation of rubbish, etc.) no further nitrate reductase activity could be induced by NO_3^-, meaning that NO_3^- was readily available and was being used.

In many cultivated plants, then, the mechanics of NO_3^- assimilation are of great importance to an understanding of development. Fortunately, much is known about the process from a biochemical point of view. To begin with, it is quite expensive in terms of both energy dissipated and carbon sequestered. The reduction of NO_3^- to the level of NH_3 requires eight electrons, or enough reductant to reduce 2 mol of CO_2 to carbohydrate (CH_2O). The first part of this process, reduction of NO_3^- to NO_2^- catalyzed by the enzyme nitrate reductase, is financed by the oxidation of CH_2O (e.g., glyceraldehyde-3-phosphate) to acid (e.g., 3-phosphoglyceric acid).[47] Because the balanced reaction for this step involves the consumption of protons (which are released as H_2O), intracellular pH maintenance considerations require than an acid accumulate, functioning as a proton donor.[48] There is direct evidence[49] that malic acid is synthesized coordinately with the reduction of NO_3^-.

A description of the remainder of the process, reduction of NO_2^- to NH_3, depends upon the site within the plant. In leaves, NO_2^- can serve directly as an acceptor for photosynthetically generated reductant,[50] and thus the cost of its reduction is not straightforward. In the traditional analysis (see Penning de Vries[51]) it is implied that CO_2 concentration is limiting to photosynthesis, and that NO_2^- reduction thus uses light energy that would otherwise be wasted. To a first approximation, if the leaf area index (LAI) is low — perhaps two or less — then every leaf can be considered to be receiving excess light (CO_2 limiting to photosynthesis). However, as a crop grows and the LAI increases, the leaves at the bottom of the canopy will no longer be light-saturated. Under these conditions, the photoreduction of NO_2^- to NH_3 uses light energy which would otherwise be productively employed in assimilating CO_2. The exact timing of the shift from "cheap" to "expensive" manufacture of reduced N obviously will depend upon crop growth rate, light extinction coefficient of the canopy, irradiance, the extent of leaf adaptation to low irradiance as the crop grows, and many other factors. Little good data are available to establish the location of NO_3^- reduction within the canopy. In a thorough study, Harper and Hageman[52,53] measured nitrate reductase activities of soybean leaves at several points during the growing season and found that the most active leaves were somewhat removed from the top of the canopy, especially during mid to late pod fill stages of growth. The question of the location of NO_3^- assimilation is obviously of great practical importance.

Penning de Vries[51] described the production of reduced N by two separate equations:

$$1.785 \text{ g glucose } (59.5 \text{ mmol } CH_2O) + 0.535 \text{ g } NO_3^- (8.6 \text{ mmol}) \rightarrow 1.00 \text{ g amino acid} + 0.567 \text{ g organic acid} + 0.447 \text{ g } CO_2 \ (10.2 \text{ mmol}) \quad (1)$$

$$49.3 \text{ mmol } CO_2 + 119 \text{ mmol NADPH} + 181 \text{ mmole ATP} + 8.6 \text{ mmol } NO_3^- \rightarrow 1.00 \text{ g amino acid} + 0.567 \text{ g organic acid} \quad (2)$$

Equation 1 describes NO_3^- assimilation at the expense of CH_2O, as would happen in darkness. The molar ratio of CO_2 produced to N assimilated is less than the long-established 2:1,[50] but there is also considerable C oxidation inherent in the production of organic acids. Equation 2 describes the process when a part of it, the reduction of NO_2^-, depends upon the direct utilization of light. For comparison, Equation 3 describes the production of glucose alone during (C_3) photosynthesis;

$$6\ CO_2 + 12 \text{ NADPH} + 18 \text{ ATP} \rightarrow 1 \text{ glucose} + 12 \text{ NADP} + 18 \text{ ADP} \quad (3)$$

Both equations 1 and 2 show a considerable transformation of C into the form of organic acids. These substances can accumulate to very high levels in the leaf, as in sugarbeet[54] or cotton,[55] or they can perhaps be transported to the roots in the phloem, there to be decarboxylated and the HCO_3^- exchanged for incoming NO_3^-.[48,56] In addition, the equations show a diversion of C into amino acids. Assuming that NO_3^- assimilation occurs at the rate of 2.2 μmol/g fr wt/hr and that leaf specific fresh wt is 30 mg/cm², then Equation 2 leads to the estimate that NO_3^- assimilation in the leaf requires a photosynthetic rate of 1.7 mg CO_2/dm²/hr. If NO_3^- assimilation occurs at this rate in limiting light, then an additional 0.3 mg CO_2/dm²/hr will not be fixed because of the competition for products of the light reactions. In this case, the CH_2O cost to the plant, 2.0 mg CO_2/dm²/hr, is the same as in Equation 1.

Because most (C_3) plants photosynthesize in the range of 10 to 40 mg CO_2 dm²/hr, this drain represents 5 to 20% of photosynthesis. To state the comparison on another basis, photosynthesizing leaves probably assimilate about 1% as many N atoms as C atoms, but the effect of the N is amplified several fold. Combining Equations 2 and 3 in the proper ratio shows that for every 1 mol of N assimilated, leaves can produce about 6 mol of C in amino acids and organic acids, and about 94 mol of C in CH_2O. Indeed, this is approximately the ratio of substances exported from the leaves in the phloem.[58]

Where, other than leaves, can NO_3^- be assimilated? In most plants, some fraction of the plant's NO_3^- is assimilated in the roots. This fraction can range from very small to very large.[59] Analysis of costs, to a first approximation, is straightforward following Equation 1. In cotton plants, Radin[60] estimated the NO_3^- assimilation rate of roots at 0.4 to 0.5 μmoles/g fr wt/hr; this was one third to one half the rate in shoots. However, the roots apparently consumed most of their own reduced N, exporting little, and they were not considered a major source of reduced N for the other parts of of the plant. In other plants, e.g., pea,[59] the roots provided most of the plant's capacity for NO_3^- reduction, with the products delivered to the leaves in the xylem sap. In such plants, the solutes must cross over from xylem to phloem before they can be delivered to the fruits. These processes are analyzed further later.

The third possible source of assimilated N for the fruits is the fruits themselves. Section II of this chapter discusses the two types of long-distance transport in plants. Because NO_3^- is carried in the xylem and is seldom if ever found in pure phloem sap,[6] it can enter the fruit only to the extent that the fruit accepts xylem sap. Fruits do transpire, albeit at greatly reduced rates compared to leaves, and a balance sheet for water showed that not enough water was delivered in the phloem to meet consumption.[61] The deduction is clear that some xylem water is delivered to fruits, bringing with it some NO_3^-. Radin and Sell[62] studied NO_3^- reduction in developing cotton bolls (fruits), and estimated that no more than 5% of the boll's N requirement was met in this way. An indirect estimate more or less corroborates this figure. Leffler and Tubertini[63] found that each boll accumulated about 5.5 mg of Ca (which, like NO_3^-, is carried exclusively in the xylem[6]) and 140 mg of N. Taking the Ca and NO_3 concentrations in the xylem sap at 3 mM and 5 to 15 mM, respectively;[14,64] one can estimate the input of NO_3^-–N at about 3 to 10 mg, or 2 to 7% of the total. By Equation 1, this costs the fruit a maximum of 0.15 g CH_2O, or 5 mmol. Of this total, about 1 mmol is respired away as CO_2, and the rest is converted to amino acids and organic acids. Although malic acid accumulates to some extent in developing cotton bolls,[65,66] probably the greatest part of the organic acid fraction is eventually converted, via transamination reactions, to amino acids. Thus, during NO_3^- assimilation in the fruit, only the C which is actually respired to CO_2 is irretrievably lost (but see below). The rest of the C is simply diverted from CH_2O to other fractions, chiefly nitrogenous.

An interesting point is that the agronomic importance of this process depends upon the nature of the desired product. For instance, in cotton bolls, most of this diversion of C from CH_2O occurs during the phase of rapid fiber growth. Because the cellulosic fiber would suffer most at the diversion of CH_2O, and because the amount diverted is

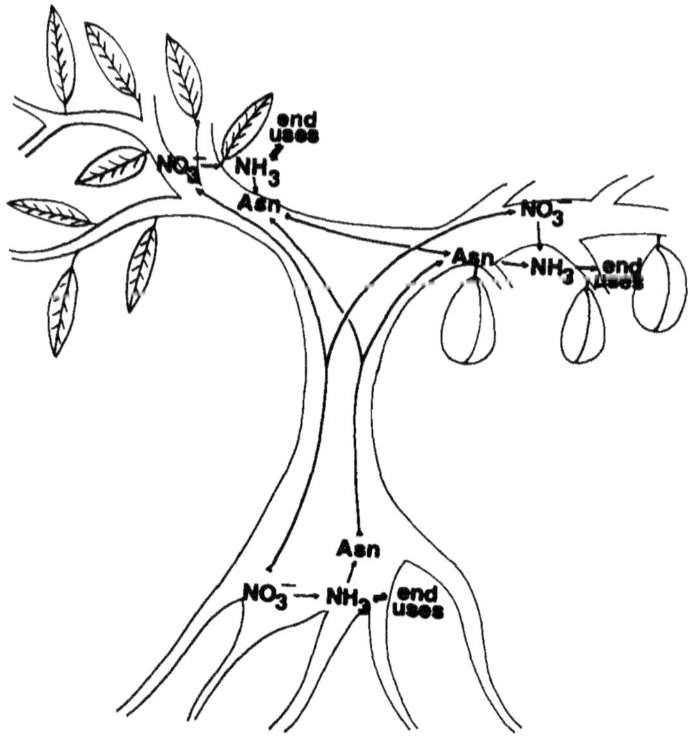

FIGURE 2. A simplified scheme for nitrogen transformations in roots, leaves, and fruits of plants. All transportable organic nitrogen is represented by asparagine (Asn).

5 to 10% of the amount eventually deposited in the fiber, NO_3^- assimilation in bolls can very well lead to important effects on boll composition.[67]

What happens to CO_2 evolved in the fruit? Because most fruits are green to some extent, the possibility exists that respired CO_2 can be recovered by photosynthesis in the carpel wall. In lupine[68] and in soybean,[69] the fruit wall was apparently very active, and it made a considerable contribution in this way. In cotton, though, the fruit wall was photosynthetically inactive, and was clearly a sink for imported photosynthate.[70] Thus, to continue our description of the cotton boll, the answer to the question posed above is "very little". However, this answer may be different for each species.

We turn now to consideration of the transport processes by which N is delivered to fruits. The pathways followed by the various nitrogenous fractions are clearly delineated. Roots take up NO_3^- and export it, along with some reduced N, in the transpiration stream. As indicated above, a small fraction moves directly to the fruits, but most is deposited in the leaves. The NO_3^- is reduced there, and the resulting amino acids, along with amino acids from the root system, are loaded into the phloem for delivery to the fruits (or other translocation sinks, such as very young leaves). We are examining here only the proximate sources of N for the fruits; thus delivery to alternate sinks and circulation between shoot and root are ignored. The error introduced by this simplification obviously depends upon the relative sink strengths of the fruits, roots, and other sites of net consumption. A schematic diagram of this simplified system is shown in Figure 2. The reduced N carriers are represented by asparagine (Asn), which in cotton[71,72] and many other plants[59] is by far the dominant amide in the translocation system. To analyze this system, we make the explicit assumption that 1 ATP molecule

TABLE 1

Reactions Which Lead to the Production of Asparagine From Aspartic Acid.

Partial reaction[a]	Enzyme	ATP produced (consumed)
α-KG + Asp → Glu + OAA	Aminotransferase	0
Glu → α-KG + NH_3	Glutamic dehydrogenase	3[b]
Glu + NH_3 → Gln	Glutamine synthetase	(1)
Asp + Gln → Asn + Glu	Asparagine synthetase	(1)
Net: 2 Asp → Asn + OAA		1

[a] Abbreviations: Asn, asparagine; Asp, aspartic acid; Gln, glutamine; Glu, glutamic acid; OAA, oxalacetic acid; α-KG, α-ketoglutaric acid.

[b] From the production of NAD(P)H.

is expended each time that one molecule of a substance is transported across a cell membrane.

In the case of reduction in the roots, NO_3^- is absorbed (1 ATP) and reduced to amino acids as discussed earlier. For simplicity, we will assume that the only amino acid is aspartic acid (Asp). By a reaction sequence involving a glutamate-asparate aminotransferase, glutamate dehydrogenase, glutamine synthetase, and asparagine synthetase, we obtain the net conversion of 2 mol Asp to 1 mol Asn plus 1 mol oxalacetic acid (OAA) (Table 1). This sequence produces 1 ATP (ignoring any intracellular cross-membrane transport, and recognizing that the energy is derived from the respiration in Equation 1.) The Asn is then secreted into the xylem (1 ATP) for transport, taken up by the leaf cells (1 ATP), removed (1 ATP), and loaded into the phloem (1 ATP). Once Asn reaches the fruit, it is unloaded (1 ATP), and Asn is hydrolyzed to Asp and free NH_3 by asparaginase[73] (0 ATP). The liberated NH_3 must be reassimilated, probably by the glutamine synthetase/glutamate synthase pathway,[74] at a cost of 4 ATP (1 ATP plus 1 NAD(P)H). The total cost of delivery, per mol NO_3^-, or per ½ mol Asn, is as follows: in the roots, 1 ATP; in the leaves, 1½ ATP; in the fruits, 2½ ATP.

When NO_3^- is reduced in the leaves, a similar kind of analysis reveals the following costs: in the roots, 2 ATP; in the leaves, 1½ ATP; in the fruits, 2½ ATP.

The simplest case of all is the direct delivery of N in the xylem to the fruit. If the N is in NO_3^-, the breakdown is 2 ATP in the roots, 0 ATP in the leaves, and 1 ATP in the fruits. This lowered cost in the fruits is occasioned by the fact that Asp can be the direct product of Equation 1 and need not be produced from Asn. If the N is delivered as Asn, we find costs of 1 ATP in the roots, 0 ATP in the leaves, and 2½ ATP in the fruits. It must be emphasized that these are estimates only of delivery costs, and are additional to the costs outlined in Equations 1 and 2, the only differences being in location. Overall the expenditure of about 6 ATP per NO_3^- for delivery represents about 14% of the rate of CH_2O use implicit in Equation 1. A similar kind of analysis could be performed for plants which predominantly transport glutamine, arginine, or any other nitrogenous compound, substituting only the costs of producing those compounds instead of Asn. It seems unlikely that the general conclusion would be greatly different.

We close this chapter by pointing out that to use this information in a simulation model of plant growth, one needs to assess the rates at which the various processes occur and to assign the energy costs accordingly. Unfortunately, this is by no means a simple task. Undoubtedly the most elegant study in this area is that of Pate et al.,[68] who developed a complete balance sheet for C, N, and water moving into developing lupine fruit. Furthermore, because they divided the fruit's 12-week development period into 2-week intervals, they were able to demonstrate the continuously changing nature of its inputs. The complexity of the different processes which contribute directly or indirectly to N accumulation in fruits make it unlikely that such a balance sheet could even be approximated by theoretical analyses. More such work is needed.

REFERENCES

1. **Barber, S. A.,** A diffusion and mass-flow concept of soil nutrient availability, *Soil Sci.*, 93, 39, 1962.
2. **Fried, M. and Shapiro, R. E.,** Soil-plant relationships in ion uptake, *Annu. Rev. Plant Physiol.*, 12, 91, 1961.
3. **Oliver, S. and Barber, S. A.,** An evaluation of the mechanisms governing the supply of Ca, Mg, K, and NA to soybean roots (*Glycine max*), *Soil Sci. Soc. Am. Proc.*, 30, 82, 1966.
4. **Barber, S. A.,** Influence of the plant root on ion movement in soil, in *The Plant Root and its Environment*, Carson, E. W., Ed., University Press of Virginia, Charlottesville, 1974, 525.
5. **Brouwer, R.,** Root functioning, in *Environmental Effects on Crop Physiology*, Landsberg, J. J. and Cutting, C. V., Eds., Academic Press, London, 1977, 229.
6. **Zeigler, H.,** Nature of transported substances, in *Encyclopedia of Plant Physiology*, Vol. 1, Zimmermann, M. H. and Milburn, J. A., Eds., Springer-Verlag, Berlin, 1975, 57.
7. **Harris, H. C.,** The effect on the growth of peanuts of nutrient deficiencies in the root and the pegging zone, *Plant Physiol.*, 24, 150, 1949.
8. **Loneragan, J. F., Snowball, K., and Robson, A. D.,** Remobilization of nutrients and its significance in plant nutrition, in *Transport and Transfer Processes in Plants*, Wardlaw, I. F. and Passioura, J. B., Eds., Academic Press, New York, 1976, 463.
9. **Menary, R. C. and Jones, R. H.,** Nitrate accumulation and reduction in papaw fruits, *Aust. J. Biol. Sci.*, 24, 531, 1972.
10. **Clarkson, D. T. and Robards, A. W.,** The endodermis, its structural development and physiological role, in *The Development and Function of Roots*, Torrey, J. G. and Clarkson, D. T., Eds., Academic Press, London, 1975, 415.
11. **Perry, M. W. and Greenway, H.,** Permeation of uncharged molecules and water through tomato roots, *Ann. Bot., (London)*, 37, 225, 1973.
12. **Pitman, M. G.,** Nutrient uptake by roots and transport to the xylem: uptake processes, in *Transport and Transfer Processes in Plants*, Wardlaw, I. F. and Passioura, J. B., Eds., Academic Press, New York, 1976, 85.
13. **Nissen, P.,** Ion uptake in higher plants and KC1 stimulation of plasmalemma adenosine triphosphatase: comparison of models, *Physiol. Plant.*, 40, 205, 1977.
14. **Anderson, W. P.,** Ion transport through roots, in *The Development and Function of Roots*, Torrey, J. G. and Clarkson, D. T., Eds., Academic Press, London, 1975, 437.
15. **Nobel, P. S.,** *Introduction to Biophysical Plant Physiology*, W. H. Freeman, San Francisco, 1974, 136.
16. **Lauchli, A.,** Symplasmic transport and ion release to the xylem, in *Transport and Transfer Processes in Plants*, Wardlaw, I. F. and Passioura, J. B., Eds., Academic Press, New York, 1976, 101.
17. **Pitman, M. G.,** Ion transport in the xylem, *Annu. Rev. Plant Physiol.*, 28, 71, 1977.
18. **Crafts, A. S. and Broyer, T. C.,** Migration of salts and water into xylem of the roots of higher plants, *Am. J. Bot.*, 24, 415, 1938.
19. **Williams, R. F.,** Redistribution of mineral elements during development, *Annu. Rev. Plant Physiol.*, 6, 25, 1955.
20. **Pitman, M. G. and Cram, W. J.,** Regulation of ion content in whole plants, in *Integration of Activity in the Higher Plant*, Jennings, D. H., Ed., Cambridge University Press, Cambridge, 1977, 391.
21. **Broyer, T. C. and Hoagland, D. R.,** Metabolic activities of roots and their bearing on the relation of upward movement of salts and water in plants, *Am. J. Bot.*, 30, 261, 1943.
22. **Glass, A. D. M.,** Regulation of potassium absorption in barley roots. An allosteric model, *Plant Physiol.*, 58, 33, 1976.
23. **Heimer, Y. M. and Filner, P.,** Regulation of the nitrate assimilation pathway in cultured tobacco cells. III. The nitrate uptake system, *Biochim. Biophys. Acta*, 230, 362, 171.
24. **Rao, K. P. and Rains, D. W.,** Nitrate absorption by barley. I. Kinetics and energetics, *Plant Physiol.*, 57, 55, 1976.

25. **Hart, J. W. and Filner, P.**, Regulation of sulfate uptake by amino acids in cultured tobacco cells, *Plant Physiol.*, 44, 1253, 1969.
26. **Maggioni, A. and Renosto, F.**, Cysteine and methionine regulation of sulfate uptake in potato tuber discs *(Solanum tuberosum)*, *Physiol. Plant.*, 39, 143, 1977.
27. **Bieleski, R. L.**, Phosphate pools, phosphate transport and phosphate availability, *Annu. Rev. Plant Physiol.*, 24, 225, 1973.
28. **Zimmermann, M. H. and Milburn, J. A., Eds.**, *Encyclopedia of Plant Physiology*, Vol. 1, Springer-Verlag, Berlin, 1975.
29. **Shih, C. H. and Currier, H. B.**, Fine structure of phloem cells in relation to translocation in cotton seedling, *Am. J. Bot.*, 56, 464, 1968.
30. **Fisher, D. B., Housley, T. L., and Christy, A. L.**, Source pool kinetics for ^{14}C-photosynthate translocation in morning glory and soybean, *Plant Physiol.*, 61, 291, 1978.
31. **Esau, K.**, *Encyclopedia of Plant Anatomy*, Vol. 2, The Phloem, Gebruder Borntraeger, Berlin, 1969, 119.
32. **Pate, J. S. and Gunning, B. E. S.**, Transfer cells, *Annu. Rev. Plant Physiol.*, 23, 173, 1972.
33. **Evert, R. F.**, Phloem structure and histochemistry, *Annu. Rev. Plant Physiol.*, 28, 199, 1977.
34. **Fondy, B. R. and Geiger, D. R.**, Sugar selectivity and other characteristics of phloem loading in *Beta vulgaris* L., *Plant Physiol.*, 59, 593, 1977.
35. **Giaquinta, R. J. and Geiger, D. R.**, Mechanism of inhibition of translocation by localized chilling, *Plant Physiol.*, 51, 372, 1973.
36. **Sij, J. W. and Swanson, C. A.**, Effect of petiole anoxia on phloem transport in squash, *Plant Physiol.*, 51, 368, 1973.
37. **Warren Wilson, J.**, Control of crop processes, in *Crop Processes in Controlled Environments*, Rees, A. R., Cockshull, K. E., Hand, D. W., and Hurd, R. G., Eds., Academic Press, London, 1972, 7.
38. **Wareing, P. F. and Patrick, J.**, Source-sink relations and the partition of assimilates in the plant, in *Photosynthesis and Productivity in Different Environments*, Cooper, J. P., Ed., Cambridge University Press, Cambridge, 1975, 481.
39. **Moorby, J.**, Integration and regulation of translocation within the whole plant, in *Integration of Activity in the Higher Plant*, Jennings, D. H., Ed., Cambridge University Press, Cambridge, 1977, 425.
40. **Wardlaw, I. F.**, The control and pattern of movement of carbohydrates in plants, *Bot. Rev.*, 34, 79, 1968.
41. **Fick, G. W., Williams, W. A., and Loomis, R. S.**, Computer simulation of dry matter distribution during sugar beet growth, *Crop Sci.*, 13, 413, 1973.
42. **Loomis, R. S. and Gerakis, P. A.**, Productivity of agricultural ecosystems, in *Photosynthesis and Productivity in Different Environments*, Cooper, J. P., Ed., Cambridge University Press, Cambridge, 1975, 145.
43. **Allison, F. E.**, *Soil Organic Matter and its Role in Crop Production*, American Elsevier, New York, 1973, 237.
44. **Russell, E. W.**, *Soil Conditions and Plant Growth*, 10th ed., Longman Group Ltd., London, 1973, 339.
45. **Rice, E. L.**, *Allelopathy*, Academic Press, New York, 1974, 77.
46. **Stewart, G. R., Lee, J. A., Orebamjo, T. O., and Havill, D. C.**, Ecological aspects of nitrogen metabolism, in *Mechanisms of Regulation of Plant Growth*, Bieleski, R. L., Ferguson, A. R., and Cresswell, M. M., Eds., Royal Society of New Zealand, Wellington, 1974, 41.
47. **Klepper, L., Flesher, D., and Hageman, R. H.**, Generation of reduced nicotinamide adenine dinucleotide for nitrate reduction in green leaves, *Plant Physiol.*, 48, 580, 1971.
48. **Raven, J. A. and Smith, F. A.**, Nitrogen assimilation and transport in vascular land plants in relation to intracellular pH regulation, *New Phytol.*, 76, 415, 1976.
49. **Ben-Zioni, A., Vaadia, Y., and Lips, S. H.**, Correlations between nitrate reduction, protein synthesis, and malate accumulation, *Physiol. Plant.*, 23, 1039, 1970.
50. **Kessler, E.**, Nitrate assimilation by plants, *Annu. Rev. Plant Physiol.*, 15, 57, 1964.
51. **Penning de Vries, F. W. T.**, Use of assimilates in higher plants, in *Photosynthesis and Productivity in Different Environments*, Cooper, J. P., Ed., Cambridge University Press, Cambridge, 1975, 459.
52. **Harper, J. E. and Hageman, R. H.**, Canopy and seasonal profiles of nitrate reductase in soybeans *(Glycine max* L. Merr.), *Plant Physiol.*, 49, 146, 1972.
53. **Harper, J. E. and Hageman, R. H.**, Seasonal and canopy variation in nitrate reductase activity of soybean *(Glycine max* L. Merr.) varieties, *Crop Sci.*, 12, 382, 1972.
54. **Houba, V. J. G., van Egmond, F., and Wittich, E. M.**, Changes in production of organic nitrogen and carboxylates (C-A) in young sugar-beet plants grown in nutrient solutions of different nitrogen composition, *Neth. J. Agric. Sci.*, 19, 39, 1971.
55. **Ergle, D. R. and Eaton, F. M.**, Organic acids of the cotton plant, *Plant Physiol.*, 24, 373, 1949.

56. **Ben-Zioni, A., Vaadia, Y., and Lips, S. H.,** Nitrate uptake by roots as regulated by nitrate reduction products of shoots, *Physiol. Plant.*, 24, 288, 1971.
57. **Canvin, D. T. and Atkins, C. A.,** Nitrate, nitrite, and ammonia assimilation by leaves: effect of light, carbon dioxide and oxygen, *Planta*, 116, 207, 1974.
58. **Pate, J. S.,** Nutrients and metabolites of fluids recovered from xylem and phloem: significance in relation to long-distance transport in plants, in *Transport and Transfer Processes in Plants*, Wardlaw, I. F. and Passioura, J. F., Eds., Academic Press, New York, 1976, 253.
59. **Pate, J. S.,** Uptake, assimilation, and transport of nitrogen compounds by plants, *Soil Biol. Biochem.*, 5, 109, 1973.
60. **Radin, J. W.,** Contribution of the root system to nitrate assimilation in whole cotton plants, *Aust. J. Plant Physiol.*, 4, 811, 1977.
61. **Pate, J. S.,** Nutrient mobilization and cycling: case studies for carbon and nitrogen in organs of a legume, in *Transport and Transfer Processes in Plants*, Wardlaw, I. F. and Passioura, J. F., Eds., Academic Press, New York, 1976, 447.
62. **Radin, J. W. and Sell, C. R.,** Some factors limiting nitrate reduction in developing ovules of cotton, *Crop Sci.*, 15, 713, 1975.
63. **Leffler, H. R. and Tubertini, B. S.,** Development of cotton fruit. II. Accumulation and distribution of mineral nutrients, *Agron. J.*, 68, 858, 1976.
64. **Anderson, W. P.,** Transport through roots, in *Encyclopedia of Plant Physiology*, Vol. 2B, Luttge, U. and Pitman, M. G., Eds., Springer-Verlag, Berlin, 1976, 129.
65. **Mauney, J. R., Chappell, J. and Ward, B. J.,** Effects of malic acid salts on growth of young cotton embryos in vitro, *Bot. Gaz., (Chicago)*, 128, 198, 1967.
66. **Dhindsa, R. S., Beasley, C. A., and Ting, I. P.,** Osmoregulation in cotton fiber. Accumulation of potassium and malate during growth, *Plant Physiol.*, 56, 394, 1975.
67. **Radin, J. W. and Sell, C. R.,** Growth of cotton plants on nitrate and ammonium nitrogen, *Crop Sci.*, 15, 707, 1975.
68. **Pate, J. S., Sharkey, P. J., and Atkins, C. A.,** Nutrition of a developing legume fruit. Functional economy in terms of carbon, nitrogen, water, *Plant Physiol.*, 59, 506, 1977.
69. **Sambo, E. Y., Moorby, J., and Milthorpe, F. L.,** Photosynthesis and respiration of developing soybean pods, *Aust. J. Plant Physiol.*, 4, 713, 1977.
70. **Elmore, C. D.,** Contributions of the capsule wall and bracts to the developing cotton fruit, *Crop Sci.*, 13, 751, 1973.
71. **Guinn, G. and Brinkerhoff, L. A.,** Effect of root aeration on amino acid levels in cotton plants, *Crop Sci.*, 10, 175, 1970.
72. **Elmore, C. D. and Leffler, H. R.,** Development of cotton fruit. III. Amino acid accumulation in protein and nonprotein nitrogen fractions of cottonseed, *Crop Sci.*, 16, 867, 1976.
73. **Atkins, C. A., Pate, J. S., and Sharkey, P. J.,** Asparagine metabolism — key to the nitrogen nutrition of developing legume seeds, *Plant Physiol.*, 56, 807, 1975.
74. **Miflin, B. J. and Lea, P. J.,** Amino acid metabolism, *Annu. Rev. Plant Physiol.*, 28, 299, 1977.

Chapter 9

THE PARADOX OF NO CORRELATION BETWEEN LEAF PHOTOSYNTHETIC RATES AND CROP YIELDS

C. D. Elmore

TABLE OF CONTENTS

I. INTRODUCTION

That plants differ with respect to leaf photosynthetic rates has long been recognized. Also recognized and implicitly stated in much of the literature is the fact that rates of leaf photosynthesis do not correlate well with yield. Statements such as "Intuitively one can easily visualize that higher rates of photosynthesis should lead to higher yields"[1] abound in the literature.[2-5] The hypothesis that higher leaf photosynthetic rates are necessary for increased yields is still popular, however, as shown by its recent inclusion in a recommendation for a research program on genetic aspects of carbon metabolism.[6] Yet, as pointed out by Evans,[3] there is little experimental evidence of any positive relationship between yield and leaf photosynthetic rate, nor is there any instance in which selection for a greater rate of leaf photosynthesis has led to increased yield. The aim of this chapter is to document what has been done so far in establishing a relationship between rates of leaf photosynthesis and crop yields, and to analyze the accumulated data in order to summarize present knowledge and establish needs for further research.

II. INTRA-SPECIES DIFFERENCES

Photosynthetic rates have been shown to differ within species on numerous occasions. Table 1 contains an extensive, though not exhaustive, listing of reports of differences. Many of the resports are of true genetic differences; however, in many cases leaf photosynthesis rates are confounded, especially when values from different experiments under different conditions are compared.

III. GROWTH-ANALYSIS APPROACH

Another chapter in this book examines the classical English growth-analysis approach and its relationship to modeling. It is also useful to consider growth analysis here in the classical sense as it pertains to the relationship between leaf photosynthetic rates and yield. Watson[93] reviewed the concepts and provided some insight into the problem in 1952, and it would be well to consider where we are now in light of his perceptions then. The problem as we have presented it is that we are trying to relate leaf photosynthetic rates, which are usually taken under optimum conditions on newly expanded leaves, and then trying to relate that to the subsequent yield of the crop. It usually does not work, although it seems that it should. At least that is the basis of the simplistic argument for the advantages of C_4 plants, and breeding for high leaf photosynthetic rates.

If we consider that the dry weight (W) of whole plants is related to photosynthesis of whole plants (NAR) and the amount of leaf area (F) present over time (LAD), we can derive relationships with yield (Y). Y is related to W, so it must also be related to NAR and LAD. NAR is related to net photosynthesis of single, expanded, well exposed leaves (that which we usually try to relate to yield), but NAR is the whole plant integral of leaf photosynthesis on a daily basis, i.e., the daily sum of net photosynthesis by all the leaves on the plant. NAR must, therefore, depend upon light interceptions, leaf area index, leaf status (wilted, diseased, necrotic, etc.), and leaf age, to name a few.

Yield is related to dry weight. In some plants it is higher than others, i.e., forage grasses vs. cotton, but it is usually a simple relationship, "the harvest index."

$$Y = f(W)$$

W, however, is a product: the product of NAR X A integrated over time. Yet, both NAR and A change during the growing season.[93] Leaf area (F) changes are covered by another term, leaf area duration (LAD), so we have another relationship: W = g(NAR, F or LAD) where LAD = h(F) T = Time). NAR, however, is related to leaf photosynthesis of all leaves and of leaf area, i.e., NAR = i(P, F, or LAI). Therefore in order to relate P and Y, we have to go through several intervening steps. As Watson pointed out in 1952[93] leaf area is probably the most important single factor. In fact, Heath and Gregory[94] once stated that NAR was a constant and leaf area was the only variable.

Carrying the growth-analysis approach still further, we can make the following analysis. As has been pointed out in Table 1, leaf-photosynthetic rate has been found to correlate well in separate instances with two other leaf parameters: specific leaf weight (SLW) and leaf N concentration. The positive relationship between differences in photosynthesis among varieties and either specific leaf weight or leaf N poses a problem, however, for plant photosynthetic production depends upon rate as well as area. New leaf area costs in terms of both dry matter and nitrogen so it cannot be produced as fast if either is limiting. Assuming N uptake is related to transpiration which in turn is related to photosynthesis, under the same vapor pressure deficit in the air,

$$dN_p/dt = a\,P_c N_s$$

where N_p is the plant nitrogen, a is the ratio of transpiration to P_c, P_c is the plant photosynthetic rate (weight per plant per time) and N_s is the soil N (weight per ground area). Using the standard rectangular hyperbola to describe the response of leaf photosynthesis (P_l) to intercepted light (I),

$$P_I = P_{max}\,k\,I/(1+kI)$$

where P_{max} and k are Michaelis-Menton constants. The standard relationship for light penetration by the leaf canopy then is,

$$I/I_o = e^{-cF}$$

where I_o is the incident light intensity at the top of the canopy, I is the intensity at some point in the canopy, and F is leaf area. Setting $P_{max} = aN_F$ where b is a constant and N_F is the nitrogen content of the leaf (weight per area), then

$$P_c = aN_F kI_o e^{-cF}/(1 - kI_o e^{-cF})$$

which must be integrated between F = 0 at the top of the canopy to total F or the LAI at the bottom of the canopy. But $dF/dt = (dN/dt)/N_F$, and also $= (dW/dt)/\Delta$, depending on whichever (N or dry matter available for leaf growth) is limiting and where Δ is the specific leaf weight (weight per leaf area).

We have two contradictory relationships here. The higher the N_F value, the higher the leaf photosynthesis rate. The lower the N_F value, the greater the leaf growth rate. The latter is not always true. The leaf growth rate is generally related directly to the plant N level; however, we are considering differences in rates among varieties. For example, among the wheat species, the photosynthetic apparatus seems to have been diluted over larger leaves as man selected for higher yields, and leaf growth rates increased.[13]

TABLE 1

Examples of intra species differences in leaf phosynthetic rate

Species	Range[a]	Comments	Ref.
Triticum aestivum	—	Varietal differences in photosynthate partitioning	7
	—	Further observations of 1950 study	8
	18—26 mg · dm^{-2} · day^{-1}	Actually NAR rates, but varietal differences	9
	28—37	Spring wheat varieties	10
	5—12	Winter wheat varieties	11,12
	25—70	Flag leaf of cultivated wheat and its ancestral parents — the parents had higher rates	13
	23—40	Differences among commercial wheat varieties of Pakistan	14
	20—40	Commercial wheat and its wild relatives	15—18
	1.40—1.52 mℓ O_2cm^{-2} · min^{-1}	Significant P rate differences among three wheat varieties	19
	28—39	Cultivated wheat and wild relatives	20
	188—341 mℓ O_2 · cm^{-2} · hr^{-1}	Variation among genotypes	21
		High CO_2 enrichment with yield increase of 0.25% per VPM of CO_2	22
Zea mays	—	Hybrid vigor effects noted	23
	—	Hybrids were 121.9% of parent lines	24
	36—58	Races of maize differed for P rates; plant dry weight did not correlate with P rates; leaf area was more important	4
	28—85	Inbreds of maizes howed differences; hybrids exhibited apparent heterosis for P rate	25
	30—45	Faster growing inbred had slower respiration rate but insignificant P rate advantage at low light	26
	29—76	Report P rate differences within inbred lines	1
	21—28	Selection for dry matter production in the juvenile maize plant resulted in 30% greater leaf area but lower P rates	27
	29—31	Further analysis of previous experiments confirm that P rates were lower but not on a chlorophyll basis	28

	26—41	Chlorophyll mutants	29
	55—68	Varietal differences, correlated P, and net assimilation rate (NAR)	30,31
	21.8—51.6	Variation in maize breeding population during vegetative growth	32
	11.4—36.9	Variation in maize breeding population during grain fill	32
Glycine max	21—31	Report of varietal differences	33
	21—31	Suggest parent-offspring relationship among varieties	34
	12—24	Yield not related to P; examined Photorespiration and found no relation to P rate	35
	18—23	Statistically significant differences among nine cultivars	36
	5—147% of parents	Genetics of P; low P rates dominant in F_1 F_2 normal distribution, typical qualitative inheritance	37
	29—43	Related P to specific leaf weight	38
	17.4—30.9	F_3 lines correlate with F_2 plants; leaf N content and SLW also correlate well with high P	39
	24—33	Leaf N content most related to P; reports a successful selection experiment for P rate	40
	24—33	Related P rates to leaf chlorophyll	41
	20—31	Related varietal differences to light and dark reaction activities	42
	26—66	Primary leaf characteristics were not closely related to P rates	43
	34—40	Correlates P rates with Specific Leaf Weight (SLW) and leaf thickness	44
	26—38	Correlated P rates, (+) with SLW but (−) with plant leaf growth rates	45
	10—40	P rates (+) correlated with chlorophyll content	46
	22—31	Response of P to high CO_2	47
Oryza sativa	5—30	An early study included two varieties	48
	36—58	Varietal differences, also cultural, age and environmental effects on P rate	49
	19—26	Leaf N content important, growth rate was determined by P rate in young plats	50
	5—12	P rates of plants at heading with two N fertilization rates; demonstrated varietal suitability to N fertilization; P increased in adapted varieties with N fertilization	51
	8—15	Leaf N content and P rates were highly correlated	52

TABLE 1 (continued)

Examples of intra species differences in leaf phosynthetic rate

Species	Range[a]	Comments	Ref.
Chlorophyll mutants	—		53
	—	Related physical parameters to yield; N at grain fill and culture conditions to promote erect leaves	54
Oryza indica	—	Varietal differences and N effects on P	55
Gossypium hirsutum	33—48	Suggested the P rate differences may not be real; differences for NAR were not detected	56
	23—40	Measured other growth analysis parameters; checked 26 species	57
	33—48	Examined races of *G. hirsutum* as well as growth analysis parameters, leaf age effects	58
	37—40	Chlorophyll mutant compared with wild type; little effect on rate/area	59
	47—53	CO_2 enrichment experiment	47
G. arboreum	23—40	Within species differences	57
Medicago sativa	26—42	Varietal differences, temperature effects	60
	20—50	P rate and SLW positively correlated ($r = 0.79$); SLW was both genetically and environmentally controlled	61,62
	20—32	P rates in breeding population unselected for P rates were not affected by selection for other characters unrelated to growth	63
Saccharum sp.	34—86	Ten cultivars differed for P rate, P was correlated with leaf thickness and porosity	64
	26—38	Field grown plants — 30 cultivars	65
	31—82	*S. spontaneum*, greenhouse	65
	26—38	*S. officinarum*, greenhouse; did not show P-rate increase with yield increase; leaf area index (LAI) was a better index	
Dactylis glomerata	20—30	P rate was not inversely related to photorespiration but to PR substrate	66
Avena sativa	22—48	Cultivar differences in P rates; correlation with SLW and internal leaf characters	67,19
Coleus spp.	2.8—6.9	Related to chlorophyll content, cultivar differences	68
Lycopersicon esculentum	—	P rate accounted for more than 60% of yield variation in a genetic experiment	69

Festuca arundinaceae	22—42	Differences of P rates in related lines were detected; genetic experiment conducted; heritability values ranged from 0.57 to 0.83 Broad Sense and 0.22 to 0.44 Narrow Sense heritability	70
	22—42	Photosynthetic enzyme activities were correlated among the lines	71
Phaseolus vulgaris	14—20	Showed P-rate difference among cultivars; genetic experiment suggested quantitative inheritance but few genes and some dominance for low P rate	72
Hordeum vulgare	2—22	Varietal differences in P rates	73
Pyrus malus	5—35	Not necessarily varietal or related to yield; literature survey	74
Prunus persica	15	Good evidence for a coupling between photosynthate supply and growth requirements	75
Nicotiana	9—13	Genetic differences consistent over 3 years	76
Brassica	10—35	P rate + related to cell surface and to leaf N content	77,78
	$m\ell O_2 \cdot cm^{-2} \cdot hr^{-1}$ 2—4	P related to number of chloroplasts at high light, but not to chlorophyll content	79,80
	$m\ell\ O_2 \cdot dm^{-2} \cdot hr^{-}$ 10—19	SLW, P rate, and Chl content were heritable 5 × 5 diallel genetic expt	81
Ipomoea	10—50	Varietal differences were present; some due to leaf area; P related to leaf N and K, (−) to carbohydrate	82,83
Populus	2.6—5.9	Differences in P rate in 16 clones, related to yield	84
	15—44	Six clones with differences in P	85
	8—17	Clonal differences existed, related P rate and enzymes devised an early selection scheme	86
Pseudotsuga	—	Showed P rates to be heritable and related to dry matter	87
	1—10	N fertility affected P rate	88
Pinus	—	Trees were shown to have genetic variability for P rate and N fertilization increased P rate	89
	—	Genetic experiment showed that P rates were heritable, but not related to seedling DW; better relationship with needle DW	90
	22—45 $mg\ CO_2 \cdot hr^{-1}\ mg\ needle^{-1}$	Estimated heritability from full sib families were 0.28—0.50; other genetic effects reported	91,92

[a] Units are mg $CO_2/dm^2/hr$ unless noted otherwise

The problem then is to maximize $dW/dt = g(P_c - hW)$ where g is the growth respiration factor and h is the maintenance rate, given N_s and I_o levels in the environment.

Assuming that flowering starts at some arbitrary F such as when $I/I_o = 0.95$, with $dF/dt = -dF/dt$ 45 days earlier (the leaf life lasting 45 days), with no leaf area increase from this point on, $dY/dt = [(dN_Y/dt)/N_Y] \times W/N$, or $= (dW/dt)/\delta_Y$, whichever (N_Y or P_c) is limiting, and with δ_Y being the partition coefficient for total **W** into the yield component of plant growth. At this point N_s will be depleted, dN_P/dt will have slowed considerably, and N_Y will have to be redistributed from N_F. P_c might be considered to build up as reserve (Starch or S) to a weight equivalent to the dry matter in leaves (the specific leaf weight × F). As N_F decreases, P_c decreases and at the same time F is decreasing through attrition. Given the above relationships and the environment, somehow the rate of usage of N and P_c or S for building Y must be optimized.

Thus, growth analysis in effect tells us that the plant has developed through evolution certain fundamental relationships between leaf area and leaf photosynthetic rate which fit it for a particular ecological niche. Man has influenced this relationship, but usually in the direction of greater leaf area, by his cultural manipulations and by his inadvertant selection.

IV. ADAPTIVE MECHANISMS

Differences in photosynthetic rates among genotypes within a species may well be related to the following adaptive mechanisms, depending upon experimental conditions. Adaptation is covered in other chapters in this book. For our purposes, consider the following adaptive relationships which may confound experiments or experimenters:

1. In a cold climate, one may well be studying how a subtropical or tropical plant adapts to cold in the field or in cool controlled environments.
2. For plants grown under artificial conditions, either in a greenhouse or growth cabinet, one might well be studying how plants adapt to weak light or to light quality.
3. For plants under intense light or high-sunlight over long periods (14- to 16-hr daylengths) but at cool temperatures typical of the temperate zone, one might well be studying adaptation to excess photosynthate supply or excess starch and sugars.
4. Plants under all conditions may well be adapting to limiting supplies of N, P, and K.[13] Values of N, P, and K decline in the plant on a percent dry weight basis, and photosynthesis is correlated with this decline. All this may be traced to root activity.
5. Plants in a desert environment may be adapting to intense light, excess photosynthate supply, high day temperatures, and low night temperatures.
6. Plants in the humid subtropics may well be adapting to high temperatures and periods of fluctuating light (periods of cloudy weather follows by periods of sunny weather).
7. In all conditions, plants may be adapting to water stress.
8. In selecting for higher-yielding cultivars, man has selected plants adapted to cultivation and larger leaves. Thus, the photosynthetic apparatus is spread out over a greater leaf surface area. This results in lower photosynthetic rates per unit leaf surface as has been shown for soybeans and wheat.[40,95]

The photosynthetic area per plant may well be an overwhelming factor in determining photosynthate supply to the plant, and it in turn can be expected to adapt to all the above factors, depending upon experimental conditions. It is likely that under most experimental conditions, several of the above adaptive mechanisms are induced, resulting in considerable interactions among the above processes. These adaptive mechanisms may be unique to the experimental conditions used and differences in photosynthesis associated with them may rarely express themselves in the field where such conditions rarely prevail or where other conditions may dominate. In view of all the above, it is no wonder that photosynthesis has rarely been associated with yield.

Photosynthesis or photosynthate supply to the plant is almost as complicated a process as the accumulation of yield itself. Cultivar differences in yield can only be established after many trials at different locations in different years. Even then there may be a big location × year × cultivar interaction term. It is naive to expect photosynthesis to behave differently than yield. From the above, it is also naive to study in great detail differences among cultivars grown under specific conditions without recognizing the adaptive mechanisms involved and studying their behavior for plants grown under a range of conditions.

Coupled with all this is a list of factors providing modulation of one process against another resulting in yield advantages. These include duration of photosynthate supply as well as ability to adapt under stress. The C_4 plants, which have highest photosynthetic rates under optimum conditions, possess drought adaptive mechanisms but lack cold tolerance.

The effort by some laboratories to convert C_3 plants to C_4 plants and thereby increase P rate by eliminating photorespiration has not been successful.[96-99] Black[5] pointed out that this was not a biologically sound approach, and Gifford[100] provides a rationale for having both C_3 and C_4 plants. Each has its own special niche and range of environment to which it is suited.

In efforts to increase yield through crop breeding, man has influenced crop evolution. As pointed out by a series of papers on wheat, however, (Table 1) man's influence to date in domesticating wheat has been to reduce the P rate and to increase the leaf area and transport capacity of the phloem, thereby increasing yield potential. If the result in wheat proves to be general among the crops man has domesticated, then further evidence is provided for the critical role and relative ease of manipulation that leaf area plays in plant productivity.

Underlying the importance of leaf area in crop growth analysis may be the results of the CO_2 enrichment experiments (Table 1). CO_2 enrichment does increase the leaf photosynthetic rate, but does not result in economical yield or dry matter increases. Some investigators have found that leaf area does not increase proportionately,[47] thus resulting in lower than expected or predicted dry matter increases. Could this be expected to happen if researchers were able to increase leaf photosynthetic rates by breeding?

It seems that in order for researchers to be able to "breed" for higher yields by selecting for higher photosynthetic rates, they must simultaneously select for (or not select against) many other important factors. Leaf area and phloem-transport capacity are but two that have been mentioned, while the specific leaf weight and leaf N seem to relate well with P rate. Do we know all of the important factors yet? Is the state of the art sufficiently advancd to permit the experiments to proceed? The answer is probably a qualified yes to both questions. Methodology is improving so that the large numbers necessary for a selection experiment can be handled adequately with sufficient precision. As early as 1955, some progress was noted in relating leaf photosynthetic rate and yield.[84] Yet followup may be lacking in utilization of the information. Recently there has been evidence of progress in a number of crop plants, notably soybean, sugarcane, maize, oat, bean, etc. (Table 1). The effort is now being made as has been suggested by research-planning conferences and programs. Only time will tell whether the effort will be successful.

ACKNOWLEDGMENT

I thank Dr. David Patterson for a critical reading of this manuscript and for helpful suggestions on the growth analysis treatment and Dr. J. D. Hesketh for suggesting the topic and for his support and advice.

REFERENCES

1. **Moss, D. N. and Musgrave, R. B.,** Photosynthesis and crop production in *Advances in Agronomy,* Vol. 23, Brady, N. C., Ed., Academic Press, New York, 1971, 317.
2. **Wallace, D. H., Ozbun, J. L., and Munger, H. M.,** Physiological genetics of crop yield, in *Advances in Agronomy,* Vol. 24, Brady, N. C., Ed., Academic Press, New York, 1972, 97.
3. **Evans, L. T.,** The physiological basis of crop yield, in *Crop Physiology,* Evans, L. T., Ed., Cambridge University Press, U.K., 1975, chap. 11.
4. **Duncan, W. G.and Hesketh, J. D.,** Net photosynthetic rates, relative growth rates, and leaf numbers of 22 races of maize grown at eight temperatures, *Crop Sci.,* 8, 670, 1968.
5. **Black, C. C.,** An assessment of C_4 photosynthesis and productivity, in *CO_2 Metabolism and Plant Productivity,* Burris, R. H. and Black, C. C., Eds., University Park Press, Baltimore, 1975, 397.
6. **Brown, A. W. A., Byerly, T. C., Gibbs, M., and San Pietro, A., Eds.,** Crop Productivity—Research Imperatives, Michigan State University, Agricultural Experiment Station and the Charles F. Kettering Foundation, East Lansing, Mich., 1975, 177.
7. **Asana, R. D. and Mani, V. S.,** Studies in physiological analysis of yield. I. Varietal differences in photosynthesis in the leaf, stem and ear of wheat, *Physiol. Plant.,* 3, 22, 1950.
8. **Asana, R. D. and Mani, V. S.,** Studies in physiological analysis of yield. II. Further observations on varietal differences in photosynthesis in the leaf, stem, and ear of wheat, *Physiol. Plant.,* 8, 8, 1955.
9. **Lupton, F. G. H.,** Varietal differences in some physiological characters of wheat, *Ann. Appl. Biol.,* 49, 557, 1961.
10. **Stoy, V.,** Photosynthesis, respiration and carbohydrate accumulation in spring wheat in relation to yield, *Physiol. Plant. Suppl.,* 4, 1, 1965.
11. **Natr, L.,** Study of the formation of grain yields produced by cereal crops. II. Effectiveness of utilization of the solar radiation by the leaf blade of winter wheat, *Separatni Vytisk.,* 149, 1965.
12. **Natr, L.,** The varietal differences in the intensity of photosynthesis, *Rostl. Vyroba,* 12, 163, 1966.
13. **Evans, L. T. and Dunstone, R. L.,** Some physiological aspects of evolution in wheat, *Aust. J. Biol. Sci.,* 23, 725, 1970.
14. **Khan, M. A. and Tsunoda, S.,** Differences in leaf photosynthesis and leaf transpiration rates among six commercial wheat varieties of West Pakistan, *Jpn. J. Breeding,* 20, 344, 1970.
15. **Khan, M. A. and Tsunoda, S.,** Evolutionary trends in leaf photosynthesis and related leaf characters among cultivated wheat species and its wild relatives, *Jpn. J. Breeding,* 20, 133, 1970.
16. **Khan, M. A. and Tsunoda, S.,** Classification of wild and cultivated wheat strains based on their leaf structures, *Tohoku J. Agric. Res.,* 21, 118, 1970.
17. **Khan, M. A. and Tsunoda, S.,** Growth analysis of cultivated wheat species and their wild relatives with special reference to dry matter distribution among different plant organs and to leaf area expansion, *Tohoku J. Agric. Res.,* 21, 47, 1970.
18. **Khan, M. A. and Tsunoda, S.,** Leaf photosynthesis and transpiration under different levels of air flow rate and light intensity in cultivated wheat species and its wild relatives, *Jpn. J. Breeding,* 20, 305, 1970.
19. **Laws, D. A. and Treharne, K. J.,** Variation in photosynthetic activity in cereals and its implications in a plant breeding programme. I. Variation in seedling leaves and flag leaves, *Euphytica,* 20, 86, 1971.
20. **Khan, M. A. and Tsunoda, S.,** Comparative leaf anatomy of cultivated wheats and wild relatives with reference to their leaf photosynthetic rates, *Jpn. J. Breeding,* 21, 143, 1971.

21. **Ghildiyal, M. C. and Sinba, S. K.**, Variation in photorespiration in wheat genotypes using Glycine-1-^{14}C decarboxylation technique, *Indian J. Exp. Biol.*, 11, 207, 1973.
22. **Gifford, R. M.**, Growth pattern, carbon dioxide exchange and dry weight distribution in wheat growing under differing photosynthetic environments, *Aust. J. Plant Physiol.*, 4, 98, 1977.
23. **Fousova, S. and Avratovscukova, N.**, Hybrid vigour and photosynthetic rate of leaf discs in *Zea mays* L., *Photosynthetica*, 1, 3, 1967.
24. **Doroyskoya, I. F.**, Formation and photosynthetic activity of assimilating surface of inbred and hybrid corn, *Fiziol. Rast.*, 9, 635, 1962.
25. **Heichel, G. H. and Musgrave, R. B.**, Varietal differences in net photosynthesis of *Zea mays* L., *Crop Sci.*, 9, 483, 1969.
26. **Heichel, G. H.**, Confirming measurements of respiration and photosynthesis with dry matter accumulation, *Photosynthetica*, 5, 93, 1971.
27. **Hanson, W. D.**, Selection for differential productivity among juvenile maize plants; associated net photosynthetic rate and leaf area changes, *Crop Sci.*, 11, 334, 1971.
28. **Hanson, W. D.**, Changes in efficiencies and numbers of chloroplasts associated with divergent selections for juvenile productivity in *Zea mays* L., *Crop Sci.*, 13, 386, 1973.
29. **Alberte, R. S., Hesketh, J. D., Hofstra, G., Thomber, J. P., Naylor, A. W., Bernard, R. L., Brim, C., Endrizzi, J., and Kohel, R. J.**, Composition and activity of the photosynthetic apparatus in temperature-sensitive mutants of higher plants, *Proc. Nat. Acad. Sci., U.S.A.*, 71, 2412, 1974.
30. **Akiyama, T. and Takeda, T.**, Studies on dry matter production in corn plant. V. Relation between the rate of leaf photosynthesis and dry matter production, *Proc. Crop Sci. Soc. Jpn.*, 44, 269, 1975.
31. **Akiyama, T., Saito, Y., and Takeda, T.**, Dry matter production in corn plant. IV. Optical environment and light extinction coefficient of corn population, *Proc. Crop Sci. Soc. Jpn.*, 44, 14, 1975.
32. **Crosbie, T. M., Mock, J. J., and Pearce, R. B.**, Variability and selection advance for photosynthesis in Iowa stiff stalk synthetic maize population, *Crop Sci.*, 17, 511, 1977.
33. **Ojima, M. and Kawashima, R.**, Studies on the seed production of soybean. V. Varietal differences in P rates of soybeans, *Proc. Crop Sci. Soc. Jpn.*, 37, 667, 1968.
34. **Ojima, M., Kawashima, R., and Sakavato, S.**, Studies on the seed production of soybean. VI. Relationship between the activity of photosynthesis of improved varieties and that of the parent ones, *Proc. Crop Sci. Soc. Jpn.*, 37, 676, 1968.
35. **Curtis, P. E., Ogren, W. L., and Hageman, R. H.**, Varietal effects in soybean photosynthesis and photorespiration, *Crop Sci.*, 9, 323, 1969.
36. **Dreger, R. H., Brun, W. A., and Cooper, R. L.**, Effect of genotype on the photosynthetic rate of soybean (*Glycine max* L.), *Crop Sci.*, 9, 429, 1969.
37. **Ojima, M., Kawashima, R., and Mikoshiba, K.**, Studies of the seed production of soybean. VII. The ability of photosynthesis in the F_1 and F_2 generation, *Proc. Crop Sci. Soc. Jpn.*, 38, 693, 1969.
38. **Dornhoff, G. M. and Shibles, R. M.**, Variety differences in net photosynthesis of soybean leaves, *Crop Sci.*, 10, 42, 1970.
39. **Ojima, M. and Kawashima, R.**, Studies on the seed production of soybean. VIII. The ability of photosynthesis of F_3 lines having different photosynthesis in their F_2 generations, *Proc. Crop Sci. Soc. Jpn.*, 39, 440, 1970.
40. **Ojima, M.**, Improvement of leaf photosynthesis in soybean varieties, *Bull. Natl. Inst. Agric. Sci. (Japan) Ser. D*, 23, 97, 1972.
41. **Watanabe, I.**, Mechanism of varietal differences in photosynthetic rate of soybean leaves. I. Correlations between photosynthetic rates and some chloroplast characters, *Proc. Crop Sci. Soc. Jpn.*, 42, 377, 1973.
42. **Watanabe, I.**, Mechanism of varietal differences in photosynthetic rate of soybean leaves. II. Variety differences in the balance between photochemical activities and dark reaction activities, *Proc. Crop Sci. Soc. Jpn.*, 42, 428, 1973.
43. **Watanabe, I. and Tabuchi, K.**, Mechanism of varietal differences in photosynthetic rate and some leaf-characters such as fresh weight, dry weight or mesophyll volume per unit leaf area, *Proc. Crop Sci. Soc. Jpn.*, 42, 437, 1973.
44. **Dornhoff, G. M. and Shibles, R.**, Leaf Morphology and anatomy in relation to CO_2-exchange rate of soybean leaves, *Crop Sci.*, 16, 377, 1976.
45. **Kaplan, S. L. and Koller, H. R.**, Leaf area and CO_2-exchange rate as determinants of the rate of vegetative growth in soybean plants, *Crop Sci.*, 17, 35, 1977.
46. **Buttery, B. R. and Buzzell, R. I.**, The relationship between chlorophyll content and rate of photosynthesis in soybeans, *Can. J. Plant Sci.*, 57, 1, 1977.
47. **Mauney, J. R., Fry, K. E., and Guinn, G.**, Relationship of photosynthetic rate to growth and fruiting of cotton, soybean, sorghum and sunflower, *Crop Sci.*, 18, 259, 1977.
48. **Yamada, N., Murata, Y., Osada, A., and Iyana, J.**, Photosynthesis of rice plant, *Proc. Crop Sci. Soc. Jpn.*, 23, 214, 1955.

49. **Murata, Y.,** Studies on the photosynthesis of rice plants and its culture significance, *Bull. Natl. Inst. Agric. Sci. (Japan) Ser. D*, 170, 1961.
50. **Osada, A. and Murata, Y.,** Varietal differences in the rate of photosynthesis of rice plant and its relation to dry matter production, *Proc. Crop Sci. Soc. Jpn.*, 33, 454, 1965.
51. **Osada, A. and Murata, Y.,** Studies on the relationship between photosynthetic and varietal adaptability for heavy manuring. III. Effects of photosynthetic characteristics on dry matter production and ripening of rice varieties, *Proc. Crop Sci. Soc. Jpn.*, 23, 460, 1965.
52. **Murata, Y.,** Physiological responses to nitrogen in plants, in *Physiological Aspects of Crop Yield*, Eastin, J. D., Haskins, F. A., Sullivan, C. Y., and van Bavel, C. H. M., Eds., American Society of Agronomy, Madison, Wis., 1969, p235.
53. **Saka, H. and Mathunaka, S.,** The growth and photosynthetic activity of artificial chlorophyll mutants in rice plant, *Proc. Crop Sci. Soc. Jpn.*, 44, 54, 1975.
54. **Tanaka, T.,** Regulation of plant type and carbon assimilation of rice, *Jpn. Agr. Res. Quar.*, 10, 161, 1976.
55. **Osada, A.,** Studies on the photosynthesis of *Indica* rice, *Proc. Crop Sci. Soc. Jpn.*, 33, 69, 1963.
56. **Muramato, H., Hesketh, J., and El-Sharkawy, M.,** Relationships among rate of leaf area development, photosynthetic rate, and rate of dry matter production among American cultivated cotton and other species, *Crop Sci*, 5, 163, 1965.
57. **El-Sharkawy, M. A., Hesketh, J. D., and Muramato, H. M.,** Leaf Photosynthetic rates and other growth characteristics among 26 species of *Gossypium*, *Crop Sci.*, 5, 173, 1965.
58. **Elmore, C. D., Hesketh, J. D., and Muramato, H.,** A survey of rates of leaf growth, leaf aging and leaf photosynthetic rates among and within species, *J. Ariz. Acad. Sci.*, 4, 215, 1967.
59. **Benedict, C. R.,** Net CO_2 Fixation in yellow-green plants, in *Net Carbon Dioxide Assimilation in Higher Plants*, Black, C. C., Ed., Cotton Inc., Raleigh, N.C., 1972, chap. 2.
60. **El-Tabbakh, A. E.,** The effect of leaf area development, leaf photosynthetic rates and temperature on growth and dry matter accumulation in certain species of forage crops, *Diss. Abstr. B*, 28, 3560, 1967.
61. **Pearce, R. B., Carlson, G. E., Barnes, B. K., Hart, R. H., and Hanson, C. H.,** Specific leaf weight and photosynthesis in alfalfa, *Crop Sci.*, 9, 423, 1969.
62. **Barnes, D. K., Pearce, R. B., Carlson, G. E., Hart, R. H., and Hanson, C. H.,** Specific leaf weight differences in alfalfa associated with variety and plant age, *Crop Sci.*, 9, 421, 1969.
63. **Chatterton, N. J.,** Photosynthesis of 22 Alfalfa population differing in resistance to diseases, insects pests and nematodes, *Crop Sci.*, 16, 883, 1976.
64. **Irvine, J. E.,** Photosynthesis in sugarcane varieties under field conditions, *Crop Sci.*, 7, 297, 1967.
65. **Irvine, J. E.,** Relations of photosynthetic rates and leaf and canopy characters to sugarcane yield, *Crop Sci.*, 15, 671, 1975.
66. **Carlson, G. E., Pearce, R. B., Lee, D. R., and Hart, R. H.,** Photosynthesis and photorespiration in two clones of orchardgrass, *Crop Sci.*, 11, 35, 1971.
67. **Criswell, J. G. and Shibles, R. M.,** Physiological basis for genotypic variation in net photosynthesis of oat leaves, *Crop Sci.*, 11, 550, 1971.
68. **Rouhani, I. and Khosh-Khui, M.,** Variations in photosynthetic rates of fourteen *Coleus* cultivars, *Plant Physiol.*, 59, 114, 1977.
69. **Rodriquez, B. P. and Lambeth, V. N.,** Photosynthetic rate-yield component relationships in winter greenhouse tomatoes, *Hortic. Sci.*, 11, 430, 1976.
70. **Assay, K. H., Nelson, C. J., and Horst, G. L.,** Genetic variability for photosynthesis in tall fescue, *Crop Sci.*, 14, 571, 1974.
71. **Nelson, C. J., Treharne, K. J., and Lloyd, E. J.,** Genetic variation in enzyme activity of tall fescue leaf blades, *Crop Sci.*, 15, 771, 1975.
72. **Izhar, S. and Wallace, D. H.,** Studies on the physiological basis for yield differences. I. Genetic variations in photosynthetic efficiency of *Phaseolus vulgaris* L., *Crop Sci.*, 7, 457, 1967.
73. **Apel, P. and Lehmann, C. O.,** Variabilität and sortenspezifitat der Photosyntheserate bei Sommergerste, *Photosynthetica*, 3, 255, 1969.
74. **Avery, D. J.,** Maximum potosynthetic rate. A case study in apple, *New Phytol.*, 78, 55, 1977.
75. **Chalmers, D. J., Canterford, R. L., Jerie, P. H., Jones, T. R., and Ugalde, T. D.,** Photosynthesis in relation to growth and distribution of fruit in peach trees, *Aust. J. Plant Physiol.*, 2, 635, 1975.
76. **Avratouscukova, N.,** Differences in photosynthetic rate of leaf discs in five tobacco varieties, *Photosynthetica*, 2, 149, 1968.
77. **Sasahara, T.,** Genetic variations in cell and tissue forms in relation to plant growth. II. Total cell surface area in the palisade parenchyma and total cell surface area: total nitrogen centent ratio in relation to photosynthetic activity in *Brassica*, *Jpn. J. Breeding*, 21, 61, 1971.
78. **Sasahara, T. and Tsunoda, S.,** Genetic variations in cell and tissue forms in relation to plant growth. I. Relationship between growth rates of the *Brassica* species and cell sizes of the shoot apex and palisada parenchyma, *Jpn. J. Breeding*, 21, 1, 1971.

79. **Kariya, K. and Tsunoda, S.**, Relationship of chlorophyll content, chloroplast area, index and leaf photosynthetic rate in *Brassica*, *Tohoku J. Agric. Res.*, 23, 1, 1972.
80. **Kariya, K. and Tsunoda, S.**, Chloroplast characters and the photosynthetic rate of cultivated *Brassica* species, *Tohoku J. Agric. Res.*, 24, 1, 1973.
81. **Waly, E. A. and Johnston, T. D.**, Genetical investigations into photosynthetic rate in *Brassica*. II. A diallel cross of inbred lines of Morrow-stem kale (*B. aleracea* L. Var. *acephala* DC), *Euphytica*, 23, 563, 1974.
82. **Tsuno, Y. and Fujise, K.**, Studies on the dry matter production of sweet potato. VI. Varietal differences of respiration and respiration photosynthetic ratio, *Proc. Crop Sci. Soc. Jpn.*, 32, 311, 1964.
83. **Tsuno, Y. and Fujise, K.**, Studies on the dry matter production of sweet potato. VII. Varietal differences of photosynthesis and factors concerning with photosynthetic rate, *Proc. Crop Sci. Soc. Jpn.*, 32, 315, 1964.
84. **Huber, B. and Polster, H.**, Zur Frage der Physiologischen ursachen der Unterschiedlichen Slofferzeugung Von Pappelklanen, *Biol. Zentralgl.*, 74, 370, 1955.
85. **Luukkanen, O. and Kazlowski, T. T.**, Gas exchange in six *Populus* clones, *Silvae Genet.*, 21, 220, 1972.
86. **Gordon, J. C. and Promnity, L. C.**, Photosynthetic and enzymatic criteria for the early selection of fast growing *Populus* clones, in *Tree Physiology and Yield*, Cannell, M. G. R. and Last, F. T., Eds., Academic Press, New York, 1976, Chap. 4.
87. **Campbell, R. K. and Rediske, J. H.**, Genetic variability of photosynthetic efficiency and dry matter accumulation in seedling Douglas-Fir, *Silvae Genet.*, 15, 65, 1966.
88. **Brix, H.**, Effects of nitrogen fertilization on photosynthesis and respiration in Douglas-Fir, *For. Sci.*, 17, 407, 1971.
89. **Reines, M.**, Photosynthetic efficiency and vigor in pines: variation, in *Forest Genetics Workshop*, Proc. Southern Forest Tree Improvement Comm. Publ. 22, Macon, Ga., 1962, 14.
90. **Robertson, C. F. and Reines, M.**, The efficiency of photosynthesis and respiration in loblolly pines: variation, in *Eight Southern Forest Tree Improvement*, Savannah, Ga., 1965, 104.
91. **Ledig, F. T. and Perry, T. O.**, Variation in photosynthesis and respiration among loblolly pine progenies, in *Ninth Southern Conference Forest Tree Improvement*, Knoxville, Tenn., 1967, 120.
92. **Burkhalter, A. P., Robertson, C. F., and Reines, M.**, Variation in photosynthesis and respiration in Southern Pines, *Ga. For. Res. Counc. Rep.*, 46, 1967.
93. **Watson, D. J.**, The physiological basis of variation in yield, *Adv. Agron.*, 4, 101, 1952.
94. **Heath, O. V. S. and Gregory, F. G.**, The constancy of mean NAR and its ecological importance, *Ann. Bot. (London)*, 2, 811, 1938.
95. **Evans, L. T.**, Physiological adaptation to performance as crop plants, *Philos. Trans. R. Soc. London Ser. B.*, 275, 71, 1976.
96. **Widholm, J. M. and Ogren, W. L.**, Photorespiratory-induced senescence of plants under conditions of low carbon dioxide, *Proc. Natl. Acad. Sci. USA*, 63, 668, 1969.
97. **Menz, K. M., Moss, D. N., Cannell, R. Q., and Brun, W. A.**, Screening for photosynthetic efficiency, *Crop Sci.*, 9, 692, 1969.
98. **Cannell, R. Q., Brun, W. A., and Moss, D. N.**, A search for high net photosynthetic rate among soybean genotypes, *Crop Sci.*, 9, 840, 1969.
99. **Goldsworthy, A. and Day, P. R.**, A simple technique for the rapid determination of plant CO_2 compensation points, *Plant Physiol.*, 46, 850, 1970.
100. **Gifford, R. M.**, A comparison of potential photosynthesis, productivity and yield of plant species with differing photosynthetic metabolism, *Aust. J. Plant Physiol.*, 1, 107, 1974.

Chapter 10

PREDICTING PRIMARY PRODUCTIVITY FOR FOREST AND DESERT ECOSYSTEM MODELS

J. F. Reynolds, B. R. Strain, G. L. Cunningham, and K. R. Knoerr

TABLE OF CONTENTS

I. INTRODUCTION

A major goal of the Analysis of Ecosystems Program of the U.S. International Biological Program (IBP) was to understand how entire ecosystems function so that their behavior, particularly in response to man-made stresses, might be predictable.[1,2] A significant part of the research was focused on increasing our knowledge of productivity in natural ecosystems. Productivity was broadly defined so as to include not only energy flux within the trophic structure of the ecosystem but also the attendant water and nutrient fluxes.[3] A major component of the research design in each of the various IBP Biome projects (i.e., grassland, desert, tundra, and deciduous and coniferous forests) was the development of ecosystem-level models of energy and material fluxes.[1,4,5]

Due to the importance of primary production as a major determinant of the rates and directions of these fluxes, autotrophs received considerable attention. In fact, assessing actual and potential primary productivity in natural ecosystems is an important and significant part of modern ecological research.[6] This has resulted in numerous computer and mathematical models of primary production which run the gamut from rough approximations to detailed physiological models. In this chapter, we describe a general carbon-balance modeling approach that we have used in our IBP-supported studies of primary production and carbon allocation in desert and forest ecosystems.

A number of reviews have appeared recently, discussing the origins and methods of simulation modeling in ecology, particularly as related in the successes and failures of the IBP modeling efforts. These include critiques of modeling at the ecosystem level[5,7-10] and comparisons of the various modeling approaches used by the different Biomes.[1,5,11-15] Although our main interest here is in models of primary production and carbon allocation, the comments below sufficiently apply to ecosystem modeling in general.

There are, of course, as many ways to model a system (e.g., plant growth) as there are views of it. Generally, the first step in model development is the choice of state variables[16] that are of interest (e.g., plant organs) and the identification of those processes (e.g., photosynthesis, translocation, respiration, etc.) that determine the time course of changes in these variables. To lessen this formidable task, the modeler attempts to select a judicious frame of reference from which to view the system, i.e., when the relationships between the state variables are most obvious, a common case being when a system is in a steady-state equilibrium with its physical and chemical surroundings.[17,18] In any case, the model is a simplified representation of the real system[19] which reflects modeling objectives, individual preferences and expertise, and constraints imposed by available data.

The choice of state variables is one of the most crucial steps in the model building process.[20] For example, to model energy flow in a forest, the vegetation could be divided into an overstory, understory, and herbaceous component, and each of these, in turn, could be separated into leaf, bole, root, and storage components. The entire modeling process which follows, i.e., the resolution of the various functions, the time-step of the model, the parsing phase[20] (the choice of relationships between state variables, e.g., energy or material), comparisons with other models, etc., is dictated by the choice of these state variables. It is obvious that the selection of state variables must be considered in light of modeling goals and, importantly, constraints imposed by the available data base. Often, simple conceptualizations of a system will provide reasonable insight into the operation of the system if sufficient data is lacking to warrant a more detailed approach; likewise, progress can be made without knowing exact relationships between system variables as long as reasonable expressions of the system emerge.[21]

II. A BASIC CARBON-BALANCE PARADIGM

A. Parsing Phase

Our modeling interests have exclusively concerned the fixation, utilization, and allocation of carbon. This is due, in part, to the fact that most IBP ecosystem models involve the flow of energy as measured by carbon (usually as total biomass).[20] The choice of carbon over other entities, e.g., minerals or protein biomass, is a logical one since carbon is intimately tied to the procurement and allocation of energy, nutrients, and water.[22,23] Also, perhaps more from a practical viewpoint, carbon procurement and allocation in ecosystems can be readily evaluated by gaseous carbon dioxide exchange measurements and incremental biomass sampling which makes the task of data collection and model validation on such a large scale more feasible.

In using only a single entity (carbon) there are, of course, limitations, since we have only a partial, albeit significant, representation of the system.[24] Thornley[25] suggests that realistic models of plant production should include a number of substrates or factors essential for growth. This allows interactions between different plant organs to be of a reciprocal nature (e.g., nitrogen from the roots, carbon from the leaves). Unfortunately, the paucity of appropriate date for natural ecosystems has severely limited the development of such models. Obtaining the necessary field data for large woody plants involves enormous logistic problems. Compounding this, Evans[26] notes other complexities presented by plants growing in the field, all of which the investigator has no control over: genetic differences between individuals, short- and long-term acclimation, the nonadditive effects of the environment, and within-plant variation. In spite of this, we feel that even preliminary modeling attempts such as ours serve to focus attention on the significant voids in our knowledge of the structure and function of certain plant species and provide guidelines for future research. This pedagogical role of model-building in ecology is extremely important.[5,20,27,28]

B. Carbohydrate Balance and Conceptual Model

The carbohydrate balance of a plant is schematically illustrated in Figure 1. This scheme depicts carbon flux as a complex phenomenon involving carbon procurement, utilization, storage, and transport[29] throughout the plant. These fluxes can be readily conceptualized as an allocation between and through various compartments, or state variables, which represent vegetative and reproduction organs of the plant; hence, the dynamics of carbon flux can be expressed as a function of the balance between inputs (e.g., CO_2 fixation, translocation) and outputs (e.g., respiration) for each compartment.

The rate-controlled flows of carbon to, from, and within an organ are illustrated in system notation in Figure 2. This is the fundamental structural unit of our carbon-balance paradigm. Two compartments per organ are utilized to represent a separation of assimilates into a structural fraction (denoted S) and a pool fraction (denoted P), corresponding to the nonlabile and labile states of carbon, respectively. The third compartment shown is the common or intermediate labile pool assimilate which serves as a "source" or a "sink" for other labile pools, depending on their respective strengths. The arrows in Figure 2 represent possible pathways and directions of carbon flow: (1) from the environment to the plant via net photosynthesis (photosynthesis rate, leaves only); (2) from the labile pool to the environment via respiration and the production of exudates, leachates, and volatiles (maintenance rate) and to the structural compartment in biomass production and growth respiration (growth rate); (3) from the structural compartment to the environment via senesence or physiological death (death

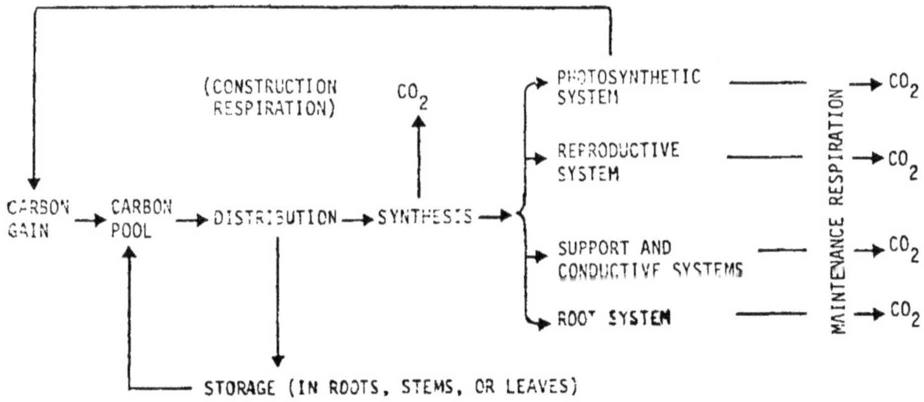

FIGURE 1. Schematic illustration of carbon flux in a plant. (From Mooney.[22] Reproduced with permission from the Annual Review of Ecology and Systematics, Vol. 3, © 1972 by Annual Reviews, Inc.)

rate); and (4) translocation of assimilates into and out of the organ (organ demand function and pool balance function).

The structural compartments are the biomass accumulations (gram dry matter) for each organ. The actual biochemical composition of this dry matter (i.e., nitrogenous compounds, carbohydrates, lipids, lignins, phenolics, minerals, etc.), if known, may be used to determine construction costs. Increases in biomass represent growth or development, whereas decreases reflect structural dieback.

Associated with each structural biomass compartment is a labile assimilate pool, the dynamics of which determine whether the organ is quiescent, actively growing, or dying at any given time. For simplicity and because of the constraints mentioned above, the labile assimilate pool is considered to be composed entirely of glucose, an intermediate product of photosynthesis. This is motivated by the work of Penning de Vries et al.[30] who calculated the number of grams of glucose required to manufacture compounds of various biochemical composition. Thus, glucose is assumed to serve as a basic substrate for the numerous anabolic and catabolic processes of the organ: (1) maintenance respiration of structural biomass (the respiratory substrate might be a hexose sugar), (2) translocation of carbohydrate into or out of an organ (mainly sucrose), (3) storage of carbohydrates (e.g., starch and various polysaccharides), (4) biomass production (the substrates will vary for polysaccharides, lignins, etc.), and (5) losses of carbon from exudates, volatiles, and leachates by the leaves and roots.

C. Mathematical Representation

The dynamics of carbon flux are modeled by writing an equation for each compartment which governs the change in the quantity of carbon through time as a function of the balance between inputs and outputs:

$$\frac{dC_i}{dt} = \sum_j I_{ji} - \sum_k O_{ik} + A_i$$

where dC_i/dt is the time derivative of compartment i, $\Sigma_j I_{ji}$ is the sum of wll inputs to compartment i from other compartments (e.g., translocation from the leaf to the root),

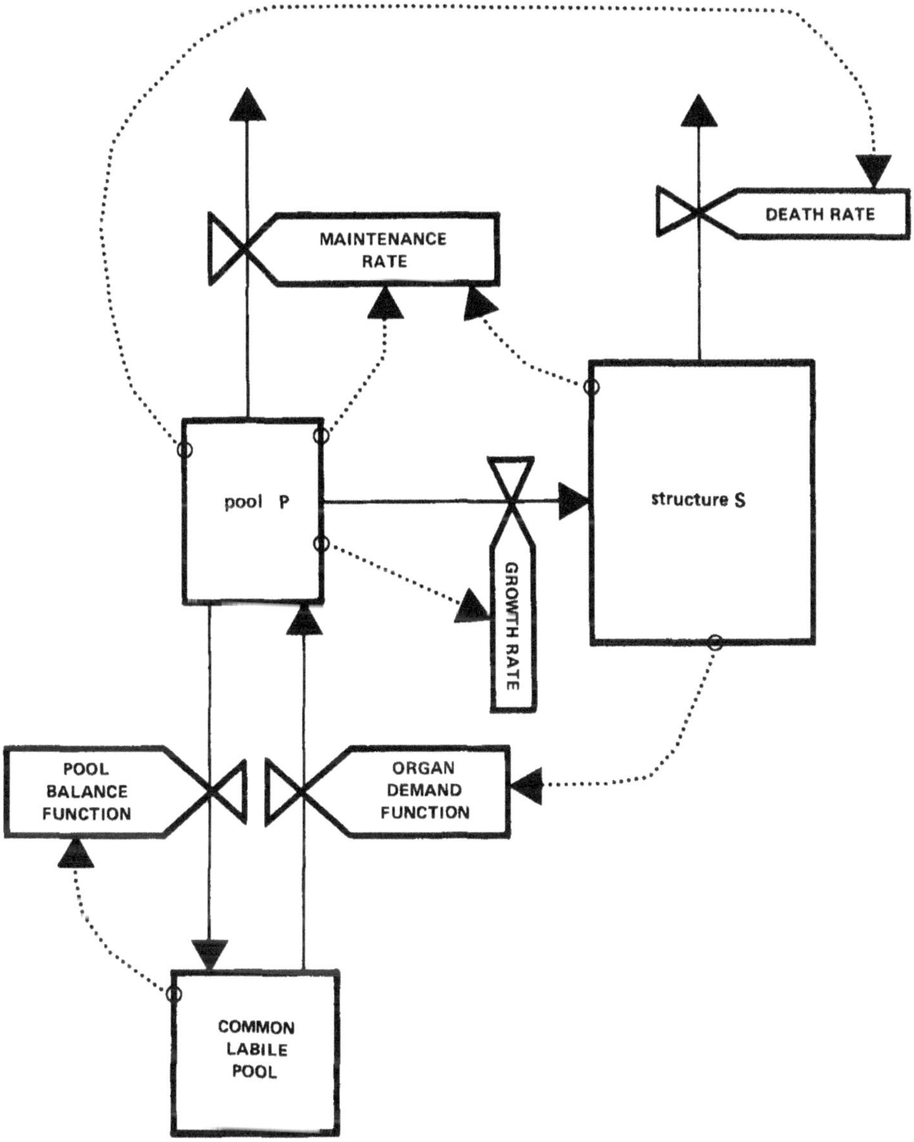

FIGURE 2. System diagram of fundamental structural unit of carbon-balance paradigm.

$\Sigma_k O_{ik}$ is the sum of all losses from compartment i to other compartments or outside the system (e.g., respiration), and A_i is the input to compartment i from any external source (e.g., atmospheric CO_2).

1. Carbohydrate Pools

The carbohydrate pool may fluctuate in size between a maximum, P** (grams glucose), and a minimum, P* (grams glucose); these limits are expressed as functions of the current structural biomass of the associated organ compartment, S (grams dry weight). The limits for a pool at time t are defined:

$$P^{**}(t) = \delta_u S(t) \tag{1A}$$

$$P^{*}(t) = \delta_l S(t) \tag{1B}$$

where the upper limit is given by δ_u (grams glucose per gram dry weight) and the lower limit by δ_l (grams glucose per gram dry weight). If these limits are passed, various consequences are possible (e.g., death of structural biomass, growth, etc.). This is explained below.

The rate of change of the carbohydrate pool is given by:

$$\frac{dP}{dt} = \min \begin{cases} \dfrac{X - TR}{1 + r_s} \\ \dfrac{Y}{1 + r_s} \end{cases} \tag{2}$$

where $X = F + I_{mp} - [dS/dt\ (\varrho + r_g) + r_m A]$; $Y = \delta_u\ (S + dS - P)\ 1/dt$; and $TR = \max\ \{X\text{-}Y, O\}$. F is net photosynthesis (grams glucose per day, leaves only) and I_{mp} is carbohydrate translocated into the pool (grams glucose per day); these are the two possible inputs to the pool. Respiration is divided into three components: maintenance, storage, and growth respiration;[30] r_m is maintenance respiration (grams glucose per dm^2 per day or grams glucose per gram dry weight per day) which is proportional to A, the living biomass (grams dry weight) or surface area (dm^2) of the organ, whereas growth respiration is given by r_g (grams glucose per gram dry weight) which is proportional to the growth rate of the organ (dS/dt [grams dry weight per day]). The respiration cost incurred in the synthesis of storage carbohydrates from soluble carbohydrates is given by r_s (grams glucose per gram glucose). Transfers of carbon from the pool to the structure compartment represent growth (dS/dt), and the accompanying substrate requirements for the synthesis of this dry matter are given by ϱ (grams glucose per gram dry weight).

2. Structural Compartments

The growth of an organ is represented by the transfer of carbon from the pool to the structure compartment. As described above, there is also a loss from the pool due to the respiration associated with growth. The general equation representing structural growth is

$$\frac{dS}{dt} = Z - D\,(S,P) \tag{3}$$

where

$$Z = \max \begin{cases} \min \begin{cases} G(S) \\ \dfrac{E}{r_g + \rho + \delta_u} \end{cases} \\ O \end{cases}$$

and

$$E = F + I_{mp} - r_m A + (P - \delta_u S)\, 1/dt$$

G(S) is some function which determines the maximum growth rate of the organ; this usually is a modified logistic (i.e.t , the growth rate is made proportional to the living biomass and is diminished as a maximum size is asymptotically approached). E is "excess" glucose (grams glucose per day) available for growth once the maintenance requirements are met and the pool, if less than P**, is brought to maximum. The denominator $r_g + \varrho + \delta_u$ determines the maximum values of dS/dt for given values of E and r_g, ϱ, and δ_u. D(S,P) is a function which determines the death rate of structural biomass (see below).

The structural growth dynamics of an organ, as defined by Equation 3, proceed as follows. The maximum growth rate of the organ, G(S), is determined. If enough assimilate is available to meet this rate, i.e., $E \geq G(S)$, then $Z = G(S)$. If $0 < E < G(S)$, then $Z = E/(r_g + \varrho + \delta_u)$, which is the realized growth rate. If $E < O$, then $Z = O$.

For the sake of brevity, all appropriate and obvious restraints have not been explicitly stated. The implementation of this paradigm is presented below.

III. A GROWTH MODEL FOR THE DESERT SHRUB *LARREA TRIDENTATA*

The central objective of the Desert Biome study of the US/IBP was to be able to predict changes that would occur in arid ecosystems due to man-caused or natural perturbation.[11] This necessarily involves an understanding of the biology of the important plant species. Mooney[22] has stressed the importance of understanding how plants gain and allocate their resources to evaluating and predicting their success in a given environment. This motivated us to develop some preliminary models of primary productivity and carbon allocation in one of the most widespread and successful species of the warm desert regions of North America, the creosotebush (*Larrea tridentata* [D.C.] COV.).

Larrea is the dominant evergreen perennial over most of its range, which in the U.S. includes the four major warm deserts: the Colorado, Mojave, Sonoran, and Chihuahuan. Much research has been devoted to understanding various aspects of its physiological ecology, particularly those aspects relative to adaptations that have allowed it to become so successful in such a variety of habitats (see review in Mabry et al.[31]).

A. Productivity Rates in Deserts

Compared to other terrestrial ecosystems, deserts have the lowest standing phytomass (1 to 40 t/ha) and the lowest net annual primary production per unit land area (10 to 250 g/m²/year).[6] However, when production is expressed per unit of maximum standing phytomass, deserts have as high or higher production than most ecosystems.[6,32] It is well established that seasonal and annual variations in water availability significantly affect net primary production in these ecosystems.[33-39] This has prompted some ecologists to label water as the "master input" to deserts.[40,41]

Noy-Meir[40] proposes that the water-controlled nature of arid ecosystems is due to the tight coupling of energy inflow and water throughflow in the soil-plant-atmosphere path; thus, plants can be regarded as converters of a water inflow to an energy inflow. Noy-Meir further states that desert ecosystems are "water limited ecosystems with infrequent, discrete, and largely unpredictable water inputs." A review of the short-term water and energy flow in arid ecosystems is given by Ludwig and Whitford.[39]

B. Productivity and Growth Patterns in *Larrea*

Rates and efficiencies of photosynthesis and annnual productivity in *Larrea* have been investigated by numerous researchers. Barbour[32] provided a recent review and critique of some of this literature. In *Larrea*-dominated communities, efficiency (production divided by incoming solar radiation per unit area) is quite low compared to, for example, certain mesic conifer-hardwood communities. An estimated efficiency of 0.017% for a *Larrea*-dominated community in southern New Mexico[39] is 1/40 of the estimated efficiency of 0.68% for a mesic forest in the eastern U.S.[42] This frequently used expression of efficiency obscures the productivity rates of the individual plants.

If productivity is expressed on a standing crop basis, *Larrea* rates fall within the range of many mesophytic forest shrubs and, in some cases, exceed them.[32] Chew and Chew [43] estimated annual productivity for *Larrea* as 1.70 grams produced per gram of dry weight of standing leaf crop. Expressed as grams produced per gram dry weight of the above- ground standing crop, this results in a relative productivity of 0.28. Peak net photosynthesis rates of *Larrea* have been estimated[44,45] to fall between 6 and 17 mg CO_2/g dw/leaf/hr.

Larrea can maintain a positive net CO_2 balance throughout the year, even under conditions of extreme air temperatures (>43°C) and water stress (<−80 bars). [46-48] Apparently, its principal adaptation to such hot arid conditions is its high level of protoplasmic tolerance or resistance to desiccation and temperature.[46,48] *Larrea* exhibits a photosynthetic temperature acclimation with season,[49] but the availability of water is of foremost importance in determining its growth responses. [39,47,48] When water becomes available, an increase in photosynthesis rate occurs, although the maximum rates obtained when water is abundant are lower than those for desert species that are only active during wet periods.[46] An adaptive advantage of this type of response has been discussed by Orians and Solbrig.[50]

Ludwig and Whitford[39] characterized the above-ground production of *Larrea* in relation to rainfall patterns over a 3-year period in southern New Mexico. Precipitation in this region is highly seasonal (rain occurring mainly in the late summer and fall), and the growth responses of *Larrea* followed this pattern very closely. They concluded that peak aboveground production occurs in the fall and that, during the interval from June to September, the production rates of green shoots were 4.6 kg/ha/day.

C. Brief Description of Model

In the model, 14 compartments are defined, two for each of seven plant organs or developmental stages of interest: (1) leaves, (2) stems, (3) roots, (4) early reproductive buds, (5) maturing reproductive buds, (6) flowers, and (7) fruits. The two compartments per organ or developmental stage represent the separation of carbon into a structural fraction (S_i) and a pool fraction (P_i), corresponding to the nonlabile and labile states, respectively. Translocation of labile assimilates between organs is achieved by two-way flows from each organ pool through a common labile pool, the leaf pool (see Figure 2). The leaf pool also receives the currently produced photosynthate.

The current version of the model allows only one cohort of reproductive biomass to progress through the maturation stages from early reproductive buds to mature fruits at any given time. Another constraint is that all members of the cohort progress at the same rate (i.e., all buds mature at the same rate). Observations[51] of *Larrea* phenology indicate that these constraints are realistic, at least for this stage of model development. The reproductive status of the plant is thus determined by the presence or absence of a cohort of structural biomass in one of the four reproductive compartments (S_4, S_5, S_6, S_7).

1. Photosynthesis Submodel

We have developed and implemented a variety of photosynthesis models to predict net gas exchange in *Larrea*. The model presented here is an analytical photosynthesis model developed by Reed et al.[52] which we have fit to CO_2 gas exchange data for *Larrea*.[53] The model describes the response of net photosynthesis to light, soil moisture availability, temperature, and ambient CO_2 concentration.

F, the net CO_2 influx (milligrams CO_2 per m^2 per second) through the stomata is given by:

$$F = \frac{C_a - C_i}{r_I} \tag{4A}$$

This is a simple flux equation describing a CO_2 concentration (milligrams per cubic meter) gradient from the ambient air (C_a) to the substomatal cavity (C_i) divided by a stomatal resistance, r_1, (seconds per meter) (boundary layer resistance is assumed negligible). C_i is solved for:

$$C_i = \frac{-b + \sqrt{b^2 - 4kc}}{2k} \tag{4B}$$

where

$$b = P_n(I,T) + k\left\{r_I P_n(I,T) - C_a - r_I W_d(I,T)\right\}$$
$$c = P_n(I,T)\left\{-C_a - r_I W_d(I,T)\right\}$$

and

$$P_n(I,T) = P_m(T,\psi)\,\alpha I \,/\, \left\{P_m(T,\psi) + \alpha I\right\}$$

$$W_d(I,T) - \left\{W_d(0,0)\exp(\eta T)\right\}\exp(-\xi I)$$

$$P_m(T,\psi) = \frac{P_m(T_o)\left\{-(T-T_I)^4 + 2(T-T_I)^2(T_o-T_I)^2\right\}}{(T_o-T_I)^4} E(\psi)$$

$P_n(I,T)$ is the Rabinowich formulation expressing the net flux of CO_2 from the substomatal cavity to the mesophyll cells (i.e., internal net photosynthesis) as a function of photon flux density (PAR) I, (μeinstein per m^2 per second), temperature, T (°C), and moisture status. $W_d(I,T)$ is the rate of dark respiration (milligrams CO_2 per m^2 per second) which is assumed to increase expontentially with temperature and decrease exponentially with irradiance. P_m (T,Ψ) is a function describing the maximum photosynthetic rate at a given temperature and moisture status with saturating photon flux density and CO_2; this is scaled (0 to 1) with the empirical function E(Ψ), which introduces the effect of soil moisture availability on net gas exchange. [46,47]

The parameters of the model that must be estimated are k (internal [mesophyll] CO_2 conductance, meters per second), α (quantum efficiency of the photochemical process, milligrams per μeinstein), W_d0,0) (dark respiration at 0°/C and I = 0, milligrams per m^2 per second), $P_m(T_o)$ (maximum photosynthetic rate at optimum temperature with saturating photon flux density and CO_2, milligrams per m^2 per second), T_0 and T_l (the optimum temperature and low temperature intercept of the temperature function, respectively,°C), η (the rate constant relating dark respiration to temperature$°C^{-1}$) and ξ (the rate constant relating the reduction of dark respiration to photon flux density [μeinstein per m^2 per second]). Stomatal resistance (seconds per meter) is calculated as a function of photon flux density using an inverted hyperbolic:

$$r_l = \frac{I + K_m}{I} a_m \tag{5}$$

where a_m (seconds per meter) is the minimum value of r_l reached at high photon flux densities, and K_m (μeinstein per m^2 per second) is the value of I such that $r_l = 2a_m$.

Nonlinear least-squares regression analysis (Marquardt's algorithm) was used to estimate the model parameters using CO_2 exchange data collected on phytotron-grown plants.[53] The range of conditions for which data were available was judged sufficient to obtain unbiased estimates of the model parameters following the procedure detailed in Reed, et al.[52] In the simulation, F is converted to grams glucose per m^2 per hour using the relationship 0.682 g glucose/g CO_2 fixed and a CO_2 to biomass ratio of 1:0.55.

The interception of radiation is calculated as a function of the leaf area index (L) and mean leaf angle (θ).[54] The projected area of the shrub canopy toward radiation (direct and diffuse) which strikes the canopy from sky elevations equal to or greater than the angle of the leaves[55] is 1−exp (−L cos θ). Hence, the intercepted radiation (I) is a fraction of incident solar radiation (I_o):

$$I = I_o [1 - \exp(-L \cos \theta)] \tag{6}$$

This assumes that most of the radiation during a day originates from high solar angles. Daily values of I_o are estimated with a sine curve equation.

The effects of physiological aging of leaves on net photosynthesis are incorporated in the model by calculating a leaf biomass weighted by the effect of physiological age. This weighting is intended to reflect effects resulting from shading by leaves produced distally as well as changes in metabolic potential resulting from senescence. Each cohort of leaf biomass (all leaf biomass produced during a given week) can progress through a maximum of 72 physiological age classes (PAGE[i], i = 1, 2, ..., 72), reflecting the maximum life expectancy of *Larrea* leaves.[43] At the end of each week, the biomass in each physiological age class is transferred to the next oldest class if,and only if, new leaf biomass has been produced during that week. Thus, physiological aging does not occur unless vegetative growth occurs. Net photosynthetic capacity is assumed to decline linearly with physiological age. This effect is introduced using a scalar, SEN(i), which assumes values from 1.0 for the first physiological age class, i.e., PAGE(1), to 0.0 for the 72nd physiological age class, i.e., PAGE(72). A weighted leaf biomass (S_1) is calculated from the product of the biomass in each age class and the associated weighting factor (SEN) reflecting its particular photosynthetic capacity:

$$\bar{S}_1 = \sum_{i=1}^{72} \text{SEN}(i) \cdot \text{PAGE}(i) \qquad (7)$$

A linear relationship between leaf dry weight and area (m^2) is assumed.

2. Respiration Submodels

For plants with large volumes of woody tissue, respiration rates are best described on a surface area rather than on a dry weight basis, because respiration is concentrated in the cambial sheath which surrounds the dead older wood.[56] For *Larrea*, the respiration rate is expressed on a dry weight basis, because this type of data is available. This is probably satisfactory since the biomass accumulations in secondary wood are relatively low in this species.

The respiratory loss of the ith organ (r^i_m) in milligrams CO_2 per gram dry weight per hour is calculated using the general formula:

$$r^i_m = \phi_1 R + \frac{\phi_2 R K}{R_L + K} \qquad (8)$$

where

$$R = R^i_{max} T^i_{mult} M^i_{mult}$$

$$R^i_{max} = f(T) \text{ or } f(T_s)$$

$$T^i_{mult} = f(T)$$

$$M^i_{mult} = f(\Psi)$$

$$K = P_i / P_i^{**}$$

$$R_L = \text{relative pool size at which } R = \frac{R^i_{max}}{2}$$

and

$$0. \leq \phi_1 \leq 1., \; 0. \leq \phi_2 \leq 1., \text{ and } \phi_1 + \phi_2 = 1.0$$

R'_{max} is the maximum respiration rate of the ith organ and is calculated as a weighted function of mean daytime and nighttime air temperature for stems and soil temperature (at various depths) for roots. For the remaining organs, it is a constant value that is moderated by T'_{mult} and M'_{mult}, which are multipliers (0 to 1) that introduce the effect of air temperature and soil water potential (Ψ), respectively, on the respiration rate of the ith organ.

It has been demonstrated that respiration rates may be influenced by the current size of the available assimilate pool.[57] To account for this, a substrate-level influence is used to modify the respiratory rate of each organ. This is accomplished by scaling the respiratory rate by a factor which is, in part, a function of the current size of a pool, P_i, compared to the possible maximum size at that time, P_i^{**}. The degree to which this dependence is important is determined by two scalars, ϕ_1 and ϕ_2, where ϕ_1 determines the base metabolism of the organ and ϕ_2 determines the substrate-dependent respiration. Under the defined constraints, if $\phi_2 = 1.0$, the rate is totally substrate dependent.

The maintenance losses due to the production of volatile and leachable compounds and the formation of nectar are calculated as donor-controlled, constant-coefficient rates. These rates are also substrate dependent as described for respiration losses. For the ith organ, the total maintenance loss, M_i (grams glucose per day), other than respiration is

$$M_i = \sum_k \left\{ \sigma_k S_i \phi_1 + \sigma_k S_i \phi_2 K / (R_L + K) \right\} \tag{9}$$

where σ_k is the constant parameter determining the kth maintenance cost (on a grams glucose per gram dry weight per day basis).

3. Carbon Allocation Submodel

A summary of the allocation submodel is provided in Table 1. Following the findings of Oechel et al.,[47] all growth is at the expense of currently produced photosynthate. Hence, allocation is simply a process of defining the proportion of F which is to be distributed to each organ. Growth can occur only if there is enough photosynthate to fulfill the maintenance requirements of all organs; assimilate beyond this amount is termed "excess" (EX, grams glucose per day). If the currently produced photosynthate is less than the total maintenance requirements of the plant, translocation of labile pool assimilate between organs may occur to satisfy these costs. These processes are shown as the "organ demand" and "pool balance" functions in Figure 2.

The organ demand function combines the maintenance requirements of an organ with a priority system of allocation. "Demand" of organ S_i denotes (1) the amount of assimilate necessary to maintain the pool at maximum (P_i^{**}) (a function of total maintenance losses) and (2) a growth and priority function which specifies the percent of excess assimilate (EX) that is to be allocated to organ S_i. There is a higher priority for allocation to reproductive organs once reproductive growth is initiated.

As shown in Figure 2, the movement of labile assimilate from each pool back to the common labile pool may occur. The pool balance function governs the occurrence of such movements, which occur when the current production of photosynthate is not great enough to meet the total maintenance demands of the plant. This may occur at any time under any growth conditions, vegetative and/or reproductive. Thus, translocation is a function of source and sink strengths among the respective labile pool compartments.

TABLE 1

Summary of Equations Governing Carbon Allocation in the *Larrea* Model

	Vegetative organs (leaves, stems, and roots)	Reproductive organs			
		Bud stages			
		Early	Maturing	Flower	Fruit
Total excess carbon in plant	$EX = F - \Sigma[r_m S - (P - \delta_u S)\ 1/dt]$				
Carbon available for organs	$EXV = EX - EXR$	$EXR = EX - \tau_1 S_1$	$EXR = \min EX \tau_2 S_i$		
Total maintenance requirements per organ	$D_{sum} = \delta_u S - [P - (r_m S + M)]$				
Carbon allocated to organ	$I_{mp} = D_{sum} + a_i EXV$ where $i = 1,2,3$	$I_{mp} = D_{sum} + EXR$			
Allocated carbon available for growth	$E = I_{mp} - D_{sum}$				
Growth rate of organ (dS/dt)	$\frac{E}{r_g + \rho + \delta_u}$	$\frac{E}{r_g + \rho + \delta_u}$	$\min \begin{cases} \beta S\left[1 - \frac{S}{S_{max}}\right] \\ \frac{E}{r_g + \rho + \delta_u} \end{cases}$	O	$\min \begin{cases} \beta S\left[1 - \frac{S}{S_{max}}\right] \\ \frac{E}{r_g + \rho + \delta_u} \end{cases}$
Death rate of organ (S,P)	$S - P/\delta_u$ and AGE(72) (chronological)	$S - P/\delta_u$			
Average weight of individual bud or fruit when formed or mature (g dw)	—	0.0047	0.0178	0.0178	0.022

TABLE 1 (continued)

Summary of Equations Governing Carbon Allocation in the *Larrea* Model

	Vegetative organs (leaves, stems, and roots)	Reproductive organs			
		Bud stages			
		Early	Maturing	Flower	Fruit
Maximum weight (g dw) attainable by organ (S_{max})	—	$\int_t^{t+\Delta t} f(t)\,dt$ where $\Delta t = 2$ days; $N_{buds} = S_{max}/0.0047$	$0.0178N_{buds}$	$0.0178N_{buds}$	$0.022N_{buds}$
Growth rate of pool (dP/dt)		$\frac{dP}{dt} = \min \begin{cases} X - TR \\ Y \end{cases}$ $X = I_{mp} \quad (dS/dt\,(\rho + r_g) + r_m S + M)$ $Y = [\delta_u (S + dS) \quad P] \quad 1/dt$			
Carbon available for growth but not used; available as short-term "storage"		$TR = \max \begin{cases} X - Y \\ O \end{cases}$			

The pool balancing procedure involves calculating the assimilate that is available (λ_{ca}, percent per day) as a proportion of the necessary assimilate level that, if available, would bring each pool to maximum size:

$$\lambda_{ca} = \frac{\text{TPOOL} - \text{LOSSES}}{\text{MAXP}} \tag{10}$$

where

$$\text{TPOOL} = \sum_i P_i + F$$

$$\text{LOSSES} = \sum_i (r_m^i S_i + M_i)$$

$$\text{MAXP} = \sum_i P_i^{**}$$

TPOOL is the current total labile assimilate (all pools plus F), LOSSES is the total maintenance requirements, and MAXP represents the labile assimilate level necessary to bring each pool to its maximum level. Each pool is then allocated from the common labile pool a portion (grams glucose per day) of the available assimilate, I_{mp} (Equation 2 and Table 1):

$$I_{mp} = \lambda_{ca} \delta_u S_i + r_m^i S_i + M_i \tag{11}$$

and $dS_i/dt = 0$. This procedure results in a pool balance since the labile assimilate levels are brought to the same percentage of P_i^{**} throughout the plant. Each organ pool either gains or loses assimilates.

a. *Growth Rates*

The growth rate of an organ is a function of growth priorities and the total assimilate available for growth; the growth equations are outlined in Table 1. Vegetative growth is a partitioning of available assimilate in a specific ratio (e.g., 1:1:2) to these organs. In the absence of reproductive growth, all excess assimilate is allocated to vegetative organs. When reproductive growth is occurring, the amount of assimilate available for vegetative growth is determined by the parameters τ_1 and τ_2 (grams glucose per grams dry weight).

Reproductive growth is initiated if (1) the mean daytime air temperature exceeds 15°C and (2) EX is greater than a certain critical percent (τ_1) of the leaf structural biomass. The amount of assimilate greater than this percent (EXR) is then allocated to initiate early reproductive buds, and the remaining assimilate is allocated to vegetative growth (EXV; see Table 1). The allocation to the early reproductive buds continues for several days, and the number of buds formed (N_{buds}) is determined from the total allocation and average bud weight at this stage (Table 1).

Once reproductive growth is initiated, all excess assimilate up to another critical percent of leaf biomass (τ_2) is available for the growth and maintenance of the buds through the flower and fruit stages; remaining assimilate is used for vegetative growth (Table 1). The buds are considered mature when their total biomass reaches a size equal to the average weight of a mature bud times the total number of buds. Once mature, the buds are considered flowers (which persist for 48 hr) and are not allocated assimilate for structural growth. Hence, only maintenance requirements (respiration and nectar production) constitute the organ demand for flowers. At the end of 2 days, the structural biomass (= flowers) is aborted, and the pool is used to initiate the structure and pool compartments of the fruit stage. The fruits are similarly considered mature when their biomass reaches a critical weight (Table 1).

b. Death Rates

The final rate to be considered in the basic structural unit of the model (Figure 2) is the death rate. This rate is assumed to be a function of (1) the chronological age distribution of biomass (in the case of vegetative structures) or (2) the reaching of a minimum level of assimilate in the labile pool (P_i*) (both vegetative and reproductive organs).

The maximum chronological age obtainable by *Larrea* leaves is approximately 72 weeks.[43] We have assumed that root and stem biomass have the same life expectancy. All structural losses due to aging are accounted for by placing vegetative biomass in 72 chronological age classes (AGE[i], i = 1, 2, . . . 72), each containing the cohort of biomass produced during a given week. The most recent cohort of biomass, AGE(1), is the total amount of assimilate allocated to vegetative structure during the latest week; the biomass, if any, in AGE(72) is dropped each week.

If the pool of any organ drops below a lower limit (Equation 1B), a structural loss of biomass occurs at a magnitude which makes the existing pool size a new maximum. Thus, the biomass lost by death in grams per dry weight is

$$D(S,P) = (S - P/\delta_u)\, U(P - \delta_l S) \qquad (12)$$

where U is a unit step (Heaviside) function:

$$U(x) = 1 \text{ if } x \leq 0$$

$$U(x) = 0 \text{ if } x > 0$$

The loss of structural biomass during the developmental stages of reproductive growth includes a recalculation of the number of reproductive buds (N_{buds}) that remain.

D. Model Behavior

To examine the behavior of the model, a 1-year simulation was conducted for an individual *Larrea* plant. The initial environmental conditions correspond to those at the starting date, March 22, in southern New Mexico; two of the driving variables used in the simulation, daily mean air temperature and soil water potential, are shown in Figure 3. The results of the simulation are presented in Figures 4 to 6 and Table 2.

The seasonal trends in net photosynthesis and respiration are shown in Figure 4. The lowest net photosynthesis rates are during May to June (due to low soil moisture; see Figure 3) and late October through February (due to low air temperatures, low irradiance, and shorter day lengths; Figure 3). The amount of assimilate available for

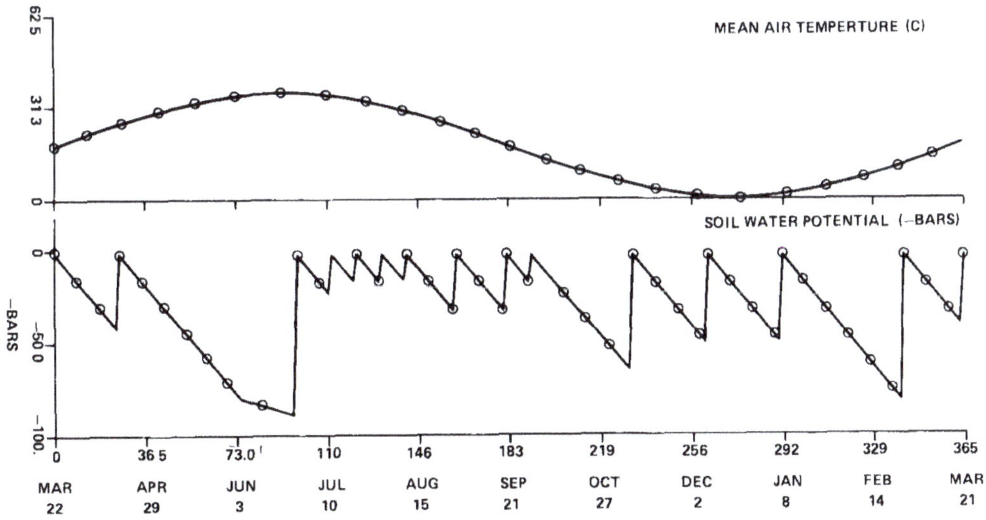

FIGURE 3. Daily values of mean air temperature and soil water potential typical of southern New Mexico (used in *Larrea* simulation).

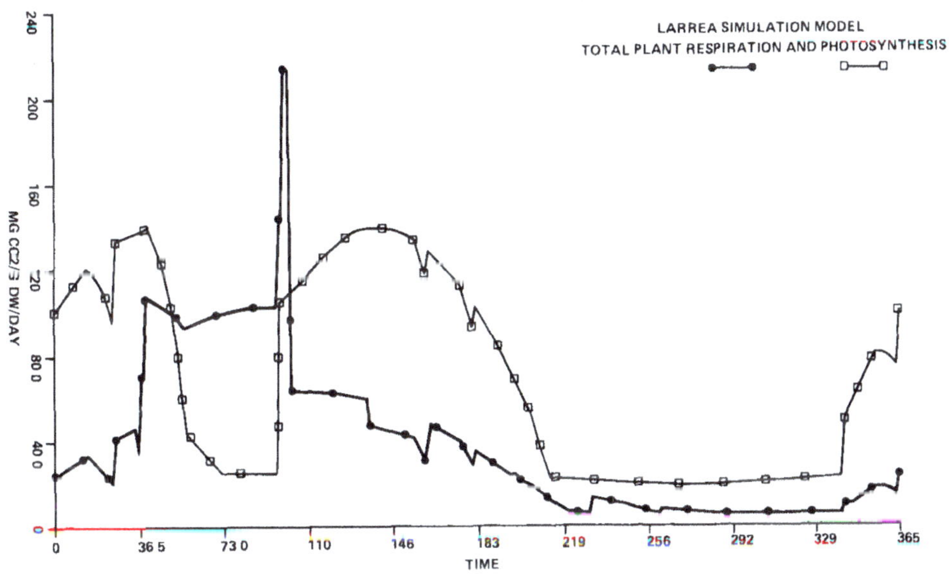

FIGURE 4. Seasonal trends in daily net photosynthesis and respiration (milligrams CO_2 per gram dry weight per day) predicted for *Larrea* during a 1 year simulation.

growth (EX; Table 1) is depicted in Figure 5 and tracks the seasonal patterns of net photosynthesis and respiration. Negative values of EX indicate that currently produced photosynthate is insufficient to meet the total respiration losses of the plant and, therefore, pool balancing is occurring. The peak respiration rates (days 99 to 101) are due to the high respiration rates of flowers which are present during this period (see Figure 6).

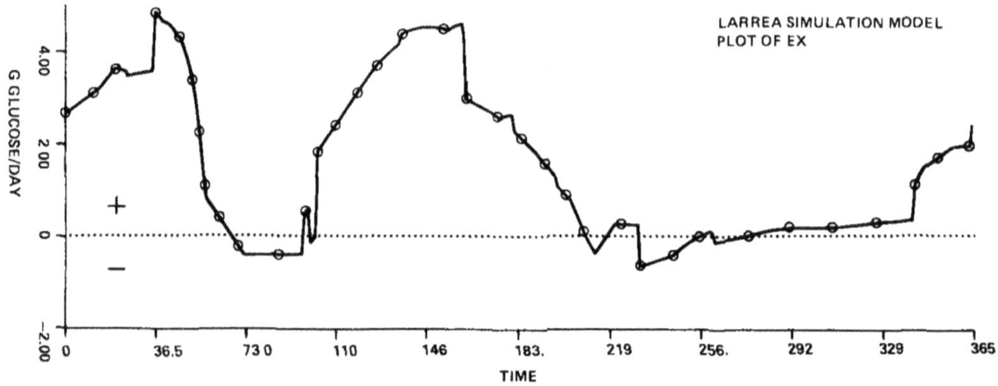

FIGURE 5. Seasonal trends of EX, the amount of assimilate (grams glucose per day) available for growth over 1-year simulation.

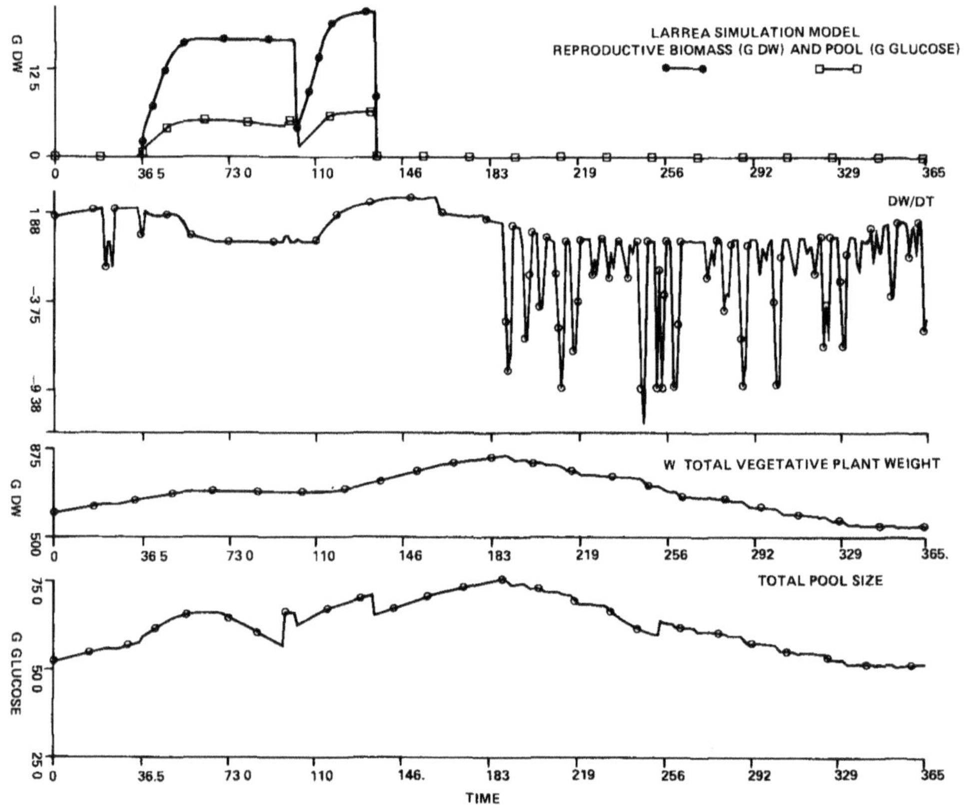

FIGURE 6. Vegetative and reproductive growth dynamics predicted for *Larrea* over a 1 year simulation.

The growth dynamics of the plant are shown in Figure 6 and summarized in Table 2. For the entire year, a total of 320 g of vegetative dry matter production was simulated using a fixed ratio for leaf to stem to root of 0.25:0.25:0.50 (i.e., $a_1 = 0.25$, $a_2 = 0.25$, $a_3 = 0.50$; Table 1). The seasonal variation in total vegetative biomass over the year (W = S1 + S2 + S3) is shown along with the daily rate of change (dW/dt); the pattern of dW/dt through the year reflects the growing season and the initial distribution of W into the 72 senescence categories. Note the decreases in the size of the total labile pool of the plant when EX is negative.

TABLE 2

Summary of Carbon Allocation During a 1-Year Simulation of a Hypothetical *Larrea* Plant

Net photosynthesis (g glucose)	1604.8
Total respiration (g glucose)	
Maintenance	1056.3
Growth	109.7
Total substrate utilization for construction (g glucose)	438.8
Vegetative growth	
Total plant weight (g dw)	
Initial	600.0
Final	559.6
Total growth (g dw)	320.7
Total senescence (g dw)	361.0

	Respiration (g glucose)		
	Maintenance	Growth	Substrate utilized for growth (g glucose)
Leaves	308.2	24.85	99.42
Stems	115.4	24.85	99.42
Roots	568.0	49.70	198.83
Totals	991.6	99.40	397.66

Reproductive growth	
Number of cohorts	1
Number of buds	937

	Weight (g dw)		Respiration (g glucose)		
	Initial	Final	Maintenance	Growth	Substrate utilized for growth (g glucose)
Early buds	0.00	4.41	0.40	1.37	5.47
Maturing buds	4.41	16.68	53.08	3.80	15.22
Flowers	16.68	16.68	3.29	0.00	0.00
Fruits	4.12	20.61	7.95	5.11	20.45
Totals			64.72	10.28	41.14

One cohort of reproductive biomass was initiated on day 36 (April 29), and final fruit maturation occurred 101 days later (August 8). Due to the dry conditions following the initiation of reproductive growth, the period for bud maturation was prolonged considerably and the associated pool can be seen to slowly decline during this period, reflecting the low assimilate levels in the plant and the restriction of no growth when $EX \leq 0$.

IV. DEVELOPMENT OF A PRIMARY PRODUCTION MODEL FOR LOBLOLLY PINE

In a recent critique of modeling activities in the U.S./IBP, O'Neill[12] discusses the philosophy of process modeling as implemented within the EDFB, detailed models of abiotic and biotic processes (e.g., water balance, decomposition, secondary production, etc.) were developed and attempts made to integrate these into comprehensive sit-specific ecosystem models. Toward this end, a number of scientists at the EDFB Research Triangle Site in North Carolina have collaborated in an intensive study of primary production and microclimate of an even-aged loblolly pine (*Pinus taeda* L.)

plantation.[58] As part of this effort, we have recently begun work on developing a model of primary productivity and carbon allocation in loblolly pine.

Loblolly pine was chosen for study because of its ecological and economic importance in the southeastern U.S. It forms pure stands over cut or burned forests and abondoned agricultural land and is the most important commercial tree species for pulpwood and timber production in the South. A plantation has unique advantages for an intensive ecosystem study: (1) the horizontal homogeneity of the stand simplifies the modeling os stand environmental parameters and (2) the uniform ages and sizes of the trees allow extropolation to the entire stand from studies of an average tree. Our work thus far in developing a primary production and carbon allocation model for loblolly pines has been at the level of this average tree.

There have been few attempts to construct physiologically based simulation models of forest tree growth. Instead, numerous empirical approaches to modeling forest tree dynamics have been utilized, e.g., yield and growth models, stochastic process models, size-class distribution models, etc. (see review in Ek[59]). However, if causal relationships within the forest ecosystem are of primary concern, a physiological approach is preferred.[60]

The major limiting factor in the evolution of sophisticated physiological growth models for forest trees has been the paucity of data. Consequently, initial attempts to develop growth models of tree species were based on data collected on tree seedlings over relatively short time periods, e.g., Ledig[61] and Promnitz.[62] Recently, however, attempts have been made to construct physiologically based ecosystem-level models of primary production in forests[60,63] and, although these models are relatively low in resolution, they are important components in any detailed study of energy and material flow in ecosystems.

A. Productivity and Growth Patterns in *P. taeda*

The growth patterns of loblolly pine have been documented elsewhere,[64-68] but a brief synopsis will be useful for purposes of the model description which follows. Growth of the main stem terminal, branch apices, and foliage occurs in distinct intervals (flushes) during the spring and summer growing period: one to seven flushes may occur during a single growing season depending upon various environmental factors and position in the live crown. The main stem bears distinct whorls of branches formed by the development of three to four axillary buds during the first flush of growth in the spring. Lateral branching may occur in subsequent flushes from the axillary buds of these new branches. Leaves are retained on the tree for two growing seasons. Thus, during any growing season, there are leaves present that have been produced during the current season as well as during the previous growing season. In addition, leaves produced during different flushes in a growing season form age classes which are morphologically and physiologically distinct from each other. In this geographical region, the branches in the upper two thirds (approximately) of the live crown produce at least two flushes of growth during a growing season, whereas the branches in the lower third never produce more than one flush.

The annual primary productivity of loblolly pine has been investigated by numerous researchers. Nemeth[69] estimated net primary production in an eastern North Carolina loblolly pine plantation to range from 4.8 kg dw/tree/year at age 4 to 18.8 kg dw/tree/year at ages 10 to 11. Kinerson et al,[70] estimated the net primary production to average about 29.1 kg dw/tree/year over a 4-year period (ages 12 through 16) for loblolly pine in the EDFB Research Triangle Site plantation. Detailed analyses of the annual production and seasonal distribution of biomass accumulation in loblolly pine have been done by Kinerson et al.[67,70] and Ralston.[71]

B. Brief Description of Modeling Approach

1. Canopy Models

The physical environment varies dramatically through the vertical profile of a forest stand. This is due, in part, to the physical and physiological processes of the plants themselves.[72] This variation can be expected to influence the processes of photosynthesis, respiration, and carbon allocation for an individual tree and, consequently, for the entire stand. Some researchers[65,73-76] have recognized zones of varying production within forest canopies, and it is clear that one height stratum affects the production of others.[77]

In a study designed to examine the influence of canopy position on net gas exchange in loblolly pine, Higginbotham and Strain[65] arbitrarily divided the live crown of trees at the Research Triangle Site into three strata such that an equal number of live nodes occurred in each (approximately six). Within each of these strata, they found differences in the physical environment, total biomass, gas exchange capacities of the photosynthetic tissue, number of flushes produced per branch during the growing season, and leaf surface area. Based on these findings, we have initiated the construction of "canopy models" of carbon allocation, i.e., Simulation Units which are defined to correspond to the different strata of the average tree in the plantation. A compartmental representation of this approach is shown in Figure 7.

Five Simulation Units have been defined to represent the average tree: (1 to 3) the leaves and branches in each of the three strata of the live crown, (4) the main stem, and (5) the roots (see Figure 7A). Each of these units contains the specific plant components necessary to model carbon flux. For example, Simulation Unit 1 (leaves and branches in the top stratum of the canopy) is composed of two age classes of needles (flush 1 and flush 2) produced during the previous growing season, two age classes of needles (flush 1 and flush 2) produced during the current growing season, and old branch wood (Figure 7B). Finally, the carbon dynamics of each individual component of the Simulation Unit is modeled using the basic carbon-balance paradigm presented previously (see Figure 2). This paradigm has been modified slightly to allow for a storage compartment for loblolly pine (Figure 7C)

The use of these Simulation Units allows us to incorporate, in a simplified fashion, spatial complexity and variation inherent in a forest stand. To illustrate this approach to modeling the carbon economy of loblolly pine, the details of Simulation Unit 1 will be presented.

For the purposes of the description which follows, Simulation Unit 1 represents six live nodes in the top stratum of the live canopy of the average loblolly pine. The characteristics of this average tree are given in Table 3. Although 44% of the total foliage biomass is found in this stratum, this Simulation Unit contains only a small percent (approximately 12%) of the total branch wood (see Table 3). It is assumed that only two flushes of growth per growing season occur at this height in the canopy (see Figure 7B) and that no reproductive growth occurs.

2. Photosynthesis Submodel

A model of net photosynthesis in loblolly pine should ideally account for differences in gas exchange rates with (1) crown position, (2) leaf age, and (3) environmental conditions. In an attempt to elucidate some of these relationships, Higginbotham[68] and Higginbotham and Strain[65] measured net photosynthesis of loblolly pine using temperature- and light-controlled cuvettes at various crown positions on shoots of different ages. These measurements were made throughout the year to determine the influence of season and leaf age on gas exchange capacities. Using these data, approximations

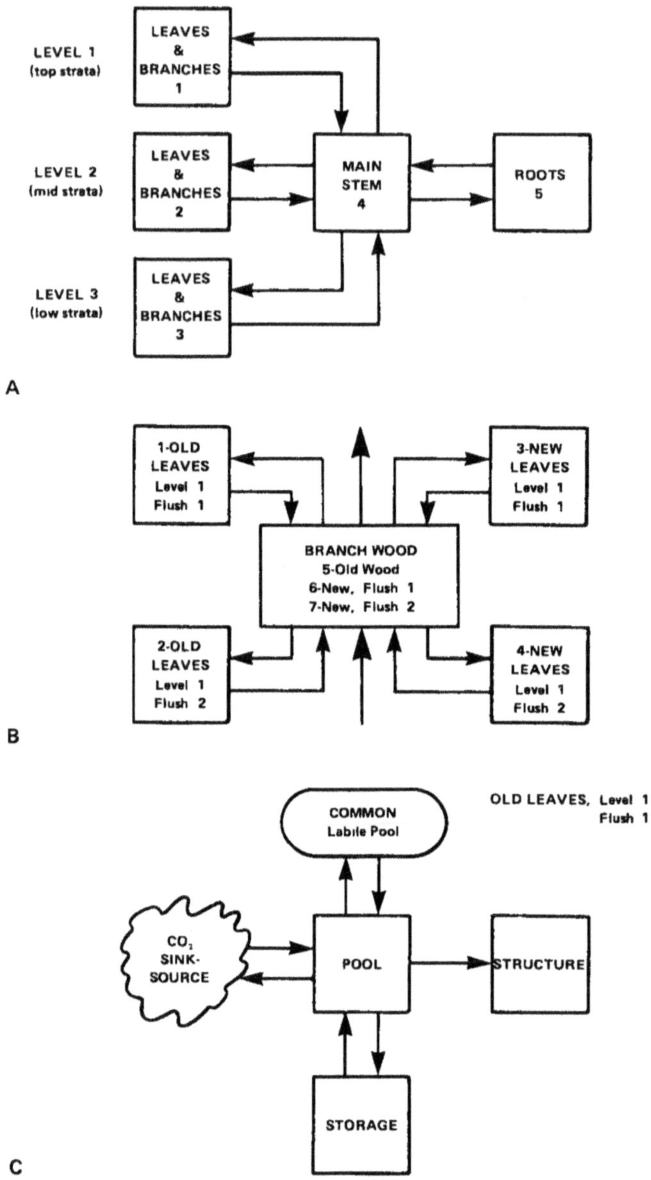

FIGURE 7. Compartmental representation of canopy modeling approach used in loblolly pine simulation model. (A) Major Simulation Units (1 to 5) used to compartmentalize the average tree in stand. (B) Details of Simulation Unit 1, the leaves and branches in the top strata of the tree. Compartments include: (1) flush 1, old needles; (2) flush 2, old needles; (3) flush 1, new needles; (4) flush 2, new needles; (5) old branch wood; (6) flush 1, new branch wood; and (7) flush 2, new branch wood. (C) Fundamental structural unit of model, such as that used for compartment 1 (old leaves, flush 1). (See Figure 2 for details of this basic structural unit.)

of CO_2 exchange rates in loblolly pine were obtained[78] using the Smith model described by Tenhunen et al.:[79]

TABLE 3

Characteristics of the Average Tree in the Loblolly Pine Plantation at the Research Triangle Site, 1974[68,71]

Age	16.0 year
Height	12.0 m
Stem diameter (at base of live crown)	11.9 cm
Stem diameter (at breast height, 1.5 m)	18.1 cm
Stem (= bole) weight	55.3 kg
Root weight	15.5 kg
Canopy depth	6.0 m

Properties of live canopy

Stratum	Height distribution in canopy (m)	Biomass[a](kg)	
		Needles	Branch wood
Top	10.0 — 12.0	1.4	0.96
Mid	8.0 — 10.0	1.8	3.70
Bottom	6.0 — 8.0	0.9	3.20

[a] From Higginbotham,[68] Ralston,[71] and unpublished data. Winter months only.

$$P_n(I, T, t) = \frac{\alpha I P_{ML}}{[P_{ML}^2 + (\alpha I)^2]^{1/2}} - R_S \qquad (13)$$

where p

$$\alpha = f(T)$$

$$P_{ML} = P_{max} V(T)$$

$$P_{max} = f(t)$$

$$V(T) = [2(T \quad T_I)^2 (T_u - T_I)^2 - (T - T_I)^4] / [T_o - T_I]^4$$

$$T_o = f(t)$$

$$T_I = T_o - 45.0$$

$$R_s = 1.05 \exp\left[9000 \ln\left(\frac{1}{293} - \frac{1}{T + 273}\right)\right]$$

P_n (I,T,t) is the rate of net photosynthesis (milligrams CO_2 per dm^2 per hour) as a function of time, t (d^{-1}), a specific photon flux density, I (μeinstein per m^2 per second, PAR) and a specific temperature, T (°C) at saturating CO_2 concentrations. P_{max} is the rate of photosynthesis at saturating photon flux density and CO_2 concentration; α is the quantum efficiency of the photochemical process (milligrams per μeinstein); V(T) is a function which gives the temperature dependence of P_m; T_o and T_l (°C) are the optimum and low temperature intercepts of the function V; and R_s is the rate of respiration (milligrams CO_2 per dm^2 per hour).[72]

Note that α, P_{max}, and T_o are computed as functions of time. The seasonal variations in these parameters for each age class of foliage was highly significant (see Higginbotham and Strain[65] for yearly trends in P_{max}). As in Equation 4, a nonlinear least-squares analysis was used to estimate the seasonal values of these parameters. In the simulation, P_m is converted to grams glucose per dm^2 per hour using the relationships given in Chung and Barnes:[80] 0.682 grams glucose per grams CO_2 fixed and a CO_2 to biomass ratio of 1:0.433.

The interception of radiation (I) at a specific height in the canopy is approximated according to the Beer-Lambert extinction law:

$$I = I_0 \exp(-gL) \tag{14}$$

where I_o is the intensity of radiation incident on the top of the canopy, g is an extinction coefficient, and L is cumulative leaf area index above the level of interest.[68] I_o is estimated using a sine-curve function.

3. Respiration Submodels

The respiration rate (milligrams CO_2 per dm^2 per hour) of the ith component of the Simulation Unit (r'_m) is calculated as:

$$r_m^i = \phi_1 R + \phi_2 RK / (R_L + K) \tag{15}$$

where all parameters, except R, are as defined in Equation 8; Higginbotham[68] computed needle respiration rates as a function of temperature using the equation:

$$R = \exp(a + bT) \tag{16}$$

There were no seasonal Q_{10} shifts in his data. On the other hand, Kinerson[81] presented data on stem respiration which exhibited marked shifts in Q_{10} before and after the initiation of cambial activity in the spring. Thus, we estimated branch respiration rates:

$$R = a \exp\left[b \frac{(T - T_b)}{T_b}\right] \tag{17}$$

where b = f(t), to account for shifts in Q_{10}, and T_b is the baseline temperature.

4. Carbon Allocation

The general patterns of photosynthate allocation in woody plants have been recently reviewed by several authors. [64,82-85] In addition, the translocation of photosynthate in loblolly pine has specifically been investigated by Chung[86] and reviewed by Murphy,[66] Barnes,[87] and Hellmers and Strain.[64] Our effort to model carbon allocation in loblolly pine is an attempt to bring these data together with those of net assimilation rates and environmental influences.

Carbon flux for each component of Simulation Unit 1 is described by Equations 1 to 3. However, the additional storage pool (see Figure 7C) must also be considered. The storage pool is defined for all components (compartments) of the Simulation Unit except shoots produced during the current growing season (compartments 6 and 7, Figure 7B). We assume that, in general, storage in needles is more of a short-term phenomenon than storage in the branches. The dynamics of each storage pool, Q (g glucose), is given by:

$$\frac{dQ}{dt} = TI \cdot U(t - t_{crit}) - OT \qquad (18)$$

where

$$TI = \min \begin{cases} (Q_{cap} - Q + OT) \cdot B(Q) \\ TR \end{cases}$$

$$Q_{cap} = Q_{max} - \delta_u S$$

$$Q_{max} = \delta_s S$$

$$B(Q) = (Q / Q_{max}) / \{K_q + (Q / Q_{max})\}$$

Input of assimilate to the storage pool, TI (grams glucose per day), is controlled by (1) the amount of available assimilate, TR (grams glucose per day) (see Equation 2), remaining after growth and maintenance demands of the organ are met; (2) a maximum storage capacity, Q_{cap} (grams glucose), determined by the parameter δ_s (grams glucose per grams dry weight); and (3) a maximum rate of storage depletion allowable, which is in turn determined by the function B(Q). In some cases, it is assumed that allocation of assimilate to storage pools is seasonal, i.e., TI = 0 when > t_{crit}.

The output of assimilate from a storage pool, OT (grams glucose per day), is a function of the total growth and maintenance demands of sinks within a defined system of allowable translocation pathways (sources → sinks). These pathways are given in Table 4. The amount of assimilate translocated from storage pool i is given by:

$$OT_i = \lambda_{nc} Q_i \qquad (19)$$

TABLE 4

Translocation pathways within Simulation Unit 1

Compartment *i*	Component	Age	Flush	Compartments for which compartment *i* serves as a: Sink	Source
1[a]	N[b]	Old[c]	1	—	3,4,5,6,7
2	N	Old	2	—	3,4,5,6,7
3	N	New	1	1,2,5	4,5,6,7
4	N	New	2	1,2,3,5	5,6,7
5	W	Old	—	1,2,3,4	3,4,6,7
6	W	New	1	1,2,3,4,5	—
7	W	New	2	1,2,3,4,5	—

Note: The strength of a sink or source is determined by current growth dynamics (see text for details).

[a] Number corresponds to those in Figure 7B.

[b] N = needles; W = branch wood.

[c] Old = produced during previous growing season of the previous year. New = produced during current growing season.

where

$$\lambda_{nc} = \min \begin{cases} \sum_{i=3}^{7} D_i \Big/ \sum_{i=1}^{5} Q_i \\ 1.0 \end{cases}$$

The ratio λ_{nc} (percent per day) is a measure of how much assimilate is needed, ΣD_i (grams glucose per day), compared to the total assimilate available in storage (ΣQ_i). The available pools in Simulation Unit 1 are compartments 1 to 5; the compartments that import assimilate from these pools are 3 to 7 (see Table 4). D_i for a given compartment is the sum of its maximum growth rate (including growth respiration), maintenance respiration, and pool deviations (if any) from P_i** minus photosynthetic gains (for leaves).

Note that the transfer of stored assimilate from compartments i to j can be calculated from Equation 19:

$$OT_{ij} = \left(\frac{D_j}{\sum_{j=3}^{7} D_j} \right) Q_i \qquad (20)$$

where i = 1, 2, . . . 5 and j = 3, 4, . . . 7. Also, the total imported assimilate to compartment j, I'_{mp}, is given by:

$$I^j_{mp} = \sum_{i=1}^{5} OT_{ij} \tag{21}$$

where j = 3, 4, 6, 7. The old branch wood compartment (j = 5) serves as common storage sink, i.e.,

$$I^5_{mp} = \sum_{i=1}^{4} OT_{i5} + \sum_{\substack{i=1 \\ i \neq 5}}^{7} TO_i \tag{22}$$

where $TO_i = TR_i - TI_i$. TO is the amount of assimilate (g glucose) actually translocated out of a compartment; this is a function of available assimilate (TR; Equation 2) and the amount of input to storage (TI; Equation 18). Finally, the amount of assimilate exported from the entire Simulation Unit is given by TO_5. This is available to the other Simulation Units (Figure 7A) in the tree; presently we assume no translocation into Unit 1 from others.

The general foliage growth equation (G[S]; see Equation 3) is

$$G(S) = \beta S \left(1 - \frac{S}{\gamma F_{max}} Z(T,\Psi) \right) \tag{23}$$

where β is the growth rate (grams dry weight per grams dry weight per day), γ is a coefficient which converts leaf area to grams dry weight (grams per m^2), F_{max} is the maximum leaf area allowable based on stand conditons, and $Z(T,\Psi)$ is a function determining the effect of temperature and soil moisture conditions on the growth rate. Branch wood growth rates are similarly determined on the basis of the structure necessary to support a given amount of foliar biomass.

In general, the following pattern of growth dynamics and assimilate allocation occurs in Simulation Unit 1 (details are given in Higginbotham[68]). During the winter months, the only needles present are those which were produced during the previous growing season (i.e., compartments 1 and 2, Figure 7B). The initiation of growth in the spring (about March 10) is controlled by photoperiod and temperature and begins with the elongation of the first flush shoots (compartment 6). This growth is at the expense of stored assimilate as well as currently produced photosynthate. First flush needles (compartment 3) start their maximum growth phase after the shoots have completed their period of maximum growth (about 4 to 6 weeks). These needles are a strong assimilate sink because they have a negative net CO_2 balance for approximately 4 to 6 weeks after emerging. During this period, the 1-year-old needles are photosynthesizing at their maximum rates.

The second flush of growth (compartment 7) is initiated about 1 1/2 months after the start of the first flush. This growth is also at the expense of stored assimilate and excess photosynthate (from compartments 1 to 3). The needles produced during the second flush (compartment 4) follow essentially the same pattern of assimilate demand as the first flush needles.

Most needles persist until the fall of their second year when decreasing temperatures

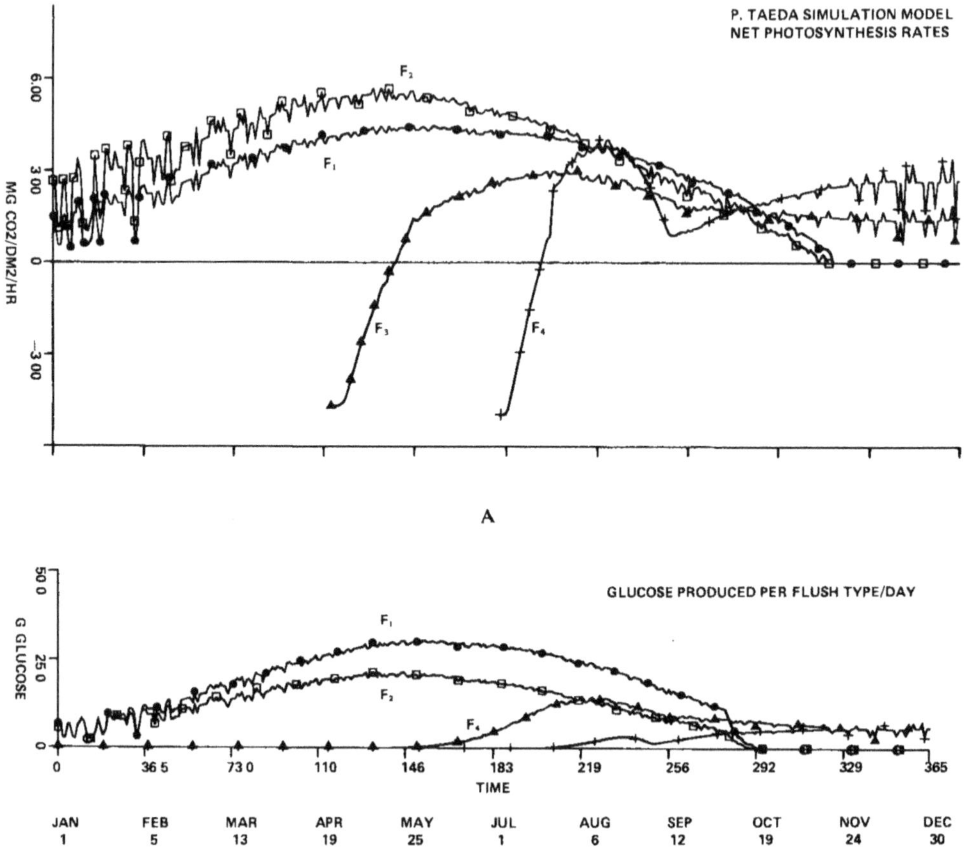

B

FIGURE 8. (A) Model output for net photosynthesis rates (milligrams CO_2 per dm^2 per hour) of top strata for average tree. F_1 and F_2 are rates of the first and second flushes of old needles, respectively, and F_3 and F_4 are rates of the first and second flushes of new needles, respectively. (B) Conversion of net photosynthesis rates to grams of glucose per flush type per day.

and plant water potential initiate and accelerate abcission. The needles do not grow during their second year on the tree. The growth rate of secondary wood is assumed to be a function of temperature and moisture status. Labile-assimilate pool balancing is done in a manner similar to that described by Equation 10.

C. Model Behavior

Model behavior is illustrated by the results of a 1-year simulation using average tree conditions (Table 3). The starting date for the simulation was January 1, and environmental conditions in central North Carolina were used as input to the model. Insight into the dynamic behavior of the model can be gained by examining seasonal variations in photosynthesis, assimilate pool sizes, structural biomass accumulations, and inputs and outputs from the various labile assimilate pools (Figures 8 to 11).

Seasonal trends in net photosynthesis of the different foliage age and flush classes are shown in Figure 8A. The magnitudes and seasonal variations of these net CO_2 exchange rates are of primary importance in determining the carbon budget of Simu-

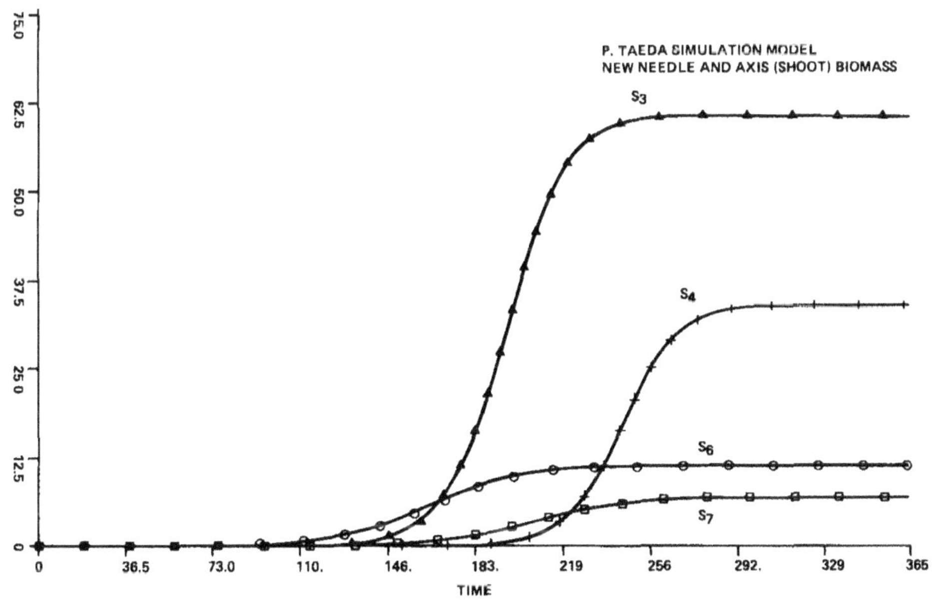

FIGURE 9. Model output showing new needle and branch growth dynamics during growing season. S_6 is the first flush of branch wood growth which supports the first flush needles, S_3, and S_7 is the second flush of branch wood growth which supports the second flush needles, S_4.

lation Unit 1. Foliage produced during the previous year (F_1 and F_2) shows a gradual increase in net photosynthesis until about the beginning of May and then a gradual decrease until these needles are lost near the end of October. Foliage produced during the current year (F_3 and F_4) initially has a negative net CO_2 balance and does not achieve net photosynthesis rates higher than those of the old needles until about the end of September. It is also interesting that foliage produced during the first flush has consistently lower net photosynthesis rates than does the second flush foliage. It is apparent that the major source of photosynthate during the course of a year is the foliage produced in the previous year (Figure 8B).

In Figure 9 are shown the growth dynamics of new shoot biomass, S_6 (first flush) and S_7 (second flush), and new foliage, S_3 (first flush) and S_4 (second flush). As can be seen from Figures 8 and 9, the initiation of growth in the spring coincides with the period of peak photosynthetic production. These new shoots and foliage are strong sinks for both stored and currently produced photosynthate.

The dynamics of the various translocation pathways are evident in Figures 10 and 11. The translocation of assimilate out of the various storage pools (as described by Equation 19) is shown in Figure 10A; this gives the total amount of assimilate translocated out of a given compartment. The specific fate of this total exported assimilate (see Equation 20) is illustrated in Figure 11.

Old foliage and branch wood are the major sources of assimilate for the first flush spring growth (Figure 11A, C, and D). The storage pools of the old foliage (i.e., compartments 1 and 2) are depleted near midsummer (Figure 11C, D); however, they are still important sources of assimilate because of their current photosynthate production. This photosynthate is shunted through the common old branch storage pool (i.e., compartment 5), which becomes the only source of assimilate for a period of time after midsummer (Figure 11). As described by Equation 20, this old branch wood storage

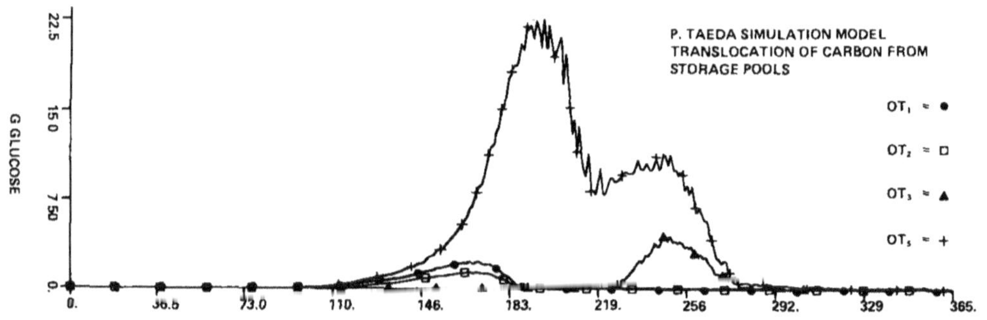

A

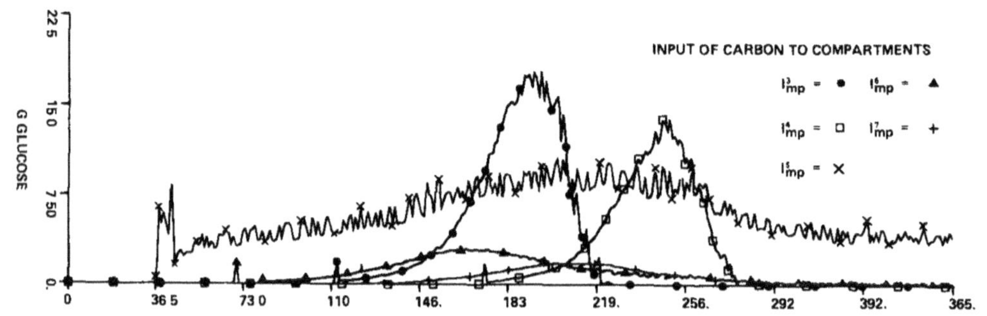

B

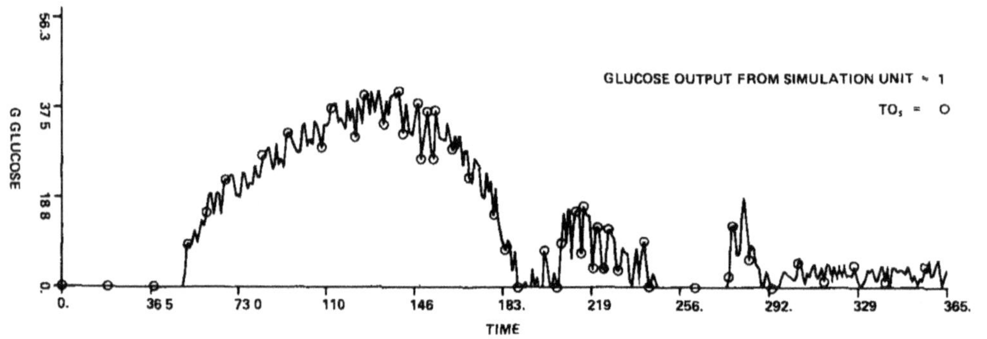

C

FIGURE 10. Model output. (A) Daily values of total carbon translocated from each compartment (see Equation 19). (B) Daily values of total carbon imported to each compartment (see Equations 21 and 22). (C) Daily values of total amount of glucose translocated out of Simulation Unit 1 (see Equation 22).

pool is the sink for excess assimilate from the foliage (see Figure 10B, I^{5}_{mp}). The second flush is, in part, supported by old foliage as well as by the first flush foliage of the current growing season (see OT_{34}; Figure 11B). The total import of assimilate by each compartment (Equations 21 and 22) is shown in Figure 10B.

In addition to supplying current growth needs of the Simulation Unit, there is a

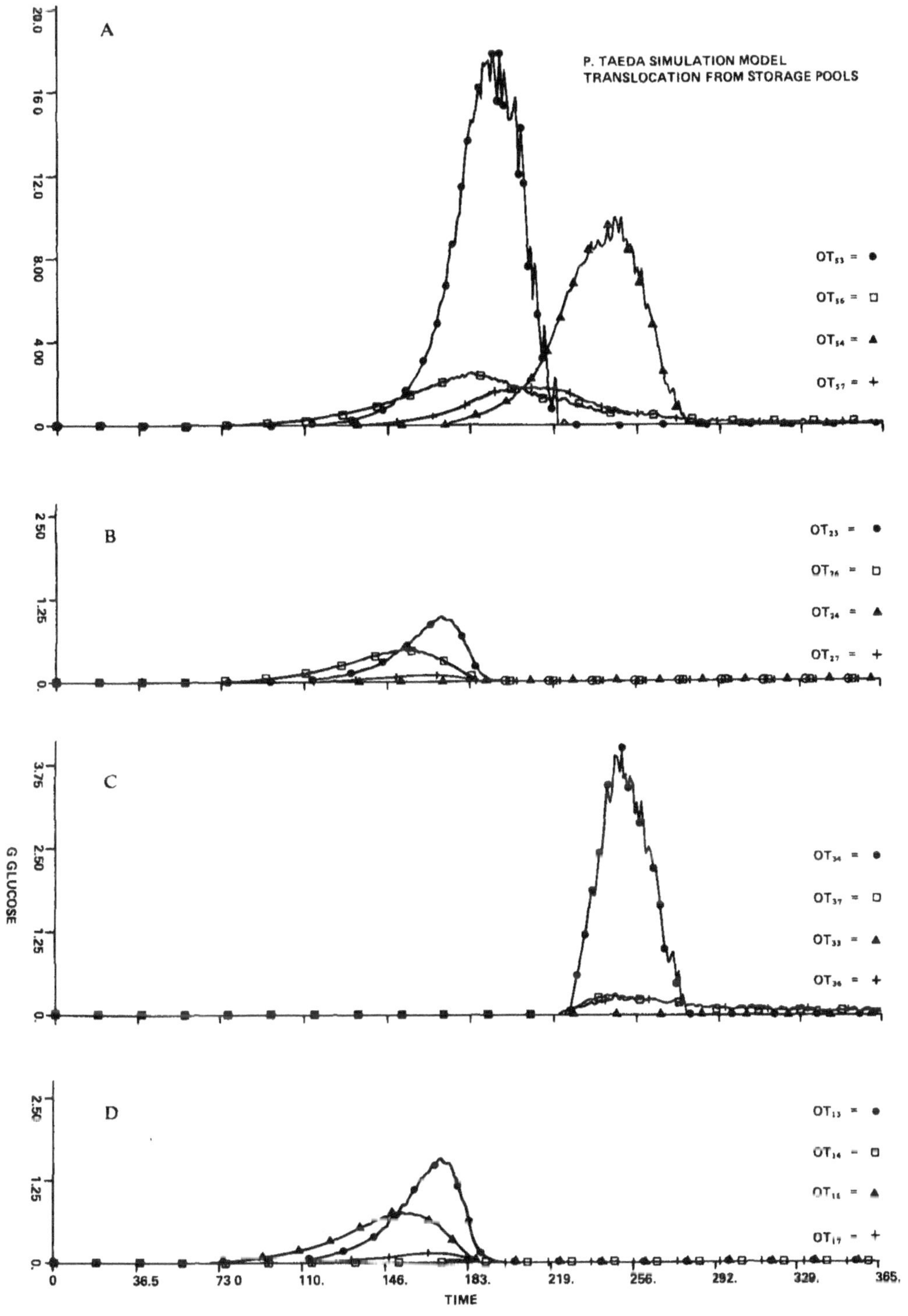

FIGURE 11. Model output. (A) Translocation of carbon from old wood (compartment 5). (B) Translocation of carbon from first flush, new needles (compartment 3). (C) Translocation of carbon from second flush, old needles (compartment 2). (D) Translocation of carbon from first flush, new needles (compartment 1).

seasonal variable export of assimilate from Unit 1 to other units (see Equation 22, TO_5). This export is shown in Figure 10C. The greatest export occurs early in the growing season when photosynthesis rates are high and growth demands are low.

V. SUMMARY AND FUTURE PROSPECTS

The modeling approach presented in this chapter was developed as part of a continuing effort to understand carbon flow and primary productivity in natural ecosystems. Ultimately, we hope to contribute to the development of models which can be used to predict the behavior of ecosystems. It is clear from this book and other recent reviews of modeling progress, however, that predictive ecosystem modeling has not yet been achieved. It is our assumption that total system modeling will not be possible until the successful development of submodels which adequately predict the behavior of critical system components. The allocation of carbon to the various biochemicals, tissues, and organs of the primary producers is obviously one critical aspect of ecosystem behavior. Any detailed study of mass and energy flux in ecosystems requires an understanding of energy procurement and allocation by the primary producing organisms. We feel that the models developed here are evidence of significant progress towards the development of predictive submodels from which predictive ecosystem models will eventually be derived.

One of the specific objectives of this study was to analyze the possibility of developing a single modeling framework which could be used to predict carbon allocation and biological productivity of different perennial growth forms in different ecosystems. A comparison of the application of our scheme to modeling growth in a desert shrub, *Larrea tridentata*, and a southeastern conifer, *Pinus taeda*, illustrates that this objective was rather successfully attained. The generality and flexibility of the model is demonstrated by its ability to accommodate these two greatly different plant species. Numerous differences in these species which must be considered include significant differences in growth form, environmental resistance, carbohydrate storage and translocation patterns, etc. It has been demonstrated that the model developed here can accommodate such extreme biological differences. Additionally, the details of internal mechanisms are sufficiently adaptable to allow an interfacing with other ecosystem-level processes. Accomplishment of this latter point is our ultimate objective as stated above. We are currently expanding the effort by the development of submodels for nutrient and micrometeorological processes.

An additional objective of this research is to use the simulations to test and formulate hypotheses on plant growth, carbon allocation, and primary productivity. Most of our effort to date has been directed toward the development of the model for use with *Larrea*. In doing this, however, we have encountered several situations in which unexpected model output has forced us to rethink our understanding of plant behavior. We have found, also, that the simulations provide a level of generality which is instructive in comparing growth behavior among other species. This type of application of the model will undoubtedly be continued and, presumably, expanded in the future.

Although the level of development of this model is low relative to that obtained in some annual crop species, it must be realized that the data base is extremely limited in natural perennial species. Models will undoubtedly be increasingly useful in directing future data-gathering efforts as ecologists attempt to "catch up" with agricultural researchers.

Although we have attempted no validation at this stage, the models appear to perform reasonably well in both a qualitative and a quantitative sense. The seasonal variations in numerous parameters have been investigated in some detail, but further examination of these patterns is necessary. In fact, a full-scale sensitivity analysis of model parameterization would seem warranted. The results of such analyses would lead to further refinement (or abandonment!) of proposed mechanisms and, hence, generate testable hypotheses.

The usefulness of developing predictive models to aid in understanding and managing natural ecosystems is well documented, particularly as an outgrowth of efforts associated with the IBP. At this stage, the major obstacle in such efforts is the paucity of data and the (often) enormous expense and difficulty in obtaining such data. However, in spite of these shortcomings, ecosystem-level models will continue to play a unique and important role in ecosystem studies.

ACKNOWLEDGMENTS

The research reported here was supported by the Desert Biome US/IBP through NSF Grant GB 15886, and by the Eastern Deciduous Forest Biome US/IBP through NSF Interagency Agreement AG-199 BMS69-01147 A09 with the Department of Energy, Oak Ridge National Laboratory. Phytotron studies were supported, in part, by NSF Grants BG 7153 and GB 19634 to the Southeastern Plant Environmental Laboratories, Duke University. We wish to thank the many researchers involved in the design and implementation of these studies and for making their data and/or results available to us. In particular, thanks are due R. L. Barnes, H. H. Chung, A. C. Dudgeon, K. O. Higginbotham, R. S. Kinerson, C. W. Ralston, and J. P. Syvertsen. Contribution Number 320 from the Eastern Deciduous Forest Biome, U.S. I.B.P.

VI. SYMBOLS

a_i	Dimensionless coefficient determining the allocation of carbon to leaves (i = 1), stems (i = 2), and roots (i = 3) (see Table 1)
a_m	Minimum value of r_l obtained at high irradiance (seconds per meter)
A	Either living biomass (grams dry weight) or surface area (dm^2) for use in determining maintenance respiration costs
AGE (i)	Vegetative biomass in the ith chronological age class (grams dry weight)
B(Q)	Function which determines the maximum rate a storage pool can be depleted; expression defined in Equation 18
C_a	Ambient CO_2 concentration in the air (milligrams per m^3)
C_i	CO_2 concentration in the internal substomatal cavity of the leaf (milligrams per m^3)
D_i	Assimilate demand of organ i to fulfill all growth and maintenance requirements (grams glucose per day)
D_{sum}	Assimilate requirement of an organ (grams glucose); expression defined in Table 1
D(S,P)	Function which determines the death rate of structural biomass; expression defined in Equation 12 and Table 1
E	Excess carbon, above that required for maintenance of an organ, available for growth, storage, or translocation (grams glucose per day)
EX	Excess carbon remaining after the total maintenance demands of the plant are satisfied (grams glucose per day) partitioned for vegetative and/or reproductive growth (Table 1)
EXR	Expression defined in Table 1

EXV	Expression defined in Table 1
$E(\Psi)$	Function which determines the effect of moisture status on P_m
F	Net photosynthesis (grams glucose per day [Equation 2] or milligrams CO_2 per m^2 per second [Equation 4])
F_{max}	Maximum leaf area allowable under a given stand density (m^2)
G(S)	Function which determines the maximum growth rate of an organ (see Table 1 and Equation 23)
g	Extinction coefficient for light penetration through the canopy
I	Photon flex density intercepted by leaves (μeinsteins per m^2 per second)
I_{mp}	Carbon translocated into the pool of an organ (grams glucose per day)
$I_{mp}{}^j$	Total imported assimilate to compartment j (grams glucose per day); expression defined in Equation 21
I_o	Incident photon flux density (μeinsteins per m^2 per second)
k	Internal (mesophyll) CO_2 conductance (meters per second)
K	The ratio P_i/P_i^{**} used in determining substrate-controlled respiration rates
K_m	The value of photon flex density at which stomatal resistance is twice its minimum value (μeinsteins per m^2 per second)
K_q	The value of Q/Q_{max} at which the rate of storage pool depletion is one half the calculated rate
L	Leaf area index (m^2 of leaf surface per m^2 of ground area)
LOSSES	Expression defined in Equation 10
MAXP	Expression defined in Equation 10
M_i	Maintenance losses of the ith organ other than respiration (grams glucose per day)
$M_{mult}{}^i$	Dimensionless scalar which introduces the relative effect of soil moisture on the respiration rate of the ith organ
N_{buds}	Number of buds initiated in reproductive growth (see Table 1)
OT	Amount of assimilate translocated out of a compartment (grams glucose per day)
OT_i	Amount of assimilate translocated out of compartment i (grams glucose per day)
OT_{ij}	Amount of assimilate translocated from compartment i to compartment j (grams glucose per day)
P	Pool fraction of carbon (labile state) associated with an organ (grams glucose)
P*	Minimum carbohydrate pool size (grams glucose)
P**	Maximum carbohydrate pool size (grams glucose)
PAGE(i)	Leaf biomass in the ith physiological age class (grams dry weight)
P_{max}	Maximum rate of net photosynthesis at saturating light and carbon dioxide concentration (milligrams CO_2 per dm^2 per hour)
P_{ML}	Maximum photosynthesis rate at saturating light and CO_2 concentration as a function of time and temperature (milligrams CO_2 per dm^2 per hour)
$P_M(T_o)$	Maximum photosynthesis rate at optimum temperature with saturating irradiance and CO_2 (milligrams CO_2 per m^2 per second)
$P_m(T,\Psi)$	Maximum photosynthesis rate at saturating photon flux density and CO_2 at a given temperature and moisture status (milligrams CO_2 per m^2 per second) expression defined in Equation 4B
$P_n(I,T)$	Internal net photosynthesis as a function of photon flux density and temperature (milligrams CO_2 per m^2 per second) expression defined in Equation 4B

$P_n(I,T,t)$	Rate of net photosynthesis at saturating carbon dioxide concentration as a function of irradiance, temperature, and time (milligrams CO_2 per dm^2 per hour)
Q	Storage pool (grams glucose)
Q_{cap}	Maximum capacity of storage pool (grams glucose)
Q_{max}	Maximum amount of labile assimilate (pool + storage) associated with an organ (grams glucose)
r_g	Growth respiration (grams glucose per grams dry weight)
r_l	Stomatal resistance (seconds per meter)
r_m	Maintenance respiration (milligrams CO_2 per hour) on a dry weight (per grams dry weight) or surface area (per dm^2) basis
r'_m	Maintenance respiration of the ith organ or plant component (milligrams CO_2 per hour) on a dry weight (per grams dry weight, Equation 8) or surface area (per dm^2, Equation 15) basis
r_s	Respiration associated with the synthesis of storage carbohydrates from soluble carbohydrates (grams glucose per grams glucose)
R_L	Dimensionless parameter which gives the relative pool size at which the respiration rate is lowered by one half
R'_{max}	Maximum respiration rate of the ith organ (milligrams CO_2 per grams dry weight per hour)
R_s	Dark respiration rate (milligrams CO_2 per dm^2 per hour) of leaves as defined in Equation 13
S	Structural fraction of carbon (nonlabile state) of an organ (grams dry weight)
S_l	Weighted leaf biomass (grams dry weight); expression defined in Equation 7
SEN(i)	Dimensionless scalar (0 to 1) which measures the relative gas exchange capacity of leaf biomass in the ith physiological age class
S_{max}	Maximum attainable weight for an organ (grams dry weight)
T	Air temperature (°C)
T_b	Baseline air temperature in Q_{10} relationship as defined in Equation 17 (°C)
T_l	Lower temperature intercept of function measuring temperature dependence of photosynthesis (°C)
T_o	Optimum temperature intercept of function measuring temperature dependence of photosynthesis (°C)
T_s	Soil temperature (°C)
TI	Excess assimilate imported to storage pool to maintain storage at capacity (grams glucose per day); expression defined in Equation 18
T_{multi}'	Dimensionless scalar (0 to 1) which determines the relative effect of temperature on the respiration rate of the ith organ
TO_i	Assimilate translocated out of compartment i (grams glucose per day); expression defined in Equation 22
TPOOL	Expression defined in Equation 10
TR	The portion of excess carbon that is available to an organ which is not used for maintenance, growth, or storage (grams glucose per day)
U	Unit step (Heaviside) function (see Equation 12)
V(T)	Function which gives the temperature dependence of P_{max}
W	Total vegetative plant biomass (grams dry weight)
$W_d(0,0)$	Dark respiration at 0°C (milligrams per m^2 per second)

$Z(T,\psi)$	Function which determines the effect of temperature and moisture status on the growth rate of an organ
α	Quantum efficiency of the photochemical process (milligrams perμeinstein)
β	Growth rate parameter (grams dry weight per grams dry weight per day) (see Table 1)
γ	Coefficient which converts leaf area to grams dry weight (grams per m^2)
δ_l	Coefficient which determines the lower size limit of a pool (grams glucose per grams dry weight)
δ_s	Coefficient which determines the maximum amount of labile assimilate (pools + storage) associated with an organ (grams glucose per grams dry weight)
δ_u	Coefficient which determines the upper size limit of a pool (grams glucose per grams dry weight)
η	Temperature coefficient of dark respiration (per °C)
θ	Mean leaf angle (degrees)
λ_{ca}	Percentage of carbon available in the plant compared to the necessary level that would bring each pool to its maximum size (percent per day)
λ_{nc}	Expression defined in Equation 19 (percent per day)
ξ	Photon flux density coefficient of dark respiration ([μeinsteins per m^2 per second])
ϱ	Substrate required for the synthesis of dry matter (grams glucose per grams dry weight)
σ_k	Parameter determining the kth maintenance loss other than respiration (grams glucose per grams dry weight)
τ_1	Parameter which governs the distribution of excess carbon (EX) for initiating reproductive growth (grams glucose per grams dry weight)
τ_2	Parameter which governs the distribution of excess carbon (EX) to reproductive growth (grams glucose per grams dry weight)
ϕ_1	Dimensionless scalar which determines the respiration of an organ due to base metabolism
ϕ_2	Dimensionless scalar which determines the substrate dependence component of respiration
ψ	Soil water potential (− bars)

REFERENCES

1. **Hammond, A. L.,** Ecosystem analysis: biome approach to environmental research, *Science,* 175, 46, 1972
2. **Blair, W. F.,** *Big Biology: The US/IBP,* Dowden, Hutchinson & Ross, Stroudsburg, Pa., 1977.
3. **Harris, W. F. and Reichle, D. E.,** Evaluating forest productivity in an ecosystem context, in *Proc. Agricultural Research Institute,* National Academy of Sciences, Washington, D.C., 1973, 57.
4. **Neuhold, J. M.,** Introduction to modeling in the biomes, in *Systems Analysis and Simulation in Ecology,* Vol. 3, Patten, B. C., Ed., Academic Press, New York, 1975, 7.
5. **Wiegert, R. G.,** Simulation models of ecosystems, *Annu. Rev. Ecol. Syst.,* 6, 311, 1975.
6. **Evenari, M., Schulze, E.-D., Lange, O. L., Kappen, L., and Buschbom, U.,** Plant production in arid and semi-arid regions, in *Water and Plant Life,* Lange, O. L., Kappen, L., and Schulze, E.-D., Eds., Springer-Verlag, Berlin, 1976, 439.
7. **Watt, K. E. F.,** Critique and comparison of biome ecosystem modeling, in *Systems Analysis and Simulation in Ecology,* Vol. 3, Patten, B. C., Ed., Academic Press, New York, 1975, 139.

8. **Walters, C. J. and Efford, I. E.**, Systems analysis in the Marion Lake IBP project, *Oecologia (Berlin)*, 11, 33, 1972.
9. **Adams, M. S., Dettmann, E. H., and Abbott, B. A.**, An ecosystem model in practice, *Environ. Conser.*, 1, 50, 1974.
10. **Cooper, C. F.**, Ecosystem models and environmental policy, *Simulation*, 26, 133, 1976.
11. **Goodall, D. W.**, Ecosystem modeling in the desert biome, in *Systems Analysis and Simulation in Ecology*, Vol. 3, Patten, B. C., Ed., Academic Press, New York, 1975, 73.
12. **O'Neill, R. V.**, Modeling in the eastern deciduous forest biome, in *Systems Analysis and Simulation in Ecology*, Vol. 3, Patten, B. C., Ed., Academic Press, New York, 1975, 49.
13. **Overton, W. S.**, The ecosystem modeling approach in th coniferous forest biome, in *Systems Analysis and Simulation in Ecology*, Vol. 3, Patten, B. C., Ed., Academic Press, New York, 1975, 117.
14. **Innis, G. S.**, Role of total systems models in the grassland biome study, in *Systems Analysis and Simulation in Ecology*, Vol. 3, Patten, B. C., Ed., Academic Press, New York, 1975, 13.
15. **Miller, P. C., Collier, B. D., and Bunnell, F. L.**, Development of ecosystem modeling in the tundra biome, in *Systems Analysis and Simulation in Ecology*, Vol. 3, Patten, B. C., Ed., Academic Press, New York, 1975, 95.
16. **Boling, R. H.**, Toward state space models for biological populations, *J. Theor. Biol.*, 40, 485, 1973.
17. **Ulanowicz, R. E.**, Mass and energy flow in ecosystems, *J. Theor. Biol.*, 34, 239, 1972.
18. **Patten, B. C.**, Ecosystem linearization: an evolutionary design problem, *Am. Nat.*, 109, 529, 1975.
19. **Goodall, D. W.**, Building and testing ecosystem models, in *Mathematical Models in Ecology*, Jeffers, J. N. R., Ed., Blackwell Scientific, Oxford, 1972, 173.
20. **Dale, M. B.**, Systems analysis and simulation in ecology, *Ecology*, 51, 1, 1970.
21. **McIntire, C. D.**, Periphyton dynamics in laboratory streams: a simulation model and its implications, *Ecol. Monogr.*, 43, 397, 1973.
22. **Mooney, H. A.**, The carbon balance of plants, *Annu. Rev. Ecol. Syst.*, 3, 315, 1972.
23. **Reichle, D. E., Dinger, B. E., Edwards, N. T., Harris, W. F., and Sollins, P.**, Carbon flow and storage in a forest ecosystem, in *Carbon and Biosphere*, Woodwell, G. M. and Pecan, E. V., Eds., No. CONF-720510, U.S. Atomic Energy Commission, Springfield, Va., 1973, 345.
24. **Thornley, J. H. M.**, A balanced quantitative model for root:shoot ratios in vegetative plants, *Ann. Bot.(London)*, 36, 431, 1972.
25. **Thornley, J. H. M.**, *Mathematical Models in Plant Physiology*, Academic Press, New York, 1976.
26. **Evans, C. G.**, A sack of uncut diamonds: the study of ecosystems and the future resources of mankind, *J. Ecol.*, 64, 1, 1976.
27. **Caswell, H.**, The validation problem, in *Systems Analysis and Simulation in Ecology*, Vol. 4, Patten, B. C., Ed., Academic Press, New York, 1976, 313.
28. **Van Dyne, G. M.**, Organization and management of an integrated ecological research program — with special emphasis on systems analysis, universities, and scientific cooperation, in *Mathematical Models in Ecology*, Jeffers, J. N. R., Ed., Blackwell Scientific, Oxford, 1972, 111.
29. **Wareing, P. F. and Patrick, J.**, Source-sink relations and the partition of assimilates in the plant, in *Photosynthesis and Productivity in Different Environments*, Cooper, J. P., Ed., Cambridge University Press, Cambridge, 1975, 481.
30. **Penning de Vries, F. W. T., Brunsting, A., and Van Laar, H. H.**, Products, requirements, and efficiencies of biosynthesis, a quantitative approach, *J. Theor. Biol.*, 45, 339, 1974.
31. **Mabry, T. J., Hunziker, J. H., and Difeo, D. R.** Creosote Bush: Biology and Chemistry of Larrea in *New World Deserts*, Dowden, Hutchinson & Ross, Stroudsburg, Pa., 1977
32. **Barbour, M. G.**, Desert dogma reexamined: root:shoot productivity and plant spacing, *Am. Midl. Nat.*, 89, 41, 1973.
33. **Rosenzweig, M. L.**, Net primary production of terrestrial communities: prediction from climatological data, *Am. Nat.*, 102, 67, 1968.
34. **Beatley, J. C.**, Dependence of desert rodents on winter annuals and precipitation, *Ecology*, 50, 721, 1969.
35. **Beatley, J. C.**, Phenological events and their environmental triggers in Mojave Desert ecosystems, *Ecology*, 55, 856, 1974.
36. **Burk, J. H. and Dick-Peddie, W. A.**, Comparative production of *Larrea divaricata* Cav. on three geomorphic surfaces in southern New Mexico, *Ecology*, 54, 1094, 1973.
37. **Bamberg, S. A., Kleinkopf, G. E., Wallace, A., and Vollmer, A.**, Comparative photosynthetic production of Mojave Desert shrubs, *Ecology*, 56, 732, 1975.
38. **Leith, H.**, Primary production: terrestrial ecosystems, *Hum. Ecol.*, 1, 303, 1973.
39. **Ludwig, J. A. and Whitford, W. G.**, Short-term water and energy flow in arid ecosystems, in *Arid Ecosystems Dynamics*, Vol. 2, Noy-Meir, I., Ed., Cambridge University Press, London, in press.

40. **Noy-Meir, I.**, Desert ecosystems: environment and producers, *Annu. Rev. Ecol. Syst.*, 4, 25, 1973.
41. **Bridges, K. W., Wilcott, C., Westoby, M., Kickert, R., and Wilkin D.**, Nature: A Guide to Ecosystem Modeling, unpublished manuscript presented at Ecosystem Modeling Symp., AIBS Meeting, Minneapolis, 1972.
42. **Jordon, C. F.**, Productivity of a tropical forest and its relation to a world patten of energy storage, *J. Ecol.*, 59, 127, 1971.
43. **Chew, R. M. and Chew, A. E.**, The primary productivity of a desert shrub (*Larrea tridentata*) community, *Ecol. Monogr.*, 33, 335, 1965.
44. **Cunningham, G. L. and Reynolds, J. F.**, A simulation model of primary production and carbon allocation in the creosotebush (*Larrea tridentata* [DC] Cov.), *Ecology*, 59, 37, 1978.
45. **Strain, B. R.**, Seasonal adaptations in photosynthesis and respiration in four desert shrubs growing *in situ*, *Ecology*, 50, 511, 1969.
46. **Odening, W. R., Strain, B. R., and Oechel, W. C.**, The effect of decreasing soil water potential on net CO_2 exchange in intact desert shrubs, *Ecology*, 55, 1086, 1974.
47. **Oechel, W. C., Strain, B. R., and Odening, W. R.**, Tissue water potential, photosynthesis, ^{14}C-labelled photosynthate utilization, and growth in the desert shrub *Larrea divaricata* Cav., *Ecol. Monogr.*, 42, 127, 1972.
48. **Oechel, W. C., Strain, B. R., and Odening, W. R.**, Photosynthetic rates of a desert shrub, *Larrea divaricata* Cav., under field conditions, *Photosynthetica*, 6, 183, 1972.
49. **Strain, B. R. and Chase, V. C.**, Effect of past and prevailing temperatures on the carbon dioxide exchange capacities of some woody desert perennials, *Ecology*, 47, 1043, 1966.
50. **Orians, G. H. and Solbrig, O. T.**, A cost-income model of leaves and roots with special reference to arid and semiarid areas, *Am. Nat.*, 111, 677, 1977.
51. **Cunningham, G. L., Syvertsen, J. P., Willson, J. M., Donahue, T., and Balding, F. R.**, Primary Production and Carbon Allocation in Creosotebush, Res. Memo No. 74-10, US/IBP Desert Biome, 1974.
52. **Reed, K. L., Hamerly, E. R., Dinger, B. E., and Jarvis, P. G.**, An analytical model for field measurements of photosynthesis, *J. Appl. Ecol.*, 13, 925, 1976.
53. **Reynolds, J. F., Cunningham, G. L., and Syvertsen, J. P.**, A net CO_2 exchange model for *Larrea tridentata, Photosynthetica*, in press, 1978.
54. **Warren, W. J.**, Stand structure and light penetration. III. Sunlit foliage area, *J. Appl. Ecol.*, 4, 159, 1967.
55. **Connor, D. J. Brown, L. F., and Trlica, M. J.**, Plant cover, light interception, and photosynthesis of shortgrass prairie: a functional model, *Photosynthetica*, 8, 18, 1974.
56. **Kinerson, R. S.**, Relationships between plant surface area and respiration in loblolly pine, *J. Appl. Ecol.*, 12, 965, 1975.
57. **McCree, K. J.**, An equation for the rate of respiration of white clover plants grown under controlled conditions, in *Prediction and Measurement of Photosynthetic Productivity*, Malek, I., Ed., Pudoc, Wageningen, The Netherlands, 1970, 221.
58. **Murphy, C. E., Sinclair, T. R., Kinerson, R. S., Higginbotham, K. O., Knoerr, K. R., and Strain, B. R.**, Modeling the primary productivity process — a coordinated effort at an intensive research site, in *Proc. Summer Computer Simulation Conf.*, Simulation Council, La Jolla, Calif., 1974, 684.
59. **Ek, A. R.**, Models for Maximizing Yield: Optimizing Spatial Patterns, Stocking, Species Composition and Harvest, presented at Workshop on Quantitative Ecology: Its Application to Forest Resources Management and Education, University of New Hampshire, Durham, December 1 to 3, 1975, 1.
60. **Sollins, P., Harris, W. F., and Edwards, N. T.**, Simulating the physiology of a temperate deciduous forest, in *Systems Analysis and Simulation in Ecology*, Vol. 4, Patten, B. C., Ed., Academic Press, New York, 1976, 173.
61. **Ledig, F. T.**, A growth model for tree seedlings based on the rate of photosynthesis and the distribution of photosynthate, *Photosynthetica*, 3, 263, 1969.
62. **Promnitz, L. C.**, A photosynthate allocation model for tree growth, *Photosynthetica*, 9, 1, 1975.
63. **Shugart, H. H., Goldstein, R. A., O'Neill, R. V., and Mankin, J. B.**, TEEM: a terrestrial ecosystem energy model for forests, *Oecol. Plant.*, 9, 231, 1974.
64. **Hellmers, H. and Strain, B. R.**, Growth of trees in general and loblolly pine trees in particular, in *Modeling the Growth of Trees*, Murphy, C. E., Hesketh, J. D., and Strain, B. R., Eds., EDFB-IBP Report No. 72-11, Oak Ridge National Laboratory, Oak Ridge, Tenn., 1972, 143.
65. **Higginbotham, K. O. and Strain, B. R.**, The influence of canopy position and age of leaves on photosynthesis in loblolly pine, unpublished manuscript, 1978.

66. **Murphy, C. E.**, TAEDA 1: a model of loblolly pine growth, in *Modeling the Growth of Trees*, Murphy, C. E., Hesketh, J. D., and Strain, B. R., Eds., EDFB-IBP Report No. 72-11, Oak Ridge National Laboratory, Oak Ridge, Tenn., 1972, 177.
67. **Kinerson, R. S., Higginbotham, K. O., and Chapman, R. C.**, The dynamics of foliage distribution within a forest canopy, *J. Appl. Ecol.*, 11, 347, 1974.
68. **Higginbotham, K. O.**, The Influence of Canopy Position and the Age of Leaf Tissue on Growth and Photosynthesis in Loblolly Pine, Ph.D. thesis, Duke University, Durham, N.C., 1974.
69. **Nemeth, J. C.**, Dry matter production in young loblolly (*Pinus taeda* L.) and slash pine (*Pinus elliottia* Engelm.) plantations, *Ecol. Monogr.*, 43, 21, 1973.
70. **Kinerson, R. S., Ralston, C. W., and Wells, C. G.**, Carbon cycling in a loblolly pine plantation, *Oecologia (Berlin)*, 29, 1, 1977.
71. **Ralston, C. W.**, Annual primary productivity in a loblolly pine plantation, in *IUFRO Symp.: Forest Biomass*, Young, H., Ed., University of Maine, Orono, 1974, 105.
72. **Murphy, C. E., Sinclair, T. R., and Knoerr, K. R.**, Modeling the photosynthesis of plant stands, in *Handbook of Vegetation Science*. Part 6, Strain, B. R. and Billings, W. D., Eds., Dr. W. Junk, B. V. Publishers, The Hague, 1974, 125.
73. **Woodman, J. N.**, Variation of net photosynthesis within a crown of a large forest-grown conifer, *Photosynthetica*, 5, 50, 1971.
74. **Helms, J. A.**, Summer net photosynthesis of Ponderosa Pine in its natural environment, *Photosynthetica*, 4, 243, 1970.
75. **Jarvis, P. G.**, Characteristics of the photosynthetic apparatus derived from its response to natural complexes of environmental factors, in *Prediction and Measurement of Photosynthetic Productivity*, Setlik, I., Ed., Pudoc, Wageningen, The Netherlands, 1970, 353.
76. **Schulze, E.-D., Fuchs, M. I., Fuchs, M.**, Spacial distribution of photosynthetic capacity and performance in a mountain spruce forest of Northern Germany. I. Biomass distribution and daily CO_2 uptake in different crown layers, *Oecologia(Berlin)*, 29, 43, 1977.
77. **Lange, O. L. and Schulze, E.-D.**, The measurement of the CO_2 gas exchange and transpiration of the beech (*Fagus sylvatica* L.), in *Integrated Experimental Ecology*, Ellenberg, H., Ed., Springer-Verlag, New York, 1971, 16.
78. **Reynolds, J. F., Dudgeon, A. C., and Knoerr, K. R.**, Models of photosynthesis and photosynthate allocation in loblolly pine *Pinus taeda* L.) (Abstr.), *Bull. Ecol. Soc. Am.*, 107, 1978.
79. **Tenhunen, J. D., Weber, J. A., Yocum, C. S., and Gates, D. M.**, Development of a photosynthesis model with an emphasis on ecological applications. II. Analysis of a data set describing the P_m surface, *Oecologia(Berlin)*, 26, 101, 1976.
80. **Chung, H. H. and Barnes, R. L.**, Photosynthate allocation in *Pinus taeda* L. I. Substrate requirements for synthesis of shoot biomass, *Can. J. For. Res.*, 7, 106, 1977.
81. **Kinerson, R. S.**, Relationships between plant surface area and respiration in loblolly pine, *J. Appl. Ecol.*, 12, 965, 1975.
82. **Kozlowski, T. and Keller, T.**, Food relations of woody plants, *Bot. Rev.*, 32, 293, 1966.
83. **Wilson, B. F**, *The Growing Tree*, University of Massachusetts Press, Amherst, 1970.
84. **Zimmermann, M. H. and Brown, C. L.**, *Trees: Structure and Function*, Springer-Verlag, New York, 1971.
85. **Kozlowski, T. T.**, *Growth and Development of Trees*, Academic Press, New York, 1971.
86. **Chung, H. H.**, Growth, Biochemical Changes, and Photosynthate Allocation in Shoots of *Pinus taeda* L., Ph.D. thesis, Duke University, Durham, N.C., 1974.
87. **Barnes, R. L.**, Some physiological considerations in the conversion of photosynthate to biomass in loblolly pine, in *Modeling the Growth of Trees*, Murphy, C. E., Hesketh, J. D., and Strain, B. R., Eds., EDFB-IBP Report No. 72-11, Oak Ridge National Laboratory, Oak Ridge, Tenn., 1972, 155.

Chapter 11

COTCROP: A COMPUTER MODEL FOR COTTON GROWTH AND YIELD

J. W. Jones, L. G. Brown, and J. D. Hesketh

TABLE OF CONTENTS

I. INTRODUCTION

Preceding chapters have considered processes and some assumptions needed for a physiological model of photosynthate production and use. The photosynthetic processes cannot be studied alone, but must be considered as part of a system, with such processes interacting with other plant physiological processes from seed germination to death. These discussions then have considered the plant as a whole and its physiology.

Familiarity with a whole plant or plant population model is helpful in studying photosynthesis as part of a system, particularly in the development of hypotheses of how photosynthesis interacts with the rest of the plant's physiology. The contents of this book were planned after considerable experience in developing a cotton crop model. It is fitting then that we conclude with a brief description of a recent version of our cotton models, with emphasis on interfacing photosynthesis with other plant physiological processes.

II. COTCROP: THE COTTON CROP MODEL

COTCROP, a cotton crop model based on photosynthate supply (photosynthesis, respiration) and on growth demand for photosynthate (morphogenesis, growth of organs) will be discussed here. The nature of this model was determined by potential users for applications in studies of pest control, irrigation, harvesting, and fertility or plant nutrition. This model was developed by an interdisciplinary research team involving engineers, economists, entomologists,and plant scientists.

Considerable research and several generations of single (average) plant models preceded the development of COTCROP, some of which provided much of the background for the current model. Two cotton plant models were devloped in the late 1960s and early 1970s, namely SIMCOT[1,2] for southeastern conditions, and COTTON[3] for western production under irrigation. SIMCOT was later revised and released as SIMCOT II.[4-6] These models grew single plants by maintaining a carbohydrate balance and later a nitrogen balance. These models predicted the production of discrete fruit numbers for an average plant, which, for plant populations in a crop, had to be multiplied by the plant population. Such models led to stepwise increases in floral buds or fruit equivalent in numbers to the plant population. For a population of 100,000 plants per hectare, the increase was 100,000 floral buds or flowers. For a model predicting the behavior of insects feeding on floral buds and fruit, the day-by-day population of floral buds or fruit was needed,as well as their respective ages. Gutierrez et al[7] expanded SIMCOT II for application to insect management for the far west and used continous densities of fruit numbers.

In COTCROP, crop growth is calculated for plants growing on one m^2 ground area. The model uses continous state variables and discrete ($\Delta t = 1$ day) time steps. It maintains carbohydrate and nitrogen balances for the plant, and water and nitrogen balances for the soil. Morphogenetic calculations determine initiation and fate of various plant organs; such calculations were based upon data from upland cotton. Figure 1 shows a simplified schematic of crop components and their interactions using notation from Forrester.[8] Here we will discuss the model structure, mathematically describe the model components, and give examples of its use.

A. Model Structure

The mathematical structure of the model is described by vectors of continuous state variables and a discrete time transition matrix, or symbolically

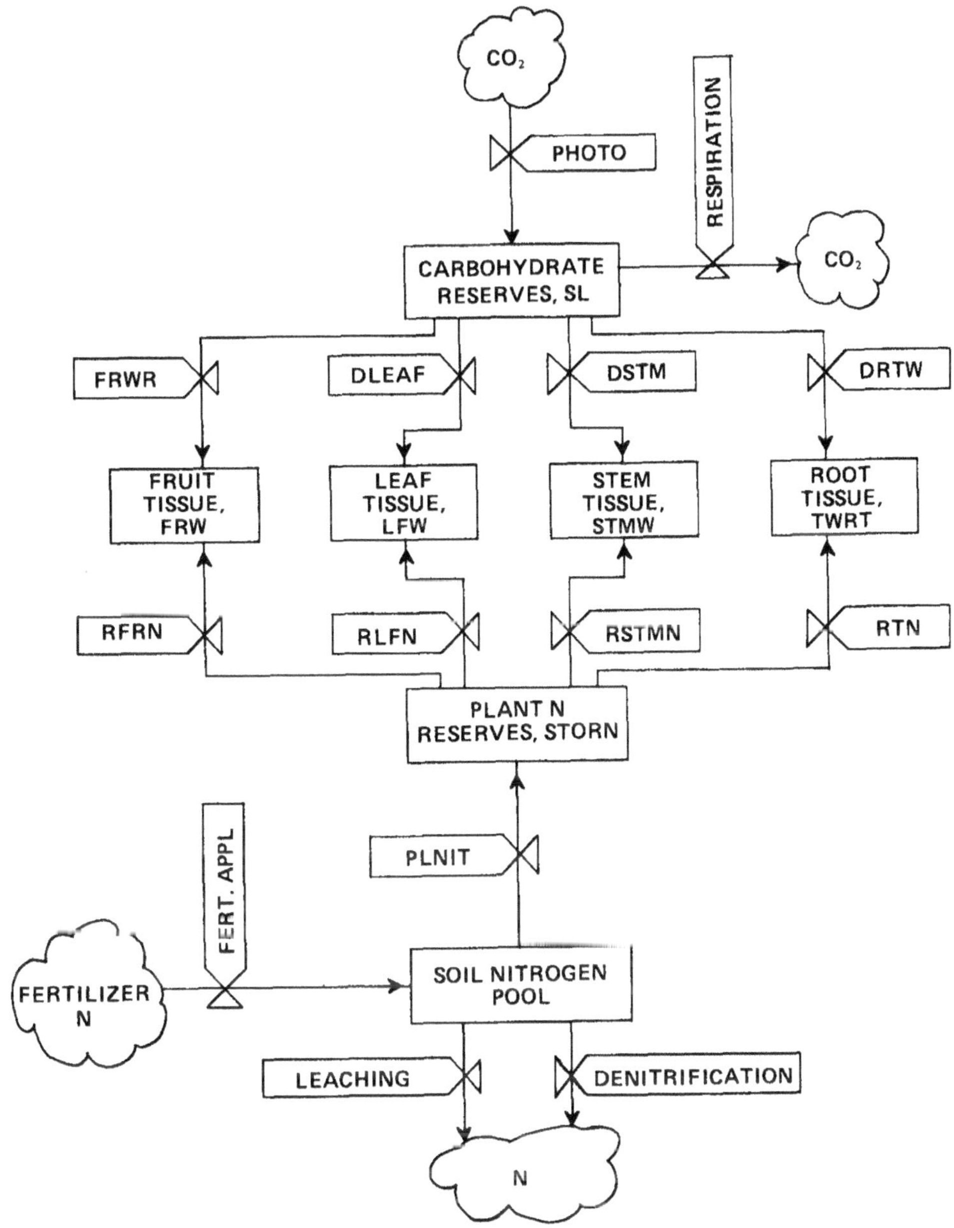

FIGURE 1. System diagram of components and rates in the cotton model COTCROP. Each crop component, such as leaves, is described by number and weight densities and nitrogen contents for daily age classes of leaves. Notation of Forrester is used although each component actually consists of more than one state variable. The soil water storage is not shown. (Adapted from Forrester, J. W., *Industrial Dynamics*, MIT Press, Cambridge, 1961. With permission.)

$$\bar{X}_t = A_t\bar{X}_{t-1} + \bar{B}_t \qquad (1)$$

where $\bar{X}_t$ = symbol for state variable vector, A_t = transition matrix for day t, and $\bar{B}_t$ = growth or organogenesis rate vector for day t.

The structure of the model is a set of simple first order difference equations, but because of the physiological detail included in the parameters and processes, solution of the model is complex and simulation is used.

The vector $\overline{X}_t$ denotes any one of a group of column vectors for describing the state of the crop, each having m elements, where m is the maximum number of days of simulation. For example, is $\overline{LFM}_t$ represents the (m × 1) vector of main stem leaf numbers, then $lfm_t(i)$ represents the remaining number of leaves on day t that were initiated on day i and are therefore (t − i) days old. Note that $lfm_t(i) = 0$ for all $i > t$, where t is the current day of simulation. The vector $\overline{LFM}_t$ then contains information about leaf numbers and their ages. Most variables in $\overline{X}_t$ are similar.

The number of new leaves to be initiated on day t is represented by $\overline{B}_{lt}$ (containing m elements) in which all elements are zero except b_{lt} (t). Since $\overline{LFM}_{t-1}$has nonzero elements only for $i \leqslant t - 1$, on day t lfm_t (t) becomes equal to b; i $l_t(t)$, the new leaf initiation rate for day t.

The matrix operation causes changes in the elements of the vector of $O \leqslant i < t - 1$ and represents senescence or abscission of organs. On day t $\overline{LFM}_t$ is calculated by adding the number of new leaves initiated on day t to $\overline{LFM}_{t-1}$after the matrix operations are performed. The state variables consist of other variables, such as branch stem leaf number, and are listed in the definition of variables. Growth will be described using vector notation. Subscripts are assumed to be t unless otherwise specified. For example, the expression $1^T \times \overline{LFM}$ is vector multiplication and represents the total number of leaves on the main stem on day t, and lfm(i) represents the number of leaves on day t initiated on day i. Lower-case letters are elements of vectors having the same upper-case letters. Cross-product and dot-product operations are denoted by X and ·, respectively.

One major assumption in the model is that the temperature effect on growth rate is linear and is the same relationship for all plant organs. The standard temperature is 26°C and one day at 26°C is referred to as one physiological day. No growth occurs below 12°C. For temperatures above 12°C, the organ growth rate at 26°C potential growth rate is multiplied by the variable TINC to calculate daily growth [2] where

$$\text{TINC} = \begin{cases} 0.0 & \text{if } T < 12.0^\circ C \\ -0.85 + 0.0714(T) & \text{if } T \geq 12^\circ C \end{cases} \quad (2)$$

TINC is calculated on an hourly basis and averaged over the day.

Physiological time is calculated by accumulating daily values of TINC and the result is stored in $\overline{CT}$. Therefore, ct(t) is equal to ct(t-i) + TINC and ct(t) is the accumulated TINC values for the first t days. On day t, organs that are (t-i) real days old are [ct(t)-ct(i)] physiological days old.

Source-sink relationships are explicitly included and are discussed as supply and demand for carbohydrates and nitrogen. Supply of carbohydrates is by processes of photosynthesis and respiration, whereas nitrogen supply rate depends on transpiration and nitrogen concentration in the soil. Demand for carbohydrate is calculated by potential growth rates of existing plant organs and initiation of new organs, using a constant requirement of carbohydrate per unit dry weight of tissue. Demand for nitrogen is calculated by specifying the minimum nitrogen concentration of growing organs. Excess carbohydrate and nitrogen are stored in reserve pools for later use. Shortage of either carbohydrate or nitrogen will reduce both growth and organ initiation rates.

Operationally, the model is a set of modular subroutines shown schematically in Figure 2. Subroutines are written in Fortran IV (ANSI standard) and the simulation language GASP IV[9] is used to provide input-output features and for scheduling dis-

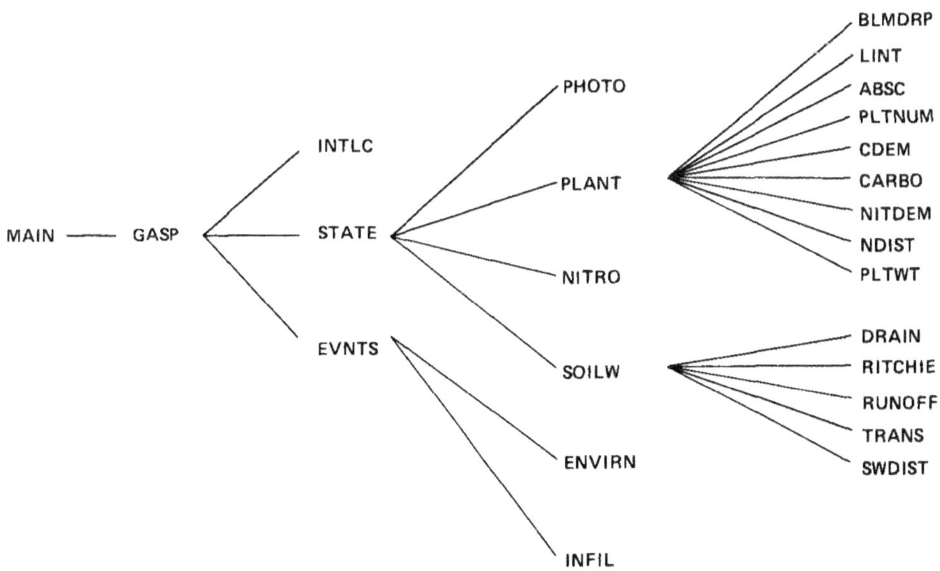

FIGURE 2. Modular subroutine structure of COTCROP. Subroutine GASP is the GASP IV subroutine that controls time and integrates the rates. The other subroutines shown are user written and provide calculations for crop growth processes.

crete events in the simulation. A small main program calls subroutine GASP, which controls input then initiates the simulation by calling other GASP IV or user-written subroutines.

The subroutine INTLC initializes parameters for the simulation. The number of plants/m^2, their leaf, stem, and root weights and N contents are specified. Soil water is set to saturation and residual N content is read. The simulation begins at crop emergence (date of 50% plant emergence) and each plant has 2 cotyledons with area 0.0009 m^2. Each plant weighs 0.5 g with 60, 20, and 20% of the weight in leaves, stems, and roots respectively. Initial leaf N content is 10% by weight, and all weights are dry weight basis.

The potential fruiting branch maximum node number (CONTR) is set depending on plant population, Figure 3. CONTR determines the maximum rate of new organ initiation per plant for fruiting branches off the main stem. To determine the maximum organ initiation rate per plastochron, vegetative branches are considered. The maximum number of vegetative branches per plant (VEGBRN) is a function of plant population, calculated by

$$\mathrm{VEGBRN} = 12.76 - 9.10(10^{-4})\,\mathrm{POP} + 2.23(10^{-8})\,(\mathrm{POP})^2 - 1.785(10)^{-13}\,(\mathrm{POP})^3 \qquad (3)$$

The vegetative branches start growth at node VB and 10 days after the first vegetative branch starts growth, all have been initiated. VB is calculated by

$$\mathrm{VB} = 10.8 - 2.2(10^{-4})\,\mathrm{POP} \qquad (4)$$

The number of vegetative branches starting growth on day t is give by

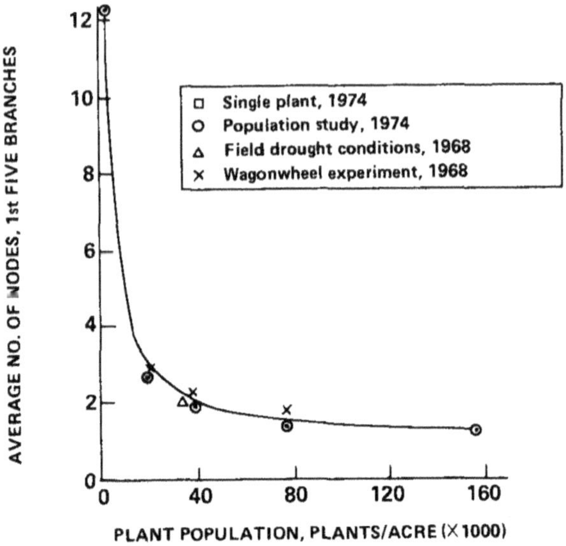

FIGURE 3. Potential number of nodes per fruiting branch vs. plant population. (From Jones, J. W., Hesketh, J. D., Colwick, R. F., Lane, H. C., McKinion, J. M., and Thompson, A. C., in *Proc. Beltwide Cotton Production Research Conf.*, National Cotton Council, Memphis, 1975, 56. With permission.)

$$vbr_t(t) = \begin{cases} VEGBRN / 10 & t_{vb} < t < (t_{vb} + 10) \\ 0 & \end{cases} \quad (5)$$

These relationships were developed from field data on four crops grown in 1972 in North Carolina and the data of Bruce and Romkens.[10] A more mechanistic relationship between vegetative branch development and physiological factors is needed.

The subroutine EVNTS is a user-written subroutine that causes discrete events to occur on specified dates. The discrete events of COTCROP are N-fertilizer application, irrigation, daily weather input, and scheduled completion of the simulation. Once daily in the simulation, EVNTS calls ENVIRN for reading values of Julian day number, daily radiation (langleys), rainfall (inches), pan evaporation (inches), and hourly temperatures (°C). (Inches are converted to cm in the model.) For our applications, we have used subroutines developed by Skeith et al.[11] to generate hourly temperature data from daily maximum and minimum temperatures.

After program initialization, GASP controls the time by calling the user-written subroutine STATE once each day. STATE then passes control to subroutines to describe processes of plant growth (PLANT) and soil water dynamics (SOILW).

Plant growth is calculated after STATE calls PLANT, wherein carbohydrate and N balances are performed. A soil-water balance is performed in SOILW and soil-nitrogen balance is accomplished in NITRO. Processes necessary for performing the material balances are calculated in various subroutines described below.

Subroutine SSAVE is a user-written subroutine for storing output data. Output will be discussed in more detail below.

B. Morphogenesis

In addition to PLANT, five subroutines are used to calculate the dynamics of plant

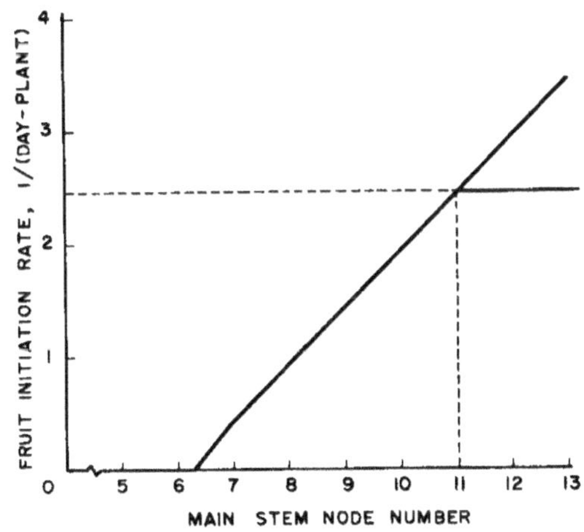

FIGURE 4. Potential fruit (square) initiation rate dependent on number of main stem nodes on the plant. At the point in time when the first fruiting branch reaches its maximum number of nodes (Figure 3), the potential rate of square initiation on fruiting branches remains constant, e.g., after node 11, the maximum fruit initiation rate is 2.45 squares per plant per day. The curve was calculated from a population model.[12] Nutrition and water stresses slow initiation rates.

organ numbers. PLTNUM calculates the numbers of new leaves, stems, and floral buds (squares) initiated on the current day on the main stem, whereas VBRNUM calculates the same numbers for vegetative branches. Subroutine ABSC causes fruit to be abscised after being selected for abscission due to stresses (carbohydrate, N, and water) and insect damage. BLMDRP causes flowering and loss of flower biomass. LINT causes opening of mature bolls and calculates lint yield.

Potential organ initiation is calculated in PLANT after TINC and $\overline{CT}$ are calculated. The plastochron concept is used, whereby organ initiation on the main stem is separated from fruiting branch organs because of morphological differences. At 26°C, data of Hesketh[2] indicate that 3 days are required between successive nodes on the main stem and 6 days are required for successive nodes within a fruiting branch. For each main stem node, a new leaf and stem internode are initiated. For every fruiting branch node, a new floral bud, a leaf, and a stem internode are initiated. (Stresses slow these rates, see below. Jones et al.[12] describe a detailed model of this process.) Therefore, a plastochron is set equal to 3 days at 26°C. For the first true leaf on the main stem, 5.2 days are required. The first fruiting branch is initiated at a node 2.5 plastochrons after the first leaf is formed at that node. Here we assumed that the first fruiting branch would occur between the 5th and 8th main stem nodes and at an average of node 7. Figure 4 shows the potential fruit initiation curve (per plastochron), depending on the stage of growth. Once plants have fruiting branches that have reached their maximum number of nodes, the potential rate for square initiation off the main stem is constant.

1. Organ Initiation

New organ numbers initiated on the current day are calculated in PLTNUM. Vectors of leaf numbers on the main stem ($\overline{LFM}_i$) and fruiting branches ($\overline{LFB}_i$), stem (internode) numbers for the main stem ($\overline{STM}_i$) and fruiting branches ($\overline{STB}_i$), and fruit numbers ($\overline{FR}_t$) are calculated. On each day of simulation, organ numbers are calculated by:

$$\overline{LFM}_t = A_{1t}\overline{LFM}_{t-1} + \overline{B}_{1t}$$

$$\overline{LFB}_t = A_{2t}\overline{LFB}_{t-1} + \overline{B}_{2t}$$

$$\overline{STM}_t = \overline{STM}_{t-1} + \overline{B}_{3t}$$

$$\overline{STB}_t = \overline{STB}_{t-1} + \overline{B}_{4t}$$

$$\overline{FR}_t = (\overline{FR}_{t-1} - \overline{FRABS}_{t-1}) + \overline{B}_{5t} \qquad (6)$$

where subscripts indicate the date and lower case symbols represent the number of new organs/m² initiated on the day t.

The $\overline{B}jt$, for j = 1,2...5, represent organ initiation. Each $\overline{B}jt$ has only one nonzero element, $b_{jt}(t)$. For example, $fr_t(t)$ is equal to $b_{5t}(t)$ for each t and the task is then to calculate that rate of new fruit initiation for each day t as it occurs in the simulation. These rates are affected by temperature, stresses, and nutritional imbalances. The A_t matrices are square (m by m) with diagonal elements representing the change of existing organ numbers due to senescence. The diagonal elements of A_1 and A_2 are found by

$$(A_{1t})_{ii} = (A_{2t})_{ii} = \begin{cases} 1.0 & 0 < ct(t) - ct(i) < 45.0 \\ 0 & ct(t) - ct(i) \geq 45.0 \end{cases}$$

$$\text{for } i = 1,t, \qquad (7)$$

such that leaves are senesced after 45 physiological days of age. The vector $\overline{FRABS}$ will be discussed in the carbohydrate stress section.

The remaining task in specifying Equation 6 is to calculate each organ initiation rate for day t. For a single plant, organ initiation rate is the inverse of days between organs, or 1/PLSTN. For plants growing on a m² soil area, the rate of main stem leaf initiation is (PLTM2) /(PLSTN) when temperature is 26°C and no stresses occur. In general, main stem leaf initiation is calculated by

$$1fm_t(t) = (TINC)(ADJ)(ADJDLY)(PLTM2) / PLSTN \qquad (8)$$

where ADJ = (ADJC) (ADJN), i.e., the product of the ratios of carbon and nitrogen (respectively) supplies to their demand, ADJDLY = cumulative water stress affect on organ initiation rate; the minimum ADJW during the last s days, where s is 3 days at emergence and increases to 20 days at flowering, PLTM2 = number of plants/m², and PLSTN = number of physiological days between successive leaves on the main stem (plastochron). Similarly, leaf initiation on the fruiting branches is calculated by

$$1fb_t(t) = (XLBR)(1fm(t)) + 1vm_t(t) + 1vb_t(t) \qquad (9)$$

where XLBR = number of new fruiting branch leaves initiated per plastochron at the current state of crop growth, Figure 4. (The model uses tabular values of XLBR as a function of main stem nodes.), $lvm_t(t)$ = new leaves initiated on vegetation main stem, and $lvb_t(t)$ = new leaves initiated on fruiting branches off vegetative structures. The rates lvm and lvb are calculated in subroutine VBRNUM by

$$1vm_t(t) = \bar{1}^T \times (\overline{VBR})\ (TINC)(ADJ)(ADJDLY)/(PLSTN) \qquad (10)$$

and

$$1vb_t(t) = 1vm_t(t) \text{ when } tdvb_t(i) > 2 \qquad (11)$$

where $\overline{VBR}$ = number of vegetative branches, and $tdvb_t(i)$ = number of plastochrons or nodes on vegetative branches initiated on day i and now (t-i) days of age.

The above relationships assume that new leaves are initiated at the same rate on vegetative branches as main stems and that no fruit or branch leaves are initiated until after 2 nodes on the vegetative branches. Also, it is assumed that no more than 1 fruiting branch node is formed off each main vegetative branch node. These vegetative branch leaves are all grouped with branch leaves since their growth characteristics are similar.[13]

The number of new internodes on the main stem is calculated by

$$stm_t(t) = 1fm_t(t) \qquad (12)$$

The number of new internodes on fruiting branches is calculated by

$$stb_t(t) = 1fb_t(t) \qquad (13)$$

The number of new fruit initiated is calculated by

$$fr_t(t) = 1fb_t(t) - 1vm_t(t) \qquad (14)$$

Subroutine PLTNUM then calculates the total number of leaves on the crop by

$$LFNUM = \bar{1}^T \times (\overline{LFM} + \overline{LFB}) \qquad (15)$$

the total number of fruit (squares plus bolls) by

$$FRNUM = \bar{1}^T \times \overline{FR} \qquad (16)$$

and the total number of squares by

$$SQNUM = \overline{SI}^T \times \overline{FR} \qquad (17)$$

where $\overline{SI}^T$ = vector whereby si_t = 1 for ct(t) − ct(i) ≤ 26 and si_t = 0 for ct(t) − ct(i) > 26, for i = 1,t.

Subroutine PLTNUM is called iteratively from PLANT and ADJ is searched until growth and organ initiation nutrient demands are balanced with nutrient supply. Nutrient or water stresses slow organ growth rates. After floral initiation, growth and organ initiation are slowed proportionally the same due to nutrient stresses. Before floral initiation, carbohydrate stresses do not slow organ initiation rates.

2. Fruit Abscission

A proportion of fruit is selected for abscission due to nutritional stresses in subroutine STRESS. In ABSC, these fruit are removed from the nutrition balance after 3 physiological days and abscised after 5 physiological days.[14] When fruit are abscised in the model, there is no mining of nutrients from fruit prior to abscission. The calculation of fruit abscission is in Equation 6.

3. Flowering

This subroutine causes squares to flower and subtracts the bloom weights from total fruit weight. Flowering occurs 26 physiological days after square initiation and the calculation is

$$frw_t(t-\tau) = frw_t(t-\tau) - \min(0.3, frw_t(t-\tau))$$

where t-τ = real age of fruit corresponding to 26 physiological days.

4. Boll Development

The subroutine LINT calculates the time required for bolls to open based on the day they were initiated.[28] The relationship is

$$TSB = 73.14 - 0.02632(DTAG) + 0.01095(DTAG)^2 - 0.0015(DTAG)^3 \quad (18)$$

where TSB = days required for boll opening if squares are initiated on day t and DTAG = date of crop emergence + t −163 + 26.

Bolls are allowed to open over a 5-day period centered about TSB. The lint percentage is allowed to vary depending on boll weights, Figure 5. The total number of open bolls (OPNBL) and total lint weight (TWLNT) are calculated by

$$OPNBL = (\overline{OPN}^T \times \overline{FR}) \quad (19)$$

$$TWLNT = (\overline{OPN}^T \times \overline{FRW} \cdot \overline{FRLNT}) \quad (20)$$

where $\overline{OPN}$ = fraction of open bolls for each age of boll, and $\overline{FRLNT}$ = fraction of boll weight that is lint for each age of bolls, calculated from Figure 5 and FR and $\overline{FRW}$.

Therefore the morphogenesis routines of PLTNUM, VBRNUM, ABSC, BLMDRP, and LINT account for numbers of plant organs being initiated, remaining on the crop, senescing, and maturing. Organs are stored in vectors whereby state variables are organ numbers of each discrete day age, $\overline{LFM}$, $\overline{LFB}$, $\overline{STM}$, $\overline{STB}$, $\overline{FR}$, and $\overline{VBR}$. Other variables above (except $\overline{FRW}$ and $\overline{OPN}$) are not state variables, but are time-varying parameters or output variables.

C. Carbohydrate Balance

In contrast to some plant models, COTCROP does not have explicit partitioning functions. Partitioning is determined by considering demand for assimilates from growing organs and from morphogenesis. Depending on the crop status, partitioning of nutrients to leaves, stems, fruit, and roots will change relative to the growing organ potential growth rates. (Relative to other organs,fruit have priority for a portion of of CH_2O.) This demand for carbohydrate (CH_2O) is calculated in CDEM and growth is adjusted to balance with available CH_2O. Available CH_2O is calculated by a photosynthesis function (PHOTO)[4] and carbohydrate subroutine (CARBO). Maintenance and growth respiration are subtracted from photosynthesis and a CH_2O reserve is calculated. Fruit abscission is calculated by using the ratios of CH_2O and of N supplied to their respective demands as well as water stress.These relationships are discussed below.

1. Carbohydrate Demand

The demand for main stem leaf growth is

$$\mathrm{DLEAF} = (\overline{\mathrm{LFM}}^{T} \times \overline{\mathrm{LFWR}})\mathrm{U} \tag{21}$$

where

$$\mathrm{lfwr}_t(i) = \begin{cases} 0.3 & \text{if } 0 \leq ct(t) - ct(i) \leq 24 \\ 0.0 & \end{cases} \tag{22}$$

and U = (TINC)(ADJ)(ADJW)/PLSTN. The demand for newly initiated leaves is $\mathrm{lfwr}_t(t)$.

The above equations cause a constant growth rate of 0.3 g per leaf per plastochron for leaves less than 24 physiological days. [13] Branch leaves were assumed to grow at half the rate, or

$$\overline{\mathrm{LFWBR}} = (0.5)\,(\overline{\mathrm{LFWR}}) \tag{23}$$

and

$$\mathrm{DLEAFB} = (\overline{\mathrm{LFB}}^{T} \times \overline{\mathrm{LFWBR}})\mathrm{U} \tag{24}$$

The potential growth for stems is calculated by

$$\mathrm{DSTM} = (\overline{\mathrm{STM}}^{T} \times \overline{\mathrm{STMWR}})\mathrm{U} \tag{25}$$

where stmwr(i) =

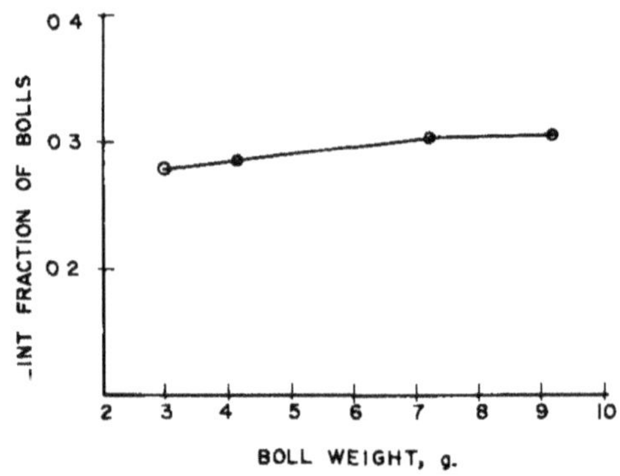

FIGURE 5. Lint fraction of bolls of different weights. (From Jones, J. W., Colwick, R. F., and Smith, D. B., unpublished data, 1971. With permission.)

$$\text{where stmwr(i)} = \begin{cases} 0.20 & \text{if } 0 \leqq ct(t)-ct(i) \leqslant 24 \\ 0.015 & \text{otherwise.} \end{cases} \qquad (26)$$

Fruiting branch growth is 0.3 times the main stem rate for 24 days, or

$$\overline{STMWBR} = (0.3)\,(\overline{STMWR}) \qquad (27)$$

and fruiting branch stem growth demand is

$$DSTB = (\overline{STB}^{T} \times \overline{STMWBR})U \qquad (28)$$

Potential fruit growth, DFW, is based on a fruit growth curve (Figure 6), denoted by $\overline{FRWR}$, or

$$DFW = (\overline{FR}^{T} \times \overline{FRWR})\,(TINC)\,(ADJ)\,/\,PLSTN \qquad (29)$$

It is assumed that water stress affects boll growth only by reduced supply of CH_2O or N. Reduced leaf water potential and turgor, causing decreased leaf growth, does not cause directly a decrease in fruit growth rates.

Root growth is treated simply as a fraction of the leaf growth, or

$$DRTW = (0.2)\,(DLEAF + DLEAFB) \qquad (30)$$

Growth demand for carbohydrate is then calculated by

$$CREQ = DLEAF + DLEAFB + DSTM + DSTMB + DFW + DRTW \qquad (31)$$

2. Carbohydrate Supply

The supply of CH_2O is calculated in CARBO as a scaler quantity in grams per m^2 per day. The total CH_2O demand (X3) for the day is calculated by Hesketh et al.[15] as

$$X3 = (1 + GR)(CREQ) + (RM)(TOTWT) \quad (32)$$

Four cases are considered.

Case 1. Total demand X3 is less than the day's photosynthesis, or X3< PR. Then the day's growth occurs at the potential rate and reserves are increased by

$$SRATE = PR - X3 \quad (33)$$

to a maximum level of B, the limit of starch accumulation. The CH_2O available for growth, QCR, is set equal to X3 and no restrictions are placed on organogenesis or growth.

Case 2. Total demand X3 is more than PR, but less than PR plus reserve CH_2O. The maximum available from reserves is

$$SRATE = -XK2\,(SL)\,(TINC)$$

In this case, QCR is set equal to CREQ, but SL reserve level is reduced. Again, Equation 33 calculates the change in reserves, but SRATE is this case is negative.

Case 3. Demand ×3 exceeds supply of PR + XK2 (SL) (TINC), and QCR is calculated by

$$QCR = (PR + (XK2)(SL)(TINC) - (RM)(TOTWT) / (1 + GR) \quad (34)$$

Therefore, QCR is less than CREQ. As a result, the factor ADJC is calculated as

$$ADJC = QCR / CREQ \quad (35)$$

Case 4. Demand for maintenance respiration causes all available carbohydrate to be used for maintenance and QCR = O.

In order that supply is balanced with use of CH_2O, the iterative procedure in PLANT is used. On successive iterations, growth and organ initiation are reduced until the supply and demand are balanced, Figure 7. For all cases, the reserve level is adjusted by

$$SL_t = SL_{t-1} + SRATE. \quad (36)$$

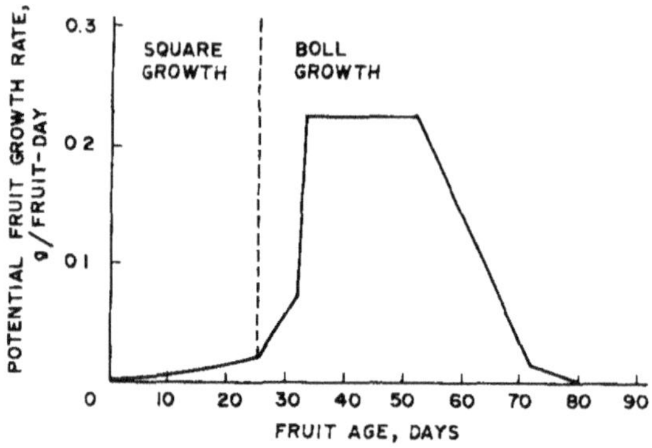

FIGURE 6. Potential fruit (square and boll) growth rates at 26°C. Square data are unpublished data; boll growth from SIMCOT II. (From McKinion, J. M., Baker, D. N., Hesketh, J. D., and Jones, J. W., in Computer Simulation of a Cotton Production System — Users Manual, ARS-S-52, 1975, 27. With permission.)

3. Photosynthesis

This subroutine is taken primarily from the plant model SIMCOT II. Light interception is calculated by

$$INT = \begin{cases} 0.05 + 0.2375\,(LAI) & 0 \leqq LAI \leqq 4.0 \\ 1.0 & LAI > 4.0 \end{cases} \tag{37}$$

Then gross photosynthesis is calculated by[4]

$$PR = [\,0.23903 + 0.137379(RAD) - 5.4136(10)^{-5} (RAD)^2\,]\,(0.68)\,(R_\psi) \tag{38}$$

The radiation (RAD) values are watts/m^2, and the 0.68 is the stochiometric conversion of CO_2 to CH_2O. R_ψ is the reduced photosynthesis potential due to high negative soil water potentials. Values from Bowen[29] are used in a table to describe the effects of soil water potential on photosynthetic rate. Plant water potential is not calculated.

4. Carbohydrate Stress

Fruit are scheduled for abscission in subroutine STRESS, depending on total physiological stresses on the crop. If there are no N, CH_2O, or water stresses, fruit abscission only occurs because of insect damage. The maximum fraction of fruit abscised depends on the fruit age; small squares and small bolls are most susceptible to physiological stresses (Table 1). Note that 11% per day abscission rate will result in 56% abscission after 7 days of severe stress.

To determine the actual fraction to abscise on a given day, the maximum rate, $\overline{RABSC}$, is multiplied by 1−(ADJ)(ADJW) so that if ADJ = ADJW = 1, no abscission will occur. If either ADJ or ADJW is 0.5 and the other is 1.0, then 50% of the maximum abscission will occur. The actual number selected for abscission 3 days later is calculated by

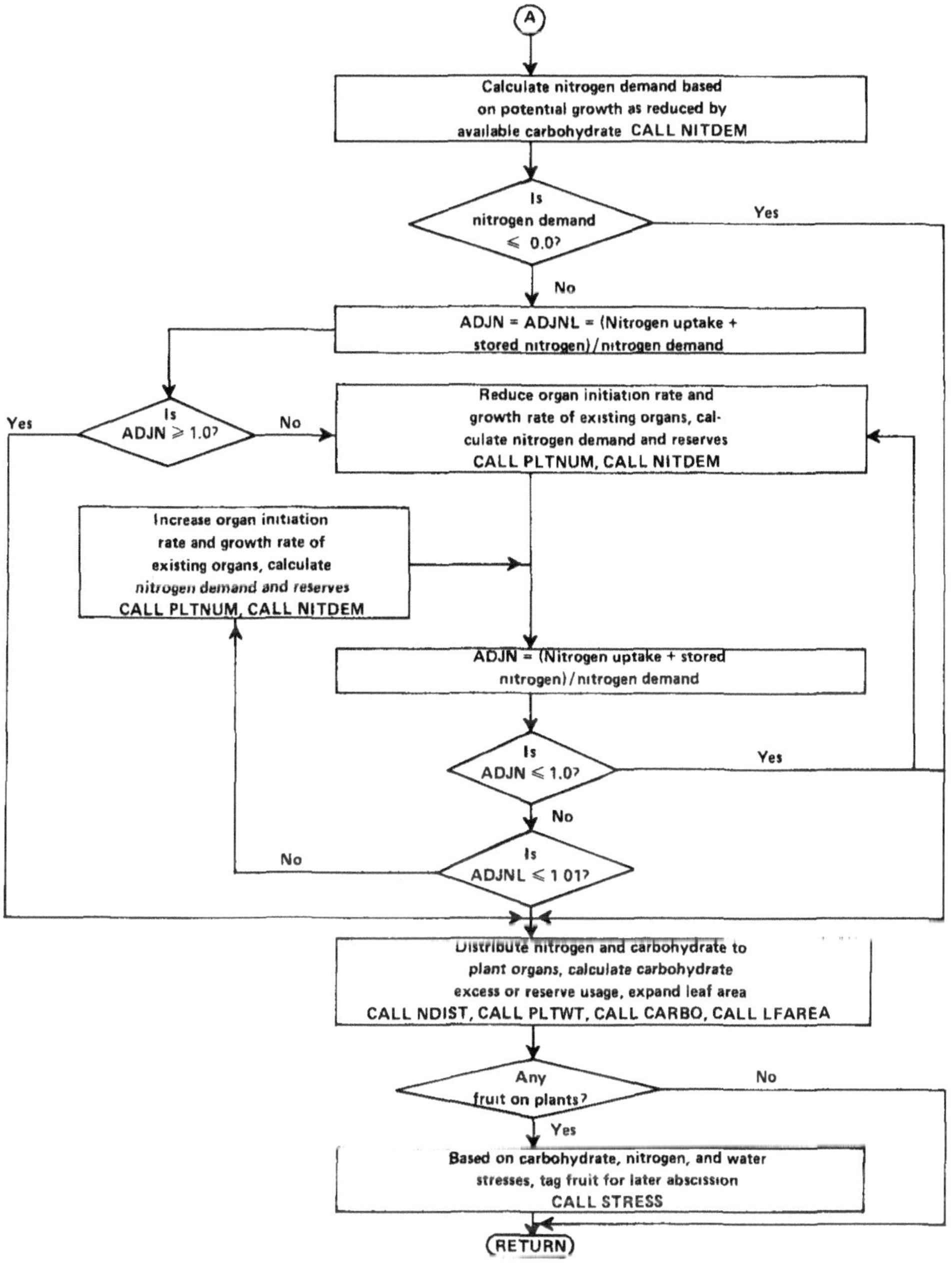

FIGURE 7. Simplified flow chart of subroutine PLANT showing iterative procedures for balancing supplies and demands of carbohydrate and nitrogen.

$$\overline{FRABS_{t+3}} = \overline{FR}_t^T \times (\overline{RABSC_t})(1 - (ADJ)(ADJW)) \qquad (39)$$

Fruit abscission is a feedback mechanism that allows the crop to produce fewer bolls based on supply and demand for nutrients.

TABLE 1.

Fraction of fruit to abscise per day for each age class of fruit for days having maximum stress; i.e., when ADJ = 0

Fruit Class	Fruit age range, days at 26°C	Fraction of fruit to abscise, maximum stress (RABSC)
Small squares	0—9	0.110
Medium squares	9—16	0.025
Large squares	16—26	0.023
Small Bolls	26—33	0.130
Medium Bolls	33—41	0.090
Large Bolls	41—	0.000

D. Nitrogen Balance

Here we discuss the nitrogen balance in the crop and its effect on growth processes. In the soil section below, we briefly discuss the operations of the soil water-nitrogen subroutines. Here, it suffices to say that the nitrogen is assumed to be nitrate and that it flows into the plant roots by mass flow with water. Therefore, knowing transpiration and soil N concentration, we can estimate the daily supply of N from the soil to the crop.

1. Nitrogen Demand

This subroutine calculates the demand for N for growth after the carbohydrate has been balanced. The data are based primarily on the study by Thompson et al.,[16] whereby N concentrations were measured for leaves and bolls of different ages. We assumed that a minimum concentration is needed to maintain normal growth and any N above that minimum is reserve. According to Miyasaka et al.,[17] a better method would be to allow growth rates to be linearly related to N concentration. We can express demand for N for leaf growth by

$$RLFN = (DLEAF + DLEAFB)\ (0.027) \tag{40}$$

where RLFN = total requirement for N for leaf growth, grams per m^2, and 0.027 = minimum leaf N fraction. Similarly, the N required for stem growth (RSTMN) is found by

$$RSTMN = (DSTM + DSTB)\ (0.01) \tag{41}$$

and N required for root growth (RTN) is found by

$$RTN = (DRT)\ (0.01) \tag{42}$$

The data of Thompson et al. [16] did not show a minimum N content for fruit; rather, N declined over time. Therefore, we took the minimum curve for fruit to describe FRNR with elements calculated by

$$frnr(t-i) = 0.03575\ EXP\ (-0.049(t-i)) \tag{43}$$

where t-1 = fruit age in days. Then total fruit N requirement XNFR was calculated by

$$\mathrm{RFRN} = \bar{1}^T \times (\overline{\mathrm{FRW}} + (\overline{\mathrm{FRWR}} \cdot \overline{\mathrm{FR}}) \cdot \overline{\mathrm{FRNR}} - \overline{\mathrm{NFR}}) \quad (44)$$

In the above, new fruit weight of a given age t (frw(t) + fr (t)· frwr (t)) is multiplied by the minimum N fraction for that age, frnr (t) and the N already in the fruit, nfr(t), is subtracted. The cross product with $\bar{1}^T$ causes addition over all ages to arrive at the total N required for fruit growth, RFRN.

The total demand for N is calculated in PLANT by

$$\mathrm{XNREQ} = \mathrm{RLFN} + \mathrm{RFRN} + \mathrm{RSTMN} + \mathrm{RTN} \quad (45)$$

2. Nitrogen Supply

The supply of N to growing organs is based on the uptake of N from the soil and on the availability of N reserves in the plant. The rate of N uptake. (PLNIT) is described below (see Equation 73).

Excess N stored in leaves is available for translocation and use if the uptake (PLNIT) was less than required by XNREQ. The plant N available for use (N reserves) is calculated by

$$\mathrm{STORN} = \bar{1}^T \times (\overline{\mathrm{NLF}} - \overline{\mathrm{RATEN}} \cdot \overline{\mathrm{LFW}}) + \bar{1}^T \times (\overline{\mathrm{NLFB}} - \overline{\mathrm{RATEN}} \cdot \overline{\mathrm{LFWB}})$$

where RATEN is a vector having elements calculated as follows

$$\mathrm{raten}(t-i) = \begin{cases} 0.027 & 0 < t-i \leq 40 \\ 0.027 - 0.0085(t-i-40) & 40 < t-i \leq 42 \\ 0.01 & 42 < t-i \end{cases} \quad (46)$$

and t-i = leaf age in days.

Thompson et al.[16] reported declining N content with age after 40 days. The above relationships allow mining of N from senescing leaves for use elsewhere. It is assumed that N cannot be removed from fruit.

In subroutine PLANT, the demand for N is compared with the availability of N, assuming that up to 80% of reserve N can be utilized within a day. The factor ADJN is calculated by

$$\mathrm{ADJN} = (\mathrm{PLNIT} + (\mathrm{STORN})\,(0.8)\,)\,/\,\mathrm{XNREQ} \quad (47)$$

If ADJN exceeds 1.0, it is set equal to 1.0 and growth is not restricted. If ADJN is less than 1.0, ADJ is calculated by considering both carbohydrate and nitrogen deficits by

$$\mathrm{ADJ} = (\mathrm{ADJC})\,(\mathrm{ADJN}) \quad (48)$$

The factor ADJ is used to reduce actual growth of organs if its numerical value is below 1.0, and represents the nutritional stress on the crop. In that case, subroutine

PLANT uses an iterative procedure to reduce growth and organogenesis so that growth just reaches the potential with existing N supply. We have assumed the same stochiometric relations for all plant parts and N contents are not added to the organ weights in the model.

E. Crop Growth

After both CH_2O and N balances have been performed individually by COTCROP, the subroutine CARBO is again called to account for changes in CH_2O balance by growth reductions from N shortages. Then three subroutines are used to accomplish growth; PLTWT, LFAREA, and NDIST. Subroutine PLTWT causes biomass growth to oocur after the nutritional balances. LFAREA causes increases in plant leaf area and NDIST causes N to be distributed to the new growth.

1. Dry Weight Growth

Subroutine PLANT calls PLTWT after the CH_2O and N balances have been performed and after the final value of ADJC, ADJN, and ADJW have been determined. PLTWT causes the weights of leaves, stems, fruit and roots to increase. New main stem leaf weights, $\overline{LFW}_t$, are calculated by

$$\overline{LFW_t} = (A_{1t}\overline{LFW_{t-1}} + (\overline{LFM_t} \cdot \overline{LFWR_t})\ U) \tag{49}$$

where U = (TINC)(ADJ)(ADJW)/PLSTN, and for fruiting branch leaves,

$$\overline{LFWB_t} = (A_{2t}\overline{LFWB_{t-1}} + (\overline{LFM_t} \cdot \overline{LFWBR_t})\ U) \tag{50}$$

Stem weights for main stems are calculated by

$$\overline{STMW_t} = (\overline{STMW_{t-1}} + (\overline{STMWR_t} \cdot \overline{STM_t})\ U) \tag{51}$$

and for fruiting branches,

$$\overline{STMWB_t} = (\overline{STMWB_{t-1}} + (\overline{STB_t} \cdot \overline{STMWBR_t})\ U) \tag{52}$$

New growth is added to fruit by

$$\overline{FRW_t} = ((A_{3t})\ \overline{FRW_{t-1}} + (\overline{FRWR_t} \cdot \overline{FR_t})\ U) \tag{53}$$

and to roots by

$$TWRT_t = TWRT_{t-1} + (0.2)(DLEAF_t + DLEAFB_t) \tag{54}$$

The total plant weight (TOTWT) is calculated by adding all plant part weights of all ages.

2. Leaf Area Growth

Potential leaf area expansion is 0.004 m^2 per plastochron for leaves on the main stem and 0.002 m^2/plastochron for branch leaves.[13] In COTCROP, leaves are allowed to expand independent of weight until a maximum leaf area ratio is reached, of 0.02 m^2/g. Also, leaves are assumed to expand linearly for 24 physiological days, then cease expansion and growth.

The equation for main stem leaf area with time is

$$\overline{LAIM}_t = (A_{1t}\overline{LAIM}_{t-1}) + (\overline{MA}_t) \cdot (\overline{LFM})\,(TINC)\,(ADJW)/(PLSTN) \tag{55}$$

and for branches

$$\overline{LAIB}_t = (A_{2t}\overline{LAIB}_{t-1}) + (\overline{MAB}_t) \cdot (\overline{LFB}_t)\,(TINC)\,(ADJW)\,/\,(PLSTN) \tag{56}$$

where elements of $\overline{MA}$ are

$$ma\,(i) = \begin{cases} 0.006 & 0 \leq ct(t) - ct(i) \leq 24 \\ 0 & \end{cases} \tag{57}$$

and elements of $\overline{MAB}$ are

$$mab(i) = ma(i)\,(0.5) \tag{58}$$

When laim(t)/lfw(t) (the leaf area per unit weight) was greater than 0.02, then ma(i) was set to zero. Total leaf area per m^2 of surface is equivalent to leaf area index and is calculated by

$$LAI = \overline{1}^T \times (\overline{LAIM}_t + \overline{LAIB}_t) \tag{59}$$

3. Nitrogen Distribution

The subroutine NDIST distributes N to leaves, fruit, stems, and roots. It is assumed that surplus N is stored in leaves and growing fruit.

The first case to consider is that there is excess N uptake and potential growth occurs with the remainder N stored in fruit and leaves. N stored in roots and stems is not a variable since concentrations are forced to remain constant in those organs. For leaves, 0.027 of the new growth must be N and

$$\overline{NLF_t} = (1 + XN3 / XNIT)(\overline{NLF_{t-1}}) + (\overline{LFWR_t} \cdot \overline{LFM_t})(0.027)\,U \tag{60}$$

and

$$\overline{NLFB_t} = (1 + XN3 / XNIT)(\overline{NLFB_{t-1}}) + (\overline{LFWBR_t} \cdot \overline{LFB_t})(0.027)\,U \tag{61}$$

$$XN3 = PLNIT - XNREQ \tag{62}$$

and

$$XNIT = \overline{1}^T \times (\overline{NLF_t} + \overline{NLFB_t} + \overline{NFR_t}) \tag{63}$$

For the case that some stored N is used in growth, the "excess" N is calculated by

$$XN3 = PLNIT + STORN - XNREQ \tag{64}$$

Fruit N calculations are expressed as

$$\overline{NFR_t} = \left(\frac{XN3}{XNIT} - 1\right)\overline{NFR_{t-1}} + [\overline{FRW_t} + (\overline{FRWR_t} \cdot \overline{FR_t})(ADJ)(TINC) / PLSTN] \cdot \overline{FRNR_t} \tag{65}$$

Therefore, any N not used in growth is added to a leaf reserve pool and to fruit, the relative amounts to each depending on the N in each. In Equation 65, XN3 is restricted to be ≥ 0 since it is the amount added to fruit.

F. Soil Processes

The soil model in COTCROP consists of water and nitrogen balances. The soil depth is divided into L homogeneous 10-cm depth layers, each having distinct volumes of water and concentrations of nitrate. Availability of water and nitrogen to the plant depend on root depth specified.[6] The following sections briefly describe the relations used to calculate soil water and nitrogen precesses.

1. Soil Water Balance

The emphasis of this soil water balance is on the overall soil volume, whereby all boundary values are calculated. The overall soil water balance is

$$W_t = W_{t-1} + I_t - D_t - E_{st} - E_{pt} \tag{66}$$

Redistribution of water within the one dimensional soil profile was calculated based on a solution to the soil water flux equation

$$q = D(\Theta) \frac{2\Theta_x}{2x} + ks \text{ and } \frac{\partial\Theta_x}{dt} = \frac{dq}{dx} - Sx \tag{67}$$

combined with the continuity equation for each soil layer. First, all boundary values, I_t, D_t, E_{st}, and E_{pt}, are calculated, as described below. Then, an iterative procedure using finite differences is used to calculate redistribution based on those four boundary conditions.

Infiltration, I_t, is calculated from irrigation or rainfall. For irrigation, the depth of water applied is set to I_t. For rainfall, the SCS[18] method is used to partition rainfall into runoff and infiltration, or

$$I_t = R_t - (R_t - 0.2S)^2 / (R_t + 0.8S) \tag{68}$$

When $I_t>0$, water is added to the soil to a depth such that soil above the depth is at field capacity. If sufficient quantity of rainfall occurs to increase the entire profile to field capacity, any excess water is added and allowed to drain.

The drainage rate is calculated using the Jackson-Whisler[19] equation. A numerical procedure is used to solve for daily drainage, D_t, based on cumulative drainage, Q_t, or $D_t = Q_{t-1}$

The Ritchie[20] model for estimating evaporation and transpiration from a row crop is used to calculate E_{st} and E_{pt}. Whereas Ritchie did not account for decreases in transpiration due to limiting soil water, we reduce potential transpiration E_{pt} as follows:

$$E_{pt} = \begin{cases} E'_{pt}\,(1 + 0.1 / \bar{h}_t) & \text{if } -3.0 < \bar{h}_t \leq -0.3 \\ E'_{pt}\,(0.6 + 0.0333\bar{h}_t) & \text{if } -15.0 \leq \bar{h}_t \leq -3.0 \end{cases} \tag{69}$$

The s_x term in Equation 69 is based on E_{pt} and the water potential profile, h_x. Water is extracted more readily from layers having higher water potentials using the following:

$$s_x = E_{pt}\,(15 - |h_x|) / \sum_{x=1}^{L} (15 - |h_x|) \tag{70}$$

The soil water potential for each layer h_x, was calculated using the Brooks and Corey[21] relation:

$$h_x = -h_b / \Theta_x^{(3/(\eta - 2))}$$
$$\text{where } \Theta_x = \frac{\Theta_x - \Theta_d}{\Theta_s - \Theta_d} \tag{71}$$

The numerical procedure used in SOILW and SWDIST subroutines calculates the volumetric water content and water potential for each soil layer. Average values of soil water content, $\bar{\theta}_t$, and soil water potential, $\bar{h}_t$, are calculated for each day. The overall soil water model was tested for Leeper Sandy Loam soil by Jones et al.[22]

:2. Soil Nitrogen Balance

Nitrogen in the soil was assumed to be nitrate. Additions of nitrogen (N) to the soil is by fertilizer and organic decomposition, N losses occur by plant uptake, leaching, and denitrification. The overall soil N balance is by

$$N_t = N_{t-1} + F_t + O_t - PLNIT_t - L_t - DN_t \quad (72)$$

The N is distributed into the L layers within the soil profile. Fertilizer N is added to the top layer on specified days in g N/m² of soil area. The rate of N leaching, L_t, is calculated from D_t and the N concentration of the bottom layer. Addition of N from organic matter decomposition, O_t, is based on percent organic matter in the soil,[6] and occurs in the top 30 cm.

Removal of N by the crop depends on the distribution of water extraction, s_x, and on concentration of N, n(x, t). Plant N uptake, PLNIT, is calculated by

$$PLNIT = \sum_{0}^{L} s_x n(x, t) \quad (73)$$

Denitrification of soil N in each layer occurs only if that layer has a volumetric water content above field capacity for more than one day.[23] Then, denitrification is assumed to be a first order reaction with rate dependent on soil carbon content and n. The rates are calculated from data of Stanford et al.[23]

Nitrogen redistribution in the soil profile is assumed to occur by convection only, and diffusion of N is not included. Dispersion of N is considered using the method outlined by Gardner.[24] A finite difference solution of the equation

$$\frac{\partial n}{\partial t} = \frac{-\partial}{\partial x}(C_D q n) - n s_x \quad (74)$$

is used to estimate redistribution of N within the profile.

III. MODEL CALIBRATION AND VALIDATION

Considerable data were available for estimating parameters of the model, such as rates of photosynthesis, leaf growth, and organ initiation. However, several parameters were not available; namely, the effects of nitrogen, carbohydrate, and water stresses on organ initiation, and fruit abscission. As in SIMCOTT II,[4] we based fruit abscission on the ratio of carbohydrate and N supply to their respective demands. Also, we included abscission of fruit and a decreased organ initiation rate due to water stress, defined by average soil water tension.

Data of Bruce and Romkens[10] were used to calibrate those stress parameters. In their 1961 experiments on a Leeper Sandy Loam soil at Mississippi State, Mississippi, Bruce and Romkens had one treatment, AAA, whereby soil water potential was main-

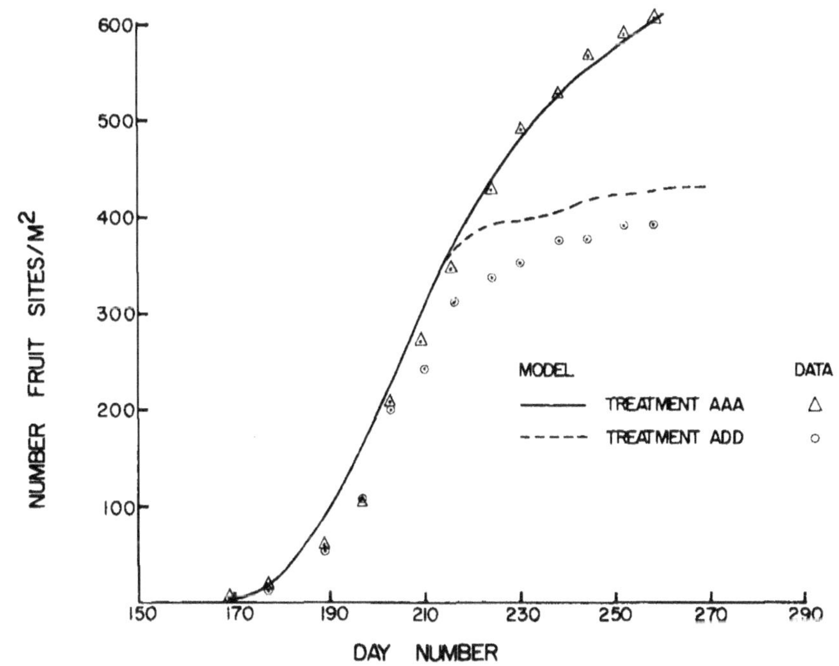

FIGURE 8. Comparison of simulated number of fruiting sites with data of Bruce and Romkens for their irrigation treatments AAA and ADD. (Data from Bruce, R. R. and Romkens, J. M., *Agron. J.*, 57, 135, 1965. With permission.)

tained above −0.3 bar throughout the season. In another treatment, ADD, they maintained the same water level until the small boll stage, then irrigated the plots when the soil water tension reached −2.4 bar. Rainfall was excluded by rainfall shelters. The detailed data they acquired on soil water status and on plant development and the corresponding weather data provided an excellent data set for model calibration. Nitrogen stress did not likely occur because of the high rate of N fertilization (300 lb/acre equivalent in the small plots).

A trial and error procedure was used to calibrate the abscission parameters using three criteria for goodness of fit: number of fruiting sites, number of fruit and final yield. Both AAA and ADD treatments were used.

Results of the calibration are shown in Figures 8 and 9. There was a considerable reduction in the number of sites produced by plants in treatment ADD. Similarly, the number of fruit on the water stressed plants was lower. Yields were 3.18 and 2.46 bales / acre for treatments AAA and ADD, respectively. Corresponding simulated yields were 3.22 and 2.69.

The percentages of fruit to abscise were 11, 2.5, and 2.5 for young, medium age, and old squares, respectively, and 13 and 9 for bolls less than 10 days old and bolls between 10 and 20 days of age, respectively. It was assumed that older bolls do not abscise. It was clear that fruit selectivity for abscission was young squares and young bolls. These percentages listed are maximum percentages per day, which occur only if ADJ is 0, or if supply of carbohydrate or N is zero, see Equation 39.

In 1977, an experiment was conducted at the same site at Mississippi State of Bruce and Ramken's experiments. However, there are two major differences in these two experiments. First, in 1977, the N application rate was 100 lb/acre. Second, water stress was allowed to occur earlier during growth. Four water treatments, 0, 3, 5, and

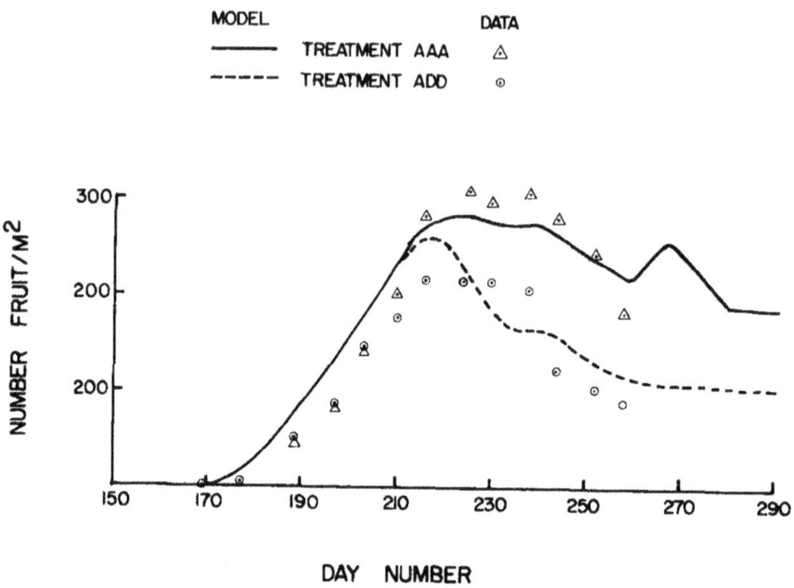

FIGURE 9. Comparison of simulated number of fruit (squares plus bolls) with data of Bruce and Romkens for their irrigation treatments AAA and ADD. (Data from Bruce, R. R. and Romkens, J. M., *Agron. J.*, 57, 135, 1965. With permission.)

8 irrigations, were established for each of four varieties. We have averaged the results from two varieties, Stoneville 213 and Deltapine 16, for comparison with COTCROP.

During early comparisons between the 1977 experiments and COTCROP, it became evident that the plants responded differently if water stress occurred prior to flowering. Regardless of the stress, organ initiation continued at about a normal rate, whereas if stress occurred first after flowering, organ initiation was sharply curtailed.[10] Therefore, at that point we changed COTCROP to reflect that behavior.

Also, because of the long water stress periods, a provision was added to the soil water model to allow recharge of the root zone with water from a water table which normally occurs at the site.

There was little difference in the 5 or 8 irrigation treatments. Figure 10 shows the simulated and actual square density curves and total main stem nodes for treatment 5. The simulated vs. actual yield data are shown in Figure 11 along with the Bruce and Romkens data. The model predicted low yield values for high water stress treatments in 1977. The yield of the 0 irrigation treatment was 18% of the 8 irrigation treatment and COTCROP predicted that it would be only 5%. The corresponding value for the 3 irrigation treatment was 76% whereas COTCROP predicted 43%. The predicted yields for the 5 and 8 irrigation treatments were within 4% of actual data. Due to differences in weather and nitrogen fertilization, yields were considerably lower in 1977 when compared with 1961 yields.

The trend in simulated results is very good, Figure 11. For the nonwater stress treatments, excellent agreement was obtained between simulated and predicted yields. The simulated results for severe water stress predict a greater reduction in yield than was realized. The ability of the crop to adapt to periods of drought through greater root proliferation or reductions in leaf osmotic potentials for turgor maintenance are concepts not included in the current model. These adaptive mechanisms may be a major reason for these variations.

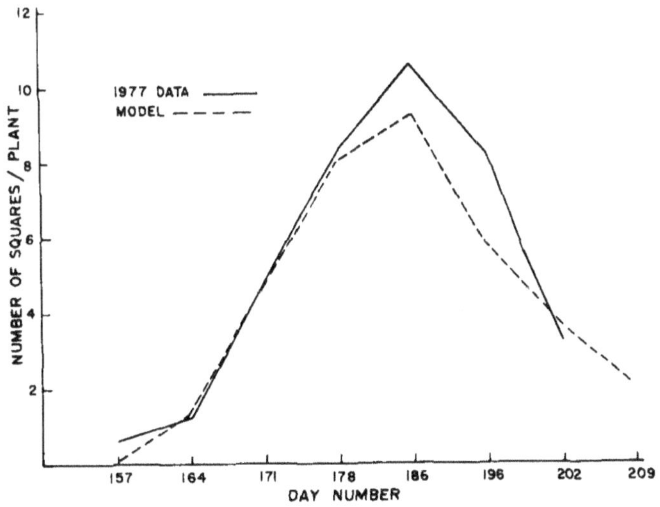

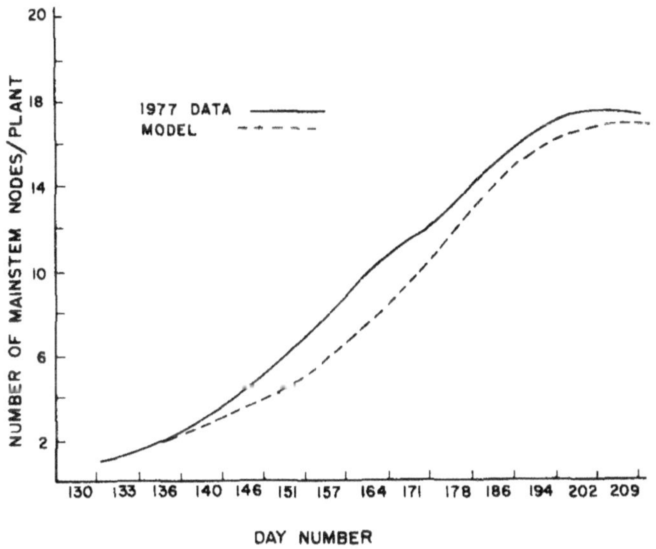

FIGURE 10. (A) Comparison of simulated squares/plant with 1977 data taken at Mississippi State University for the five irrigation treatment. (B) Comparison of the simulated mainstem nodes/plant with the 1977 data.

IV. APPLICATIONS

A major application of COTCROP is for study of insect pest management. Models of major cotton pests, the boll weevil,[25] *Heliothis zea* and *H. virescens*,[26] have been developed. These models have been interfaced with COTCROP[27] for studying conflictive cotton pest management strategies. Early season applications of insecticide for boll weevil control destroy predators that regulate *Heliothis* populations and *Heliothis* are becoming resistant to approved insecticides. Studies are being conducted to deter-

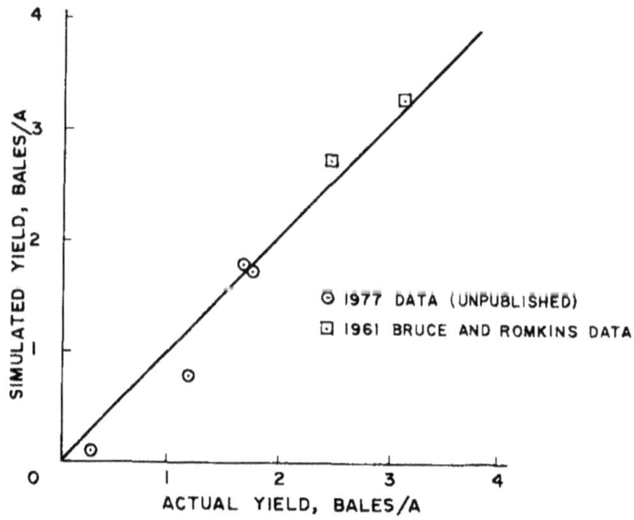

FIGURE 11. Comparison of simulated and actual yields for 1961 and 1977 irrigation treatments.

mine the possible benefits for various strategies, including the effects of timing insecticide applications on insect populations and on crop productivity.

To demonstrate the response of COTCROP to insect damage, 16 simulations were performed. The 1961 Mississippi State weather data were used with two water treatments: (1) the irrigation treatment ADD of Bruce and Romkens[10] and (2) natural rainfall. Nitrogen fertilizer was kept the same (300 lb/acre) for both cases. Superimposed on each water treatment were eight insect damage treatments. The simulated crops were maintained insect free except for one 20-day period in each of the eight runs. An insect treatment consisted of each crop being exposed to insect damage from 10,000 *Heliothis* larvae per hectare for a discrete time period of 20 days. The time that each 20-day period started were on days 170, 190, 210, 230, 270, 290, 310. These treatments were designed to determine the susceptibility of the crop to insect damage to fruit at different stages of growth and to determine if water treatment affected that susceptibility.

The treatment ADD had adequate water most of the season and produced (simulated) 2.69 bales (500 lb) /acre. The natural rainfall treatment experienced erratic rainfall thoughout the season and produced only 1.12 bales /acre. The results of the simulations are presented in Figure 12. The greatest reduction in yield occurred when insect damage was initiated early, on day 170. For the natural rainfall treatment, the yield reductions continually decreased until day 270 and damage occurring after day 270 caused no yield reduction. For the ADD irrigation treatment, maximum yield loss also occurred when insect damage started on day 270. However, if the insect damage occurred during the period 210 to 230, a slight yield increase occurred. In earlier experiments whereby we periodically pruned fruit from cotton, yield increases occurred for some treatments.[28]

Figure 13 shows a plot of the absolute yield losses simulated for both water treatments. The results indicate that the magnitude of yield loss is about the same for the low-yielding cotton as for the high-yielding cotton. The cotton is responding to damaged fruit and the nonirrigated cotton produced fewer fruit, but had the same number damaged. Based on these simulated results, crops having differing yield potentials have different relative yield losses for the same level and timing of insect damage. The ex-

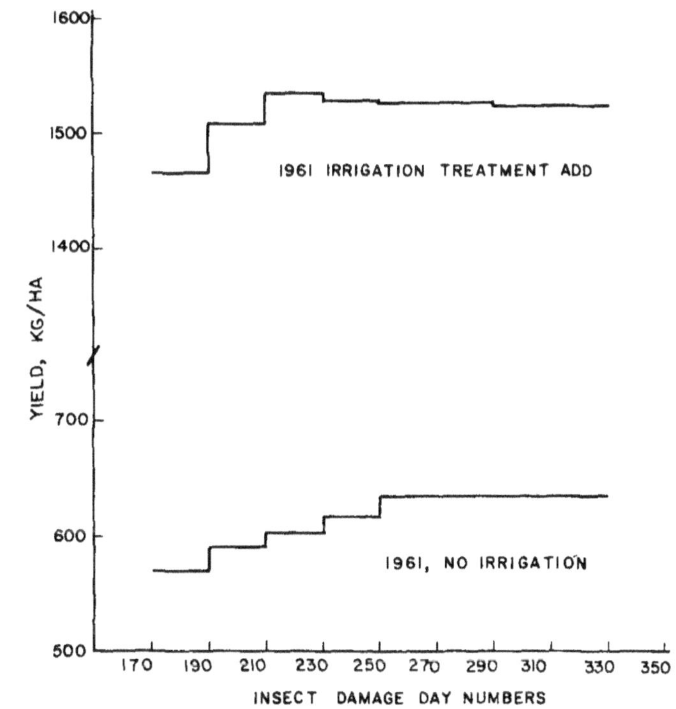

FIGURE 12. Simulated yield values for eight different insect damage periods and two water treatments using 1961 Mississippi State University weather data. During each insect damage period, 10,000 larvae per hectare were allowed to damage fruit for a 20-day period.

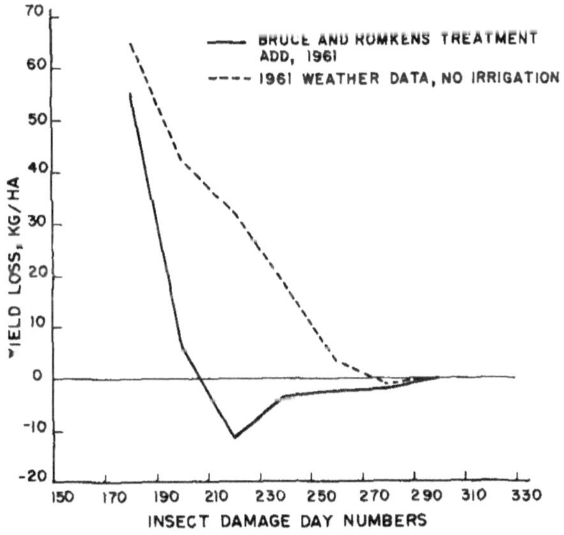

FIGURE 13. Absolute yield losses for simulated trials described for Figure 12.

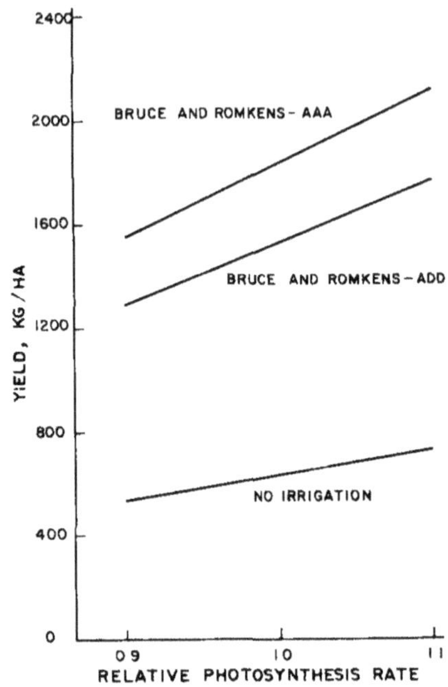

FIGURE 14. Sensitivity of COTCROP to relative changes (± 10%) in gross photosynthesis rates using 1961 weather data and irrigation treatments. (Irrigation treatments from Bruce, R. R. and Romkens, J. M., *Agron. J.*, 57, 135, 1965. With permission.)

tension of such results to include economics of pest management offers exciting opportunities.

In this book, emphasis is on predicting photosynthesis. In Figure 14 we show the results of a sensitivity analysis of COTCROP to varying photosynthesis rates. Gross photosynthesis was increased or decreased by 10% for each of Bruce and Romkens's[10] 1961 AAA and ADD treatments, and for 1961 weather with no irrigation. For each case, a 10% decrease in photosynthesis resulted in 13 to 16% decrease in lint yield, whereas a corresponding increase caused 16 to 18% increase in yield. The largest relative increase occurred for the nonirrigated treatment.

It is not our purpose here to discuss a thorough sensitivity analysis. Rather we want to point out some potential applications. The model is currently being tested or validated for use in a large scale cotton pest management program. Other possible applications include irrigation management, harvest scheduling, and various research applications. However, critical tests or validation must preceed any new application. Also, COTCROP is very sensitive to soil parameters, including the soil water content-potential relationship and root depth. It is anticipated that experience must be gained for each new soil type and even specific field sites before the model can depict growth for those conditions.

The experience gained by the authors thus far in applying COTCROP to study cotton production and insect pest management indicates that it is a powerful research tool.

V. SYMBOLS

A. State Variables

$\overline{FR}_t$ — vector of fruit numbers per m^2 on day t, elements $fr_t(i)$, where t-i is the age of fruit.

$\overline{FRABS}_t$ — vector of fruit numbers per m^2 to be abscised on day t, elements $frabs_t(i)$, where t-i is the age of fruit.

$\overline{FRW}_t$ — vector of fruit weights on day t, elements $frw_t(i)$, where t-i is the age of fruit; g/m^2.

$\overline{LAIB}_t$ — vector of leaf areas for fruiting branch leaves on day t, elements $laib_t(i)$, where t-i is the age of leaves; m^2/m^2.

$\overline{LAIM}_t$ — vector of leaf areas for main stem leaves on day t, elements $laim_t(i)$, where i is the age of leaves; m^2/m^2.

$\overline{LFB}_t$ — vector of leaf numbers per m^2 on fruiting branches on day t, elements $lfb_t(i)$, where t-i is the age of leaves.

$\overline{LFM}_t$ — vector of leaf numbers per m^2 on main stems on day t, elements $lfm_t(i)$, where t-i is the age of leaves.

$\overline{LVB}_t$ — vector of leaf numbers per m^2 on vegetative fruiting branch on day t, elements $lvb_t(i)$ where t-i is the age of leaves.

$\overline{LVM}_t$ — vector of leaf numbers per m^2 on vegetative main stems on day t, elements $lvm_t(i)$, where t-i is the age of leaves.

$\overline{LFW}_t$ — vector of leaf weights for leaves on the main stem on day t, elements $lfw_t(i)$, where t-i is the age of leaves; g/m^2.

$\overline{LFWB}_t$ — vector of leaf weights for leaves on fruiting branches on day t, elements $lfwb_t(i)$, where t-i is the age of leaves; g/m^2.

$\overline{NFR}_t$ — vector of nitrogen weight in fruit on day t, elements $nfr_t(i)$, where t-i is fruit age; g/m^2.

$\overline{NLF}_t$ — vector of nitrogen weight in main stem leaves on day t, elements $nlf_t(i)$, where t-i is leaf age; g/m^2.

$\overline{NLFB}_t$ — vector of nitrogen weight in fruiting branch leaves on day t, elements $nlfb_t(i)$, where t-i is leaf age; g/m^2.

$\overline{OPN}_t$ — vector of fractions of bolls open day t, elements $opn_t(i)$, where t-i is boll age.

SL — total carbohydrate reserves, g/m^2.

$\overline{STB}_t$ — vector of fruiting branch stem numbers per m^2 on day t, elements $stb_t(i)$, where t-i is stem age.

$\overline{STM}_t$ — vector of main stem numbers per m^2 on day t, elements $stm_t(i)$, where t-i is stem age.

$\overline{STMW}_t$ — vector of main stem weights on day t, elements $stmw_t(i)$, where t-i is stem age; g/m^2.

$\overline{STMWB}_t$ — vector of fruiting branch stem weights on day t, elements $stmwb_t(i)$, where t-i is stem age; g/m^2.

STORN — total nitrogen reserves, g/m^2.

TWRT — total root weight, g/m^2.

$\overline{VBR}_t$ — vector of number of vegetative branches per m^2 on day t that were initiated on day i, elements $vbr_t(i)$, where t-i is branch age.

B. Secondary Variables

A_{1t}, A_{2t}, A_{3t} — transition matrices for causing abscission of organs on day t, for main stem leaves, fruiting branch leaves, and fruit, respectively.

ADJ — total reduction in growth caused by carbohydrate and nitrogen stresses = (ADJC)(ADJN).

ADJC —	ratio of carbohydrate supply to demand.
ADJN —	ratio of nitrogen supply to demand.
ADJDLY —	fraction between 0 and 1 for reducing morphogenesis due to water stress history.
ADJW —	reduction in leaf expansion due to current day's soil water tension, between 0 and 1.
$\overline{B}_{jt}$ —	organ initiation on day for main stem leaves, branch leaves, main stem internodes, branch stem internodes, and fruit for j = 1,2,3,4,5, respectively, number /m²-day.
B —	maximum fraction of carbohydrate reserves that can be used within a day.
$\overline{CT}_t$ —	vector of elements ct(i), cumulative physiological time on day i, organs of age t-i have physiological age ct(t) − ct(i) on day t.
CREQ —	total carbohydrate demand for fruit growth on day t, g/m^2-day.
DFW —	carbohydrate demand for fruit growth on day t, g/m^2-day.
DLEAF —	carbohydrate demand for leaf growth for leaves on the main stem, g/m^2-day.
DLEAFB —	carbohydrate demand for leaf growth for leaves on fruiting branches, g/m^2 -day.
DRTW —	carbohydrate demand for root growth on day t, g/m^2-day.
DSTB —	carbohydrate demand for fruiting branch growth on day t, g/m^2-day.
DSTM —	carbohydrate demand for main stem growth on day t, g/m^2-day.
DTAG —	day number of crop emergence plus t minus 137, used to calculate days required for boll opening.
$\overline{\text{FRLNT}}$ —	vector of fraction of boll weight that is lint for each age of bolls.
$\overline{\text{FRNR}}$ —	vector of minimum nitrogen concentrations for fruit of each age.
FRNUM —	total number of fruit (squares and bolls) on the crop on day t per m^2.
$\overline{\text{FRWR}}$ —	carbohydrate demand vector for fruit of all ages, g/ fruit -day.
GR —	growth respiration, g respired /g growth.
LAI —	leaf area index of crop on day t.
LFNUM —	total number of leaves on the crop on day t, leaves per m^2.
$\overline{\text{LFWBR}}$ —	vector of leaf growth rates for leaves of different ages on fruiting branches, g/ leaf-day.
LFWR —	vector of leaf growth rates for leaves of different ages on the main stem, g/ leaf-day.
OPNBL —	total number of open bolls per m^2 on day t.
PLNIT —	daily total nitrogen uptake from the soil for day t, g/m^2-day.
PLSTN —	number of physiological days between successive leaves on the main stem, or plastochron.
PLTM2 —	plant population, plants per m^2.
POP —	plant population, plants /acre.
PR —	daily rate of photosynthesis for day t, g/m^2-day.
QCR —	total carbohydrate available for growth on day t, g/m^2-day.
$\overline{\text{RABSC}}$ —	vector of maximum fractions of fruit to abscise on a given day due to stresses, 1/day.
R_ψ —	photosynthetic reduction factor due to soil water potential.
RAD —	daily total radiation, langleys /day.
$\overline{\text{RATEN}}$ —	vector of mininum leaf nitrogen concentrations for each leaf age.
RFRN —	total nitrogen demand for fruit growth, g/m^2-day.
RFLN —	total nitrogen demand for leaf growth, g/m^2-day.

RM —	maintenance respiration rate, g carbohydrate/g crop-day.
RSTMN —	total nitrogen demand for stem growth, g/m^2-day.
RTN —	total nitrogen demand for root growth, g/m^2-day.
$\overline{SI}^T$ —	row vector for calculating numbers of floral buds.
SQNUM —	total number of floral buds (squares) on the crop on day t per m^2.
SRATE —	rate of addition or depletion of carbohydrate reserves, g/m^2-day.
$\overline{STMWBR}$ —	vector of carbohydrate demands for fruiting branch growth for branches of all ages, g/stem-day.
$\overline{STMWR}$ —	vector of carbohydrate demands for main stem growth for stems of all ages, g/stem-day.
t —	real time from start of simulation, days.
T —	temperature, °C.
t_{vb} —	real time at which vegetative branches start growth, days.
TINC —	number of physiological days on day t; at 26°C, TINC = 1.0 and 1 real day equals 1 physiological day.
$\overline{TOTWT}$ —	total crop weight on day t, g/m^2.
TSB —	number of real days required for bolls to open for floral buds initiated on day t.
TWLNT —	total lint weight on day t, g/m^2.
VB —	number of plastochrons at which time vegetative branches start growth.
VEGBRN —	maximum number of vegetative branches per plant, a function of plant population.
X3 —	total growth and respiration carbohydrate requirement, g/m^2-day.
XK2 —	maximum fraction of reserve carbohydrate that can be used within a day, 1/day.
XLBR —	number of new fruiting branch leaves per plastochron at the current crop growth stage, reflecting morphogenetic potential.
XN3 —	excess nitrogen added to reserves on day t, g/m^2-day
XNIT —	total nitrogen in the crop on day t, g/m^2.
XNREQ —	total nitrogen demand for growth on day t, g/m^2-day.

C. Soil Variables

C_D —	dispersion coefficient for soil water transport
D_T —	volume of water drained from profile on day t, cm per day
DN_t —	denitrification rate of nitrate N on day t, grams per m^2 per day
$D(\Theta)$ —	soil water diffusivity, cm^2 per day
E_{pt} —	transpiration on day t, cm per day
E'_{pt} —	potential transpiration on day t calculated from Ritchie[20], cm per day
E_{st} —	soil evaporation on day t, cm per day
F_t —	amount of fertilizer N applied on day t, grams per m^2 per day
h —	h(x, t), or the soil water tension at depth x on day t, bar
h_b —	bubbling pressure of soil, bar
$\overline{h}_r$ —	average soil water tension of the root zone, bar
I_t —	irrigation water applied on day t, cm
k_s —	saturated hydraulic conductivity, cm/day.
L_t —	soil N leached on day t, g/m^2.
L —	number of homogenous 10-cm layers in soil profile.
n —	n (x, t) or nitrogen concentration in soil water at depth x and on day t, gN/cm^3 water.

N_t —	total N in soil on day t, g/m^2.
O_t —	N added to soil on day t from organic matter decomposition, g/m^2.
q —	flux of water, cm/day.
R_t —	rainfall on day t, cm.
S —	empirical parameter of SCS runoff equation.
s_x —	rate of water extraction from layer x on day t, cm/day,
η —	parameter for the Brooks and Corey[21] relation.
Θ_d —	volumetric air dry soil water content.
Θ_s —	volumetric saturated soil water content.
Θ_x —	volumetric soil water content at depth x on day t.

REFERENCES

1. **Duncan, W. G.**, SIMCOT: A simulator of cotton growth and yield, in *Proc. Workshop Modeling Tree Growth,* Murphy, C. M. and Hesketh, J. D., Eds., Oak Ridge National Laboratory, Oak Ridge, Tenn., 115, 1972.
2. **Hesketh, J. D., Baker, D. N., And Duncan, W. G.**, Simulation of growth and yield in cotton: II. Environmental control of morphogenesis, *Crop Sci.*,12,436, 1972.
3. **Stapleton, H. N., Buxton, D. R., Watson, F. L., Nolting, D. J., and Baker, D. N.**, Cotton: a computer simulation of cotton growth, *Tech. Bull. 206,* Agricultural Experimental Station, University of Arizona, 1973.
4. **McKinion, J. M., Baker, D. N., Hesketh, J. D., and Jones, J. W.**, SIMCOT II: A simulation of cotton growth and yield, in *Computer Simulation of a Cotton Production System — Users Manual,* U.S. Department of Agriculture ARS-S-52, 1975, 27.
5. **McKinion, J. M., Jones, J. W., Hesketh, J. D.**, Analysis of SIMCOT: photosynthesis and growth, *Proc. Beltwide Cotton Production Research Conf.*, National Cotton Council, Memphis, 1974.
6. **Jones, J. W., Hesketh, J. D., Kamprath, E. J., and Bowen, H. D.**, Development of a nitrogen balance for cotton growth models: a first approximation, *Crop, Sci.*, 14, 541, 1974.
7. **Gutierrez, A. P., Leigh, T. F., Wang, Y., and Cave, R. D.**, An analysis of cotton production in California: *Lygus Hesperus* injury — an evaluation, *Can. Entomal.*, 109, 1375,1977.
8. **Forrester, J. W.**, *Industrial Dynamics,* MIT Press, Cambridge, 1961.
9. **Pritsker, A. A. B.**, *The GASP IV Simulation Language,* John Wiley & Sons, New York, 1974.
10. **Bruce, R. R. and Romkens, J. M.**, Fruiting characteristics of cotton in relation to soil moisture tension, *Agron. J.*, 57, 135, 1975.
11. **Skeith, R. W. and Ballard, J. L.**, An Industrial Engineering Approach to Ecosystem Analysis, *Proc. Am. Inst. Ind. Eng.*, Spring Conference, New Orleans, La., 1974.
12. **Jones, J. W., Hesketh, J. D., Colwick, R. R., Lane, H. C., McKinion, J. M., and Thompson, A. C.**, Predicting population of squares, flowers and bolls at different stages of organogenesis for a stand of cotton., *Proc. Beltwide Cotton Production Research Conf.*, National Cotton Council, Memphis, 1975, 108.
13. **McKinion, J. M.**. Simulation of plant growth: morphogenetic control of leaf expansion, *Proc. Beltwide Cotton Production Research Conf.*, National Cotton Council, Memphis, 1975, 56.
14. **Bacheler, J. S., Jones, J. W., Bradley, J. R., and Bowen, H. D.**, Influence of temperature on abscission of cotton squares infested with boll weevil eggs, *J. Econ. Entomol.*, 68, 298, 1975.
15. **Hesketh, J. D., Baker, D. N., and Duncan, W. G.**, Simulation of growth and yield; respiration and the carbon balance, *Crop Sci.*, 11, 394, 1972.
16. **Thompson, A. C., Lane, H. C., Jones, J. W., and Hesketh, J. D.**, Nitrogen concentrations of cotton leaves, buds and bolls in relation to age and nitrogen fertilization, *Agron. J.*, 68,617, 1976.
17. **Miyasaka, A., Murata, Y., and Iwata, T.**, Leaf area development and leaf senescence in relation to climate and other factors, in *JIBP Synthesis, Crop Productivity and Solar Energy Utilization in Various Climates in Japan,* Murata, Y., Ed., University of Tokyo Press, Tokyo, 1975, chap, 3.2, 72.
18. **Schwab, G. O., Frevert, R. K., Edminster, T. W., and Barnes. K.K.**, *Soil Water Conservation Engineering,* John Wiley & Sons, New York, 1966, 103.

19. **Jackson, R. D. and Whisler, F. D.,** Equations for approximating the vertical nonsteady-state drainage of soil columns, *Soil Sci. Soc. Am. Proc.*, 34, 715, 1970.
20. **Ritchie, J. T.,** Model for predicting evaporation from a row crop with incomplete cover, *Water Resour. Res.*, 8, 1204, 1972.
21. **Brooks, R. H. and Corey, A. T.,** Hydraulic properties of porous media, *Hydrology Papers*, Colorado State University, Fort Collins, Colorado, 3, 1, 1964.
22. **Jones, J. W., Whisler, F. D., and Hartsog, J.,** Soil water model for crop level simulations, *Proc. Beltwide Cotton Production Research Conf.*, National Cotton Council, Memphis, 1975, 116.
23. **Stanford, G., Vanderpol, R. A., and Dzienia, S.,** Denitrification rates in relation to total and extractable soil carbon, *Soil Sci. Soc. Am. Proc.* 39, 284, 1975.
24. **Gardner, W. R.,** Movement of nitrogen in soil, in *Soil Nitrogen*. Bartholomew, W. V., and Clark, F. E., Eds. American Society of Agron., Madison, Wis., 1965, 557.
25. **Jones, J. W., Bowen, H. D., Stinner, R. E., Bradley, J. R., Jr., and Bacheler, J. S.,** Simulation of boll weevil population as influenced by weather, crop status, and management practices, *Trans. ASAE*, 20, 121, 1977.
26. **Brown, L. G., Jones, J. W., and Harris, F. A.,** A simulation of insect pest management alternatives by integration of a *Heliothis* spp. model and a cotton crop model. *Am. Soc. Agric. Eng.*, Paper No. 765025, St. Joseph, Mich., 1976.
27. **McClendon, R. W., Brown, L. G., Jones, J. W., Hesketh J. D., Hartsog, J., Harris, F. A., and Parvin, D. W.,** Modeling crop-insect pest ecosystems for studying conflicting alternatives caused by species interactions, *Proc. Int. Fed. Auto. Control, Symp. Control Mech. Bio- and Ecosystems*, Leipzig, GDR, 5, 121, 1977.
28. **Jones, J. W., Smith D. B., and Colwick R. F.,** Mississippi State, Miss., Unpublished data.
29. **Bowen, H. D.,** North Carolina State Univesity, Raleigh, N.C., Unpublished data.

INDEX

A

D

E

F

G

H

I

J

K

L

M

N

O

P

Q

R

S

T

U

V

W

X

Y

Z

For Product Safety Concerns and Information please contact our EU representative GPSR@taylorandfrancis.com Taylor & Francis Verlag GmbH, Kaufingerstraße 24, 80331 München, Germany

Batch number: 10731074

Printed by Printforce, the Netherlands